Historical Dictionary of Data Processing

Historical Dictionary
of
Data Processing

TECHNOLOGY

James W. Cortada

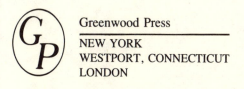

Greenwood Press
NEW YORK
WESTPORT, CONNECTICUT
LONDON

004.03
C82h

Library of Congress Cataloging-in-Publication Data

TH

Cortada, James W.
    Historical dictionary of data processing.

    Bibliography: p.
    Includes index.
    1. Electronic data processing—Dictionaries.
2. Electronic data processing—History.   I. Title.
QA76.15.C67   1987   004'.03'21   86-22751
ISBN 0-313-25652-7 (lib. bdg. : alk. paper)

Library of Congress Catalog Card Number: 86-22751
ISBN: 0-313-25652-7

First published in 1987

Greenwood Press, Inc.
88 Post Road West, Westport, Connecticut 06881

Printed in the United States of America

The paper used in this book complies with the
Permanent Paper Standard issued by the National
Information Standards Organization (Z39.48-1984).

10   9   8   7   6   5   4   3   2   1

Contents

Preface

This historical dictionary is one of three volumes being published by Greenwood Press to provide basic, important research tools for the study of data processing. The three dictionaries will cover biographies (*Historical Dictionary of Data Processing: Biographies*), institutions such as companies, societies, and laboratories (*Historical Dictionary of Data Processing: Organizations*), and technology, including both software and hardware achievements (*Historical Dictionary of Data Processing: Technology*). Taken together, the three books present nearly 400 entries on all aspects of data processing, from the earliest beginnings to recent years. These range from short, 500-word essays to others of several pages. The attempt is to provide basic factual material and, where appropriate, historical interpretation. The entries conclude with bibliographic references to lead the reader to additional material.

When an entry in any of the three volumes mentions a topic that has its own entry, a cross-reference is indicated by one of three symbols: * indicates that the subject entry is in this dictionary; ** points to the biographical dictionary; and † means that the topic appears in the volume on organizations. Space limitations, economics, and the availability of good hard data made it necessary to limit the entries to those included. As the field of data processing history matures, a more ambitious publication can be considered.

This dictionary covers computers, electronic components, early calculating machines, software, special applications and projects, and general computer concepts. Terms of historical significance are generally omitted, as they are covered in numerous dictionaries of computer terms. While this work suggests key issues as reflected in current historical literature, it does not claim to provide a definitive history of data processing or to be the ultimate guide for the specialist. An attempt has been made to provide a balance between the needs of the general reader and those of the specialist.

For the general user, the book reviews major subjects based on the assumption that the reader has only limited knowledge of data processing technology. For the specialist, it is a starting point and a reference tool, providing details on technological and economic issues. Narrative histories of individual computers, for example, are written with the nonspecialist in mind, but for those interested in more details tables present technical features. Each entry indicates the topic's significance and presents up-to-date facts.

The topic of data processing is broad, complex, and little understood. It is too soon to determine the relative importance of various factors, industries, technologies, and economies of scale, to define properly the merit of seemingly conflicting and even contradictory characteristics of data processing. Nonetheless, historians have already learned a great deal. For example, they know how specific computers were developed, they have begun writing biographies of the key people, and they are beginning to express concern about the role of various institutions (such as government agencies and universities as well as private companies) as agents of change in the industry.

Daniel J. Boorstin has noted that the production of new technology is one of the nation's great strengths. His *The Republic of Technology* (1978) pictures a nation on a path of spiralling technologies that have spawned additional innovations. In his monumental *The Americans* (1974), Boorstin was already cognizant of the influence of data processing technology, observing that when current data became available "about the quantities of everything, businessmen and consumers could not help thinking quantitatively."[1] The lion's share of the industry and its technology has developed in the United States, and these three dictionaries, therefore, focus on American developments.

As with other technologies in America—steam, railroads, electricity, photography, nuclear energy—the problem facing students of technology is the fundamental one of how it happened. Are there indeed technological imperatives, as Boorstin hints, by which developments build on each other? Elsewhere, I have argued that technology speeds itself on, building an intensity that results in yet more change. Certainly, the material presented in this book lends credence to that view.[2] Certain trends are readily noticeable but have not been fully described. In other words, some obvious answers exist based on what we know today, but more evidence must be generated before those answers can be validated. As suggested in this dictionary and in its two companion volumes, eight factors have contributed to the development of data processing.

Modern data processing—which for the purpose of this book may be said to have begun with the birth of electronic digital and analog computers—grew out of the demands in World War II for unique information handling. These requirements led to the development of flight simulation equipment, cryptoanalysis for decoding enemy communications, and, by 1942, artillery and bomb fire control systems. ENIAC,* developed at the University of Pennsylvania, is one project that emerged from these needs.

Technological innovations, both evolutionary and radical, have spurred the industry forward each year. In the 1930s, for instance, the Massachusetts Institute of Technology's work with analog computers* led to evolutionary improvements over earlier technologies, whereas in the late 1940s, after Bell Laboratories't breakthrough to the transistor, new technologies emerged quickly and dramatically. These, in turn, led to profound and rapid changes in the construction of computers.

Improvements in price performance of hardware have been continuous. In many cases it has become less expensive to perform a function with a computer than to have people do it or to use older technologies. As a result of this phenomenon, the less expensive a function becomes, the more people demand to use that function. Bold product announcements, such as the S/360* or the Applet and International Business Machines Corporation's (IBM's)† PC micros, have lowered the cost of technologies and have therefore expanded our dependence on data processing at a dizzying pace.

Improvements in software as well as in hardware have encouraged increased use of data processing. As software has become easier to use, more people have become users and hence dependent on data processing. In order to use a computer in the 1930s and 1940s, one had to be an engineer, have an intimate working knowledge of the machine, and be able to program with complicated and cumbersome languages. In the 1980s a child of ten, working with a PC, could write powerful programs in BASIC.*

The declining need to know data processing in general has also meant that people can use the technology more easily. In addition to the widespread use of microcomputers, workers can work with terminals, using human commands, to use large processors. Today we frequently see terminals on desks or on the floor of a factory; five out of every one hundred workers use terminals today; and the industry expects that by the turn of the century, there will be one terminal for each member of the labor force in the industrialized world. Fewer than 10 percent of all workers had any knowledge of or need to use data processing in 1960; by 1985 over half the people employed in the United States needed computers to do their jobs.

Competitive pressures, caused by both established firms and emerging new ones, made rivalry frenetic. This competition led to the rapid creation of new products, offered customers more selection, and motivated companies to drive down prices while making their products easier to use.

The number of scientists working in the twentieth century is greater than in all other periods combined, and has supported the growth of data processing and sped up the development of new technology. The results achieved by today's scientists are broadcast faster than ever before, and new products emerge more rapidly than in earlier centuries. That phenomenon has an impact on data processing as it does on physics, medicine, or space-related sciences.

Increasingly, *psychological stimulation* is being recognized as contributing to the explosive growth in the use of data processing, particularly by children. Recent studies suggest that the intellectual interplay between mind and computer helps to define how people think and how they identify themselves with nature and society. Ultimately, it could provide a method for describing how the mind works.[3]

A number of constants are also apparent in this fifty-year-old industry, four in particular, as follows.

Change is the most obvious. There have been five generations of technology, each with different types of physics at work and obvious price/performance and economic considerations. Profound changes are characteristic from one decade to another. Such changes are seen as important movements to observers outside the industry, whereas they are viewed simply as normal day-to-day business to those within the industry. Therefore, observers outside of data processing are usually best fitted to interpret the role of change within the industry.

In each decade, the successful company's *willingness to commit* itself to the industry totally and to change constantly is an observable trend. Those companies willing to risk their whole future on data processing have been leaders in the introduction of new technologies and thus have won handsome profits (IBM, Wang, and Apple, for instance). Those that split their financial and managerial loyalties lost (Radio Corporation of America,† or Philco,† concerned with television and computers). IBM's total commitment to the S/360, for instance, with its major change from the computers of the past, contributed dramatically to the development of necessary technologies. The opposite illustration is provided by Remington Rand,† which was slow to change products.

Phenomenal growth and expansion have always characterized this industry. In the 1950s once data processing became active outside of research laboratories and a few isolated commercial data centers, growth curves in usage became impressive. Annual growth, as measured in total industry revenues, was 27 percent or better in the 1960s, 10 to 12 percent in the 1970s, and better than 14 percent in the 1980s—rates that far exceeded those of industries such as oil, automobiles, and electricity, particularly in the late 1970s and the 1980s. By 1985 and 1986 executives within the industry expected data processing to be the world's largest industry in the late 1990s.[4] The only other industry likely to rival it in size could be the medical care complex, which has been constantly growing owing to the increase in age of people in North America and Europe.

The need for *better information handling* has coincided with the emergence of technologies designed to facilitate data management through the use of electronic means. It is no accident that companies and universities have become bigger or that their requirement to handle more knowledge (data or information) has increased. In fact, a number of technologies have sped up the process—transportation has become faster with automobiles, trucks, and airplanes; communications with the telegraph, telephone, and satellite; and the printing of books and newspapers with the growing rate of literacy around the world. All have led to a demand for more information and, consequently, to better methods of managing it.

This first edition of a dictionary on relatively new technology does not claim to be more than an initial attempt to gather historically relevant data in one convenient source. The views and contents of this book are solely mine; any suggestions for improving this book or either of its two companion volumes are exceedingly welcome.

NOTES

1. Daniel J. Boorstin, *The Americans: The Democratic Experience* (New York: Vintage Books, 1974): 205.

2. James W. Cortada, *Strategic Data Processing: Considerations for Management* (Englewood Cliffs, N.J.: Prentice-Hall, 1984): 19–21, 28–37.

3. Two useful introductions to this concept, with considerable data and thought already assembled, are Sherry Turkle, *The Second Self: Computers and the Human Spirit* (New York: Simon and Schuster, 1984) and a classic study, Joseph Weizenbaum, *Computer Power and Human Reason: From Judgment to Calculation* (San Francisco: W. H. Freeman and Co., 1976).

4. For a quick perspective on the industry's prospects, see Ulric Weil, *Information Systems in the 80's* (Englewood Cliffs, N.J.: Prentice-Hall, 1982), passim.

Acknowledgments

Many individuals contributed their skills at various stages of this project. I am particularly indebted to the librarians at Vassar College, the U.S. Library of Congress, Smithsonian Institution, University of Virginia, Vanderbilt University, Georgia Tech and the International Business Machines Corporation's library in Poughkeepsie, New York.

A number of other individuals helped along the way, too many to allow a listing of all their names here. However, several deserve particular consideration. Hank Tropp suggested entries, led me to information, pointed out other contacts, and served as all-round supporter throughout the project. Nancy Stern edited my original list of proposed entries. Paul Ceruzzi was also supportive at various stages. Just as exciting was the pleasure and honor of talking to living "pioneers," such as Mina Rees, all of whom were helpful as I chased down information and confirmed facts. Jean E. Sammet set a high standard of workmanship for me to emulate along with Emerson Pugh and Bernard A. Galler.

At Greenwood Press, home for a half dozen of my earlier books, friends and professionals worked as a team in getting all three dictionaries out smoothly. Cynthia Harris, editor at Greenwood, was an outstanding manager and an intellectual asset. The Greenwood staff, always efficient and effective, was fun to work with again. I would like to thank those publishers who allowed me to use material quoted or otherwise used in this dictionary.

A special thanks goes to my family for their patience and understanding, especially to my wife Dora.

TECHNOLOGY

A

A–2, A–3. These programming languages* were developed at Remington Rand†
for use on their computers in the 1950s. This company developed the first
automatic coding systems that were compilers (the A–0 and A–1), yet the A–2
and A–3 were more widely used. The A–2 appeared in 1955 and was the
programming result of work headed by Grace B. M. Hopper** as part of the
support given to the company's UNIVAC* computers. A–2 used the twelve-
character word common to the UNIVAC by relying on a three-address code.
A–3, sometimes also called ARITH-MATIC, evolved from A–2. It was hardly
used because it appeared at the same time as MATH-MATIC* (AT–3) which
Remington Rand sold as its answer to the International Business Machines
Corporation's (IBM's)† FORTRAN* in the late 1950s. Even so, A–2 and A–3
represented significant advances in programming languages over tools available
in the late 1940s and contributed to the overall attractiveness of the UNIVAC
computers in the 1950s.

If Remington Rand had been able to maintain its marketing advantages of the
early 1950s with its UNIVACs into the 1960s, the A series of programming
languages might today be viewed as precursors to other more widely used
software tools. But by the late 1950s IBM overtook Sperry and forced the
UNIVACs out of the limelight, along with their software* packages. The A
series had sent a clear signal throughout the industry that no computer vendor
could offer to sell computers without also providing programming tools such as
the A–2 and A–3.

For further information, see: Jean E. Sammet, *Programming Languages: History and
Fundamentals* (Englewood Cliffs, N.J.: Prentice-Hall, 1969).

ABACUS. This device, an ancestor of the modern calculator and computer, was
one of the first, if not the earliest, calculating machines that could conduct
mathematics and store data. It is the most long-lived calculator in history, being
still used extensively in the Far East and to a lesser extent in the Middle East.

For many centuries, it was the only major calculator in many parts of the world. The abacus is a widely used symbol within the data processing industry in advertisements.

The need for a calculating device grew slowly as humankind began to establish numbering systems. Some of the earliest aids to counting simply included coiled rope techniques, collections of stones, and marks on objects (especially on wood). Counting stones were still prevalent in Europe as late as the sixteenth century and in the Middle East in the early twentieth century. The leap from rows of stones gathered on the ground to stone-like objects strung together in rows within a wooden frame (the abacus) was thus not a major one. Groups of five or ten stones could then be moved back and forth in patterns in order to perform addition, subtraction, multiplication and division functions.

Although the Greeks had abacus-type devices as early as 500 B.C., the earlier Babylonians are believed to have been the first to perfect it. Although we do not know whether others may have been doing the same thing in China or elsewhere, we do have evidence that the abacus existed prior to the flowering of Greek civilization. The name *abacus* is of Phoenician origin, *abak*, which means sand spread on a surface for writing. The first abacus may have simply consisted of stones on a flat surface which were moved around; recordkeeping was done on sand and later between printed or etched lines on the surface of a board. The wires used in the modern abacus could have been, to use today's computer language, "a technical enhancement" achieved centuries later to improve the overall design of the abacus. Herodotus, the father of history, described rows of stones on sand beds in the fifth century B.C. and explained calculations done on such a device, including one application for calculating interest payments on a loan. Two centuries later Eytocious of Ascalon showed how to calculate accurately the square of 3,0133/4 on an abacus, whereas in the previous century Demosthenes, Diogenes, and Polybius described the principle of the abacus in their writings.

The abacus therefore evolved over the centuries from pebbles on a sand board to wax-covered plates for stones to ruled boards using stones, pieces of bone, glass, or metal. Finally, either grooves on a wooden board were used, or, as is common today, beads were strung on wire in rows. Each line or wire could represent units of one, tens, hundreds, and so on, or groups of values such as for currency. Each type of abacus and counting method described above was well known in Roman times. Over subsequent centuries there were variations on the number of beads on a string and the number of rows.

Through their use of the abacus the Romans contributed a major term to mathematics—*calculus*, a word meaning small stones. Thus, *calculi* (stones) represented values of money in units of one, tens, hundreds, and so forth, and provided a counting method. The excess of ten was carried over by sliding a higher value pebble over from the next row up and, when reaching a 100, one bead from the third row up, and so forth. Thus, when counting up to eleven, the user pushed ten one-unit pebbles to one side on the first row of beads and at ten pushed one pebble worth ten units over on the second strand in the same direction. The user started counting eleven forward by going back to the first row of beads, each worth one unit, and starting again by pushing to the side one bead—just as we do today.

This principle has remained essentially unchanged to the present regardless of whether stones (Roman and Greek method) or bamboo rods (early Chinese technique) were used. The bamboo type, which was described as early as the sixth century B.C., was still evident in Korea as late as the nineteenth century. In the sixth century these devices made their way into Japan and were called *sangi* or *sanchu*. The introduction of the abacus to China and Japan was an important event because its use spread rapidly and has continued in these two countries to the present. By the sixteenth century the Japanese abacus was commonly called the *soroban*. Mathematical uses of the abacus became more sophisticated over time. Thus, the first use involved simple counting, but in time, the Japanese, for example, figured out how to determine algebraic coefficients on this device, "displaying" the results. Variations in the basic design of the abacus, designed to handle more complex functions, kept on appearing throughout the Orient and Europe. For major transactions requiring large calculations, whole counting tables were used that simply had more beads. That innovation was common even in Europe in the early years of the nineteenth century.

In Western Europe the abacus began to go into decline when people accepted the Arabic numerals and the 0—all of which were easier to write than Roman numerals—while the 0 opened up a vast new area of complex mathematics. The device almost disappeared in Spain and Italy in the late 1400s and early 1500s, in France during the late 1500s, and in Germany and England a century later. Yet it has not completely disappeared either in Europe or in America. The billiard counters (beads on a long wire usually hung over a pool table) still exist, as do poker chips, both of which have originated from the abacus.

The abacus was popular because it had many applications for mathematics, particularly in societies with awkward written numbering systems. This device also facilitated larger amounts of counting than could be done in one's head, while it could "save" and remember numbers for "recall" later. The abacus was also easy to learn and could be mastered relatively quickly. As late as the 1940s competitions between experienced users and mechanical adding machines sometimes resulted in the victory of the older device. In the hands of a skilled user, the abacus can beat simple written mathematical calculations easily and, frequently, even some calculations done with the aid of a hand calculator.

Through the abacus, modern computer scientists, as well as earlier scientists working on mechanical calculators, learned many of the problems that any new computational device had to address, including how to identify numbers in a system and how to carry values, for example, from single digits to tens, tens to hundreds, and running totals (if not final totals) until removed from the device's "memory." The abacus was capable of performing all of these functions, and so any device designed to replace it had to carry out these functions more efficiently.

For further information, see: Japan Chamber of Commerce and Industry, *Soroban* (Tokyo: Chamber of Commerce and Industry, 1967); J. M. Pullan, *The History of the Abacus* (London: Hutchinson, 1968).

ABC. See ATANASOFF, JOHN VINCENT (in *Biographies*)

ACE. The Automatic Computing Engine (ACE) was an early British electronic digital computer* employing many of John von Neumann's** concepts regarding the stored program. It reflected Great Britain's growing interest in computers immediately following World War II by scientists who, during the war, had developed machines to break enemy communication codes, primarily at Bletchley Park.† Often known as the Pilot ACE, this machine was the product of the Electronics Division of the National Physical Laboratory (NPL). Built in the late 1940s, it became operational on May 10, 1950.

The design was based on the work of Alan M. Turing** who had studied the role and architecture of computers since the mid-1930s. In a widely read paper on the subject, published in 1936, he argued that a particular automation could always be explained by using a specific collection of instructions, and, when given to a computer, that device would imitate or recreate a sequence of calculations. That concept influenced the logic design of computers throughout the 1940s and beyond, including the work of an influential American scientist, John von Neumann, who in turn used his knowledge and government contacts to encourage the development of various American computers throughout the 1940s, including EDVAC* and earlier, ENIAC*—two of the most important of the early digital electronic computers.

The key architects of the ACE in the 1940s were J. H. Wilkinson, a mathematician, and E. A. Newman, an expert in electronics. ACE became the fastest of the early British computers, functioning at 1 megacycle, while instruction times varied from 64 microseconds to 1,024 milliseconds. Mercury delay lines were used for main memory* which could house 128 32-bit words. By the end of 1951 this memory had grown to 352 words, and in 1954 it added a 4,096-word drum device. Input/output was based on eighty-column card technology along with printed hard copy output. ACE employed 800 thermionic valves in its original version, and all components were mounted on specially built racks (see Table 1).

The first ACE program was run in May 1950, and ACE was publicly demonstrated in December. That event drew the first public attention to computers in Great Britain and, therefore, for the first time raised the possibility that computers could be of interest outside of what had been a small circle of scientists and engineers. Even the specialists found ACE interesting, especially because of the machine's high reliability and its use of fewer components than many

Table 1
Technical Features of ACE (NPL's Pilot ACE)

Word length (in bits)	32
Instruction length	32
Instruction format	(2+1)-address
Instruction set	(15)
Store size, fast	352 full words
Store type, fast	Delay (mercury)
Store size, backing	0
Average add time	0.54 ms
Average multiply time	2 ms
Input medium	5-track paper tape reader
Output medium	Teleprinter, 5-track paper tape or hard copy
Digit period	1 microsecond
Main valve type	ECC81[a]
Approximate number of valves	800

[a]ECC81 was an English Electronic Company valve.

SOURCE: Reprinted with permission from *Early British Computers*, Simon Lavington, First Edition. Copyright © Digital Press/Digital Equipment Corporation (Bedford, MA), 1980.

other devices. The computer broke down less frequently than many of its contemporaries. Unlike numerous other computers, most of which had been designed based on the model of the American EDVAC,* ACE (sometimes also called Turing's ACE) used different components, such as British-designed valves. It was modified continuously throughout the early 1950s. Work on ACE-type machines did not cease at the NPL until 1957. The final ACE was a 48-bit word machine that continued to use delay-line memory but had been sped up to the point where, for example, a multiplication could be done in 448 microseconds. The last machine had some 7,000 valves in its makeup. The last machine was constructed with the help of the English Electric Company and was called the DEUCE,* introduced in 1955.

For further information, see: B. V. Bowden, *Faster Than Thought* (London: Sir Isaac Pitman, 1953); Simon Lavington, *Early British Computers* (Bedford, Mass.: Digital Press, 1980); A. M. Turing et al., *The Automatic Computing Engine* (Shrivenham, England: Military College of Science, 1947).

ALGOL. This second-generation language, which was developed during the last half of the 1950s and which evolved continuously during the 1960s, spawned many languages. It reflected a growing desire within the young data processing community to standardize programming conventions. Although ALGOL was never used as widely as FORTRAN* or COBOL,* it nonetheless made important technical contributions to the design of programming languages.* It also generated considerable debate among scientists in the United States and Europe regarding the future evolution of languages.

The first major spurt in the development of uniform languages occurred in the mid-1950s. Between the end of 1955 and the start of 1958, many new tools became available to programmers with such functions as the ability to write sentence-like code, use more compilers, and greatly shift programming administration to languages so that a programmer could spend more time on defining and solving a problem. The change could even be seen in the vernacular of the day: programmers wrote less "code" and instead began to "write" in particular "languages." New and relatively powerful languages emerged, including FORTRAN in 1957 as the most widely used language of the late 1950s and early 1960s.

Considerable attention was being paid to the development of new languages (particularly those based on algebraic expression), which would be run on many new computers appearing in the late 1950s. The potential development of so many languages caused scientists to try to standardize programming with a universal language. Committees were formed made up of users, vendors, government agencies, and academic research groups to study the problem. Both ALGOL and COBOL were developed in this manner.

Work on what would become ALGOL began in Germany in 1955 when a European association of applied mathematics and mechanics called GAMM (*Gesellschaft für angewandte Mathematik und Mechanic*) established a committee to devise a new language. On May 9-10, 1957, representatives of SHARE,† DUO, USE,† and the Association for Computing Machinery (ACM)† met in Los Angeles to do the same thing. In the United States alone, there were already nearly two dozen major projects at various locations to develop languages; thus, the threat of many idioms existed. In October GAMM contacted ACM to determine whether both efforts could be combined to produce one language which all could support. The president of ACM, John W. Carr III, formed a committee to study the matter of a universal language with representatives from the Massachusetts Institute of Technology (MIT), International Business Machines Corporation (IBM),† Remington Rand,† Philco,† Bendix, Westinghouse, the universities of Pennsylvania, Chicago, California, Carnegie Tech, Johns Hopkins, and the National Bureau of Standards.† All were already well-established major forces within the data processing community. Work began on January 25, 1958, and after three committee meetings, the group settled on an algebraic format for the language, designed parts of it, and described its overall organization. Additional efforts with GAMM at a meeting held in Zurich between May 27 and June 2, 1958, led to a consolidated report.

One of the most interesting results of these efforts was the name of the language. The initial working title used since the mid-1950s had been International Algebraic Language (IAL), but it was later changed to the more exact ALGOL 58 (Algorithmic Language). The report, which described this new language and its features, almost immediately generated controversy both in Europe and the United States. People on both sides of the Atlantic debated its strengths and weaknesses. Nonetheless, as a result various organizations developed ALGOL-like languages in subsequent years and described programming issues in ALGOL-like terms. The new language also led to other industrywide meetings to define more precisely the new language which various supporters felt should become *the* universal language. Because ALGOL was very close to FORTRAN in form, one might ask why the industry did not simply adopt FORTRAN as the universal language. At the time the members of each committee were concerned that FORTRAN was IBM's language and thus was not in the public domain. By not being public, it was felt that the language might not be readily available on non-IBM computers as it had to be if it were to be a universal language. Today FORTRAN is available on most non-IBM computers and is legally in the public domain.

ALGOL 58 was an elegant language for its day, graceful in form, and a relative of the familiar FORTRAN. Hence, it was more suitable for scalar mathematical processing than for straight commercial applications. For many, it was simple. ALGOL introduced a new concept in programming in which the language would be at three levels—reference language, publication language, and hardware language—making ALGOL far more flexible for a programmer than earlier tools.

No sooner was ALGOL 58 out than work began to refine it, leading to ALGOL 60. At an international data processing conference hosted by UNESCO in Paris during June 1959, John Backus** of IBM presented a paper defining how syntax might be handled in ALGOL. His ideas soon became known as the Backus-Naur Form (BNF), and this paper, a technical milestone in the development of programming languages, showed the way to a more disciplined approach to the design of programming languages. In 1960 a committee made up of Europeans and Americans worked on revisions to ALGOL, the result of which was ALGOL 60. Revisions continued throughout the 1960s, and various bulletins on the language were published primarily by the ACM. Yet ALGOL 60 formally came into existence when it was first described in a report published in May 1960 and when at the same time the language entered the public domain.

ALGOL 60 received mixed reviews, as had its predecessor. The language had more support in Europe than in the United States. In North America, the primary sponsor of ALGOL 58, SHARE, chose not to support ALGOL 60 as enthusiastically as before. In addition to technical reasons inherent in the functional characteristics of ALGOL 60, members of this IBM user group had increasingly been using FORTRAN and thus were reluctant to give up an enormous investment already made in programming by converting to ALGOL. IBM in turn did not want to support ALGOL when it became obvious that its customers were not interested in the language. Yet, as in Europe, ALGOL became a base of reference by which to describe and measure future languages.

ALGOL's contributions were technical, but they are worth reviewing briefly inasmuch as so many future languages relied on the work done by the committees preparing ALGOL. ALGOL helped to define block structures and the scope of variables in languages; syntax for languages was thereafter more formalized;

recursive procedures became common afterwards; the way was shown for greater simplicity in languages; it introduced the idea of a language being detailed at various levels; it created the demand for better methods to implement languages; and it encouraged the development of other languages. Some of the better known languages, which were either variations of ALGOL or which emerged from designs influenced by it, included LISP* 2, DIAMAG, GPL, NELIAC, MAD, and JOVIAL.*

Although ALGOL was less frequently used for day-to-day programming than many others were, particularly in the United States, it became a useful basis and format for discussing other languages. It was probably the most discussed language of the 1960s among those interested in the design of systems software and idioms. Certainly, an enormous body of literature was generated by the debate over ALGOL. Ironically, ALGOL was one of the least used programming tools, especially in the United States.

For further information, see: Peter Naur, "The European Side of the Last Phase of the Development of ALGOL 60," in Richard L. Wexelblat, ed., *History of Programming Languages* (New York: Academic Press, 1981): 92–139; Alan J. Perlis, "The American Side of the Development of ALGOL," in ibid.: 75–91; Jean E. Sammet, *Programming Languages: History and Fundamentals* (Englewood Cliffs, N.J.: Prentice-Hall, 1969).

ALGY. ALGY was developed in the 1950s and was perhaps the oldest programming language* designed to do formal algebraic manipulation. It expressed notations in a format similar to FORTRAN* as input and ran on a Philco† 2000 computer. ALGY represented one of the first attempts to process formal mathematical expressions independent of any reliance on their numeric values. Although other researchers had worked on the problem as early as 1953 using computers, ALGY was the first successful and practical contribution to this kind of computing.

The language had no definitions for arithmetic, and it lacked features to handle loop controls or transfers. Yet it had a small collection of commands to handle such functions as input, renaming of files, searches, manipulation of notations, and expansion of sine and cosine functions. It was a language that was used only by a few researchers. Although the literature on ALGY is limited, descriptions of this language illustrate how to build various functions for a general class of expressions within one language. ALGY is considered a direct forerunner of FORMAC.*

For further information, see: M. D. Bernick et al., "ALGY—An Algebraic Manipulation Program," *Proceedings, WJCC* 19 (1961): 389–392; Jean E. Sammet, *Programming Languages: History and Fundamentals* (Englewood Cliffs, N.J.: Prentice-Hall, 1969).

ALTRAN. This particular programming language, designed to perform formal algebraic manipulations, was one of many programming languages* developed in the 1960s to enhance the calculation of algebra through the use of computers. It was developed at Bell Laboratories† at Murray Hill, New Jersey, and was based largely on another software package called ALPAK, which was a collection of subroutines to handle polynomials and rational functions. ALTRAN's strength lay in its ability to deal with rational functions. It was created in 1964 and ran on an IBM 7090/94 and later on a 7040/44.

The language represented an extension of the widely used FORTRAN* for scientific computing. Most of the elements drawn from FORTRAN came out of FORTRAN II, with other features extracted from FORTRAN IV. ALTRAN had features that led a user to employ substitution as the main algebraic statement. Its functions otherwise were limited and appear not to have contributed significantly to the art of designing programming languages. ALTRAN was hardly used outside of Bell Laboratories.

For further information, see: W. S. Brown et al., "The ALPAK System for Non-numerical Algebra on a Digital Computer," *Bell System Technical Journal* 42, no. 5 (September 1963): 2081–2219, 43, no. 2 (March 1964): 785–804, 43, no. 4, pt. 2 (July 1964): 1547–1562; Jean E. Sammet, *Programming Languages: History and Fundamentals* (Englewood Cliffs, N.J.: Prentice-Hall, 1969).

AMBIT. AMBIT was an obscure, little-used string and list processing language of the early 1960s. It may have run on only two computers and, even then, not efficiently. It is an example of a genre of programming languages* developed primarily in the United States to do string and list processing in the late 1950s and early 1960s. By 1968 string and list programming languages became especially useful for research in artificial intelligence* and were used primarily at American universities.

AMBIT was created in 1964 by the Massachusetts Computer Associates in ALGOL* in an attempt to create a language that could manipulate mathematical symbols. Although similar in approach to the better known list processors COMIT* and SNOBOL,* little was known about this language. It appeared that it was not a fully matured programming tool, lacking a wide set of functions. Perhaps for that reason it was never well received.

For further information, see: C. Christensen, "On the Implementation of AMBIT, A Language for Symbol Manipulation," *Communications, ACM* 9, no. 8 (August 1966): 570–573.

AMTRAN. Also known as the Automatic Mathematical Translation system, this programming language was designed at the National Aeronautics and Space Administration (NASA) to do numerical scientific problems. In the early 1960s NASA needed a language that scientists and engineers could use to solve problems quickly online. At this time there were too many scientists and engineers and not enough professional programmers to act as intermediaries with the computers. Taking many ideas first developed in the creation of JOSS* (another programming language) and other research conducted at the Hudson Laboratories of Columbia University, AMTRAN was developed by the mid-1960s. In addition to being an early online programming language for scientific problems, it was one of the first to be used in connection with the exploration of space through the development of flight planning and the construction of space ships.

The first known implementation of the language was as an interpreter on NASA's IBM 1620* computer. Although a complex system using what even by the standards of the time was a small computer, AMTRAN worked nonetheless. Programs were entered into the computer by way of a typewriter keyboard specially equipped with 224 push buttons. Approximately half of these buttons could be "programmed" to perform specific operations in addition to the normal ones provided by the language itself. Later, a small subset of the language, called the *Sampler*, relied on a standard console typewriter and a normal card reader on a 1620 to do data entry and programming. AMTRAN also worked with 5-inch display screens, then known as scopes (today as cathode ray terminals) to display programs that could be printed out on the system's printer.

Users likened the functions of AMTRAN to FORTRAN* II. What distinguished it from the International Business Machines Corporation's (IBM's)† widely used FORTRAN were such features as the push buttons and the lack of significant notation. FORTRAN's wide range of formatting functions was also not available on AMTRAN. But even the lack of FORTRAN's fixed-point and variable-precision arithmetic did not dissuade users from seeing the language as fulfilling a similar need as FORTRAN.

AMTRAN was used in the 1960s on a variety of NASA's second-generation computers. Among the specific machines it ran on were a Burroughs B5500* and an IBM 1130. By the 1970s more useful languages for scientists were available and thus AMTRAN was relegated to history.

For further information, see: S. E. James, "Evolution of Real-Time Computer Systems for Manned Spacecraft," *IBM Journal of Research and Development* 25, no. 5 (September 1981): 417–428; Jean E. Sammet, *Programming Languages: History and Fundamentals* (Englewood Cliffs, N.J.: Prentice-Hall, 1969); James E. Tomayko, "NASA's Manned Spacecraft Computers," *Annals of the History of Computing* 7, no. 1 (January 1985): 7–18.

ANALYTICAL ENGINE. This machine, designed by Charles Babbage** in the early 1830s, marked an important step in the evolution of technology and computers specifically. It proved to be a quantum leap beyond any device designed in Western Europe in centuries, and it incorporated some of the features that would characterize computers of the twentieth century.

The analytical engine was an outgrowth of Babbage's experiences in designing and partially constructing the difference engine* during the 1820s. He began work on the design of the analytical engine in the mid-1820s and worked out its essential elements by the early 1830s. He continued work on this device until his death in 1871. Although the machine was never built owing to Babbage's lack of funds and constant redesign, he left hundreds of design drawings documenting the essential principles of this sophisticated calculator.

Babbage designed the analytical engine to perform all types of arithmetical operations and to be able to string them in any order to solve mathematical problems. This ability represented a major advancement over the difference engine which had been designed only to execute polynomial calculations. Herman H. Goldstine,** himself an important developer of the computer in the 1940s and 1950s, called it "in concept a general purpose computer" in his history of computers. Some sixty mathematical operations were to be executed concurrently in this design, all mechanically using gears, cranks, and probably steam power.

Babbage's machine was to consist of four parts. The first, called the "store," would warehouse numerical data on columns of wheels, each capable of handling ten digits. His conception of a store function had the capacity to house 1,000 fifty-digit numbers. The second part he named the "mill" and would equate with today's processor within a computer, or that part of the device that actually performs calculations. These calculations would be accomplished through various rotations of the gears and wheels. The third part consisted of more gears and levers to carry numbers to and from the mill to the store. He designed the fourth to move data in and out of his system (in today's language, input/output devices).

Babbage designed a series of machines which in modern parlance would be called a computer system. Many of his ideas for such a system were best described by a close friend of his who understood mathematics. Lady Lovelace,** daughter of the poet Lord Byron, worked with him for a number of years and explained the concepts of his analytical engine in publications and to visitors. She likened the design of his machine to a device that "weaves *algebraic patterns* [her italics] just as the Jacquard-loom weaves flowers and leaves." The analogy was an important one because Babbage, like Joseph Marie Jacquard** before him, intended to use cards with holes in them, representing data, as his method for programming instructions for the machine. The same method would be used to instruct data processing equipment in the twentieth century, and as of this writing it is still a widespread technique for sending data into a computer. Babbage also wanted to use cards to generate output. Generating readable results would not be limited to cards, however, as he also designed printers attached to his system, just as computer systems have today.

In addition to the parallels with modern hardware, some characteristics of programming were evident in his thinking. For instance, Babbage saw the wisdom of storing data in devices which, on command, the processor could draw on. He thought that tables of data could be stored in the system and used for calculations. Using tables rather than reconstructing calculations each time from formulas would save time. (Using tables of data is a common technique in modern programming.) This thinking also led Babbage to add a condition function to his calculator such as the IF statement in FORTRAN.* "If a certain step was taken, then another had to come next" was the type of logic inherent in Babbage's sense of design. After he had worked out the essential elements by the end of 1836, Babbage continued to modify his machine into the 1860s, and most of these modifications involved programmable characteristics.

After Babbage's death in 1871, his son Henry continued to do some development work on the analytical engine, especially with a printer which he ultimately built and demonstrated. In 1910 he demonstrated a second printer and a revised mill to the Royal Astronomical Society. Thereafter, no additional work was done on Babbage's engine, and so it slipped into history.

For further information, see: A. G. Bomley, "Charles Babbage's Analytical Engine, 1838," *Annals of the History of Computing* 4, no. 3 (July 1982): 196–217; B. Collier, "The Little Engines That Could've; The Calculating Machines of Charles Babbage" (Ph.D. diss., Harvard University, 1971); J. M. Dubbey, *The Mathematical Work of Charles Babbage* (Cambridge: Cambridge University Press, 1978); A. Hyman, *Charles Babbage: Pioneer of the Computer* (Princeton, N.J.: Princeton University Press, 1982); A. Lovelace, Countess of, "Sketch of the Analytical Engine Invented by Charles Babbage, by L. F. Menabrea of Turin, Officer of the Military Engineers, with Notes Upon the Memoir by the Translator," *Taylor's Scientific Memoirs* 3 (1843): 666–731.

ANTIKYTHERA DEVICE. This analog astrolabe* was of Greek origin and represented an advanced design from earlier types, incorporating a drive mechanism not found in other aids to astronomical calculations. Until the discovery of this device, historians believed that the use of gearing did not begin until the middle of the sixteenth century, with the first possible example dating from about 1575. Discovery of the Antikythera device pushes back that date to the dawn of Christianity. While other drive mechanisms may have been invented earlier, they were not in common use in machinery in the intervening 1,500 years, particularly in Europe.

Greek sponge fishermen first discovered the gadget in 1900 in the wreck of a ship sunk off the coast of Antikythera Island in the Aegean Sea, probably during a storm. Decades later scientists speculated that the ship, laden with marble statues and other items, had been sailing from Rhodes to Rome. Based on the statues on board, all of which were raised from the sea's bottom by the Greek government, the ship was dated to the time of Christ. The device found aboard was a mass of encrusted bronze which remained unstudied at the Greek National Archaeological Museum until Professor D. De Sola Price of Yale University took an interest in it during the 1950s. De Sola Price, a specialist on Greek machinery, had the device X-rayed, cleaned, and restored, and performed a number of tests on the remains. He found that Cicero describes a gadget similar to the Antikythera device. It appears that the device was originally housed in a wooden box, perhaps 1 foot high, 8 inches wide, and 4 inches deep, with a protruding crank on one side to turn gears. Dials in the front and back of the box would have turned, representing positions of the sun and moon for every day in the lunar calendar. Yet the remaining physical evidence was not exact enough to make other definitive statements. However, based on the remains found on the ship in 1900 and based on Cicero's comments, the device appeared to have had thirty gears made out of a bronze alloy. These gears, utilizing an epicyclic differential turntable, consisted of one large gear, which rotated in various directions, and two smaller ones, which activated the large one and meshed and rotated at different rates.

For further information, see: D. De Sola Price, *Gears from the Greeks* (New York: Science History Publications, 1975).

APL. Originally a language intended for mathematical applications in the 1960s, by the end of the 1970s A Programming Language (APL) was also being utilized for a wide variety of business and academic applications. It had a unique set of characters, and its users discovered that it could solve problems rapidly, problems which, with other languages, required far more extensive programming.

APL was the brainchild of Kenneth E. Iverson during the 1950s. Later, he and Adin D. Falkoff turned it into a fully developed programming language. The idea for APL originated in 1956 when Iverson, a Harvard student, began searching for a way to describe basic data processing concepts. The partnership with Falcoff, which developed after Iverson joined International Business Machines Corporation (IBM)† in 1960, accelerated the development of what would become APL. While at Harvard, Iverson had studied subroutines that would allow a program to experiment with various mathematical methods in a convenient way. During a second phase in the evolution of the language (1961–1963), he and Falkoff wrote machine descriptions of APL as a byproduct of their interest in the forthcoming IBM System 360* family of computers. During this period they developed a collection of programs describing systems and worked out the concepts necessary for shared variables as a way to communicate among programs.

Until this point their work had been theoretical. Next, they had to implement the language, which they did between 1964 and 1968 when they accumulated sufficient descriptions of APL. In addition, by then interest in the possibilities of the language had grown within IBM. The two authors and those who worked with them made two basic decisions which helped give birth to APL. First, as characters in the notation of APL they elected to use only those that were already available on an IBM SELECTRIC® printing element. Since that set of eighty-eight characters was available on a variety of terminals (e.g., the IBM 2741 which was widely used for online input/output to computers), it conformed to existing technologies. Second, they made APL easy to use by the standards of the day, thereby increasing its popularity. Ease-of-use was translated into a language that required fewer lines of programming than others to accomplish the same task.

One of the earliest versions ran on the IBM 7090 computer in batch mode, with data entry in the form of punched cards. The following year, in the fall of 1966, it was moved to the System 360. With its successful operation, broader support for the language developed within IBM, and by the late 1970s it was aggressively marketed by the company. The first important publication on APL did not appear until November 1966, and a complete operating manual was not issued until 1968. Since then, it has run on each of the new operating systems and computers introduced by IBM during the 1970s and 1980s. Today it is a widely used language in government agencies, businesses, and academic environments, performing mathematical work, spreadsheet analysis, and "what if" computing relating to business statistics. It is even available on microcomputers.

Visually, the language was quite different from its contemporaries. Iverson's description of its characteristics has been quoted frequently. "The language is based on a consistent unification and extension of existing mathematical notations, and upon a systematic extension of a small set of basic arithmetic and logical operations to vectors, matrices, and trees." Nearly two decades later, he observed that APL "remained much closer to mathematical notation, retaining (or selecting one of) established symbols where possible, and employing mathematical terminology."

Because of APL's popularity, it is continually being refined. As of this writing, IBM is still releasing new versions, supporting it under its new computers (both large and small). APL today has few rules to bind users, a limited set of generalized functions that are extremely powerful, and familiar symbols that make users comfortable with APL quickly, and it stresses brevity—all fundamental design criteria established nearly two decades ago. As popular as this language became, throughout its greatest period of evolution (1960s and 1970s), only five people controlled its development. Its future evolution appears to be focused on keeping up with new computers and operating systems,* on preserving the original intents of its designers, and on providing programming support for those interested in working with set theory or vector calculus. For the later versions, additional operators (symbols that define the actions a programming language can perform) will have to be developed. As always with APL, pressure exists to expand its functions. Because of its simple organization, it has played little role in advancing the science of designing computer languages, despite its wide acceptance.

For further information, see: A. D. Falkoff and K. E. Iverson, "The Design of APL," *IBM Journal of Research and Development* 17, no. 4 (July 1973): 324–334 and their "The Evolution of APL," in Richard L. Wexelblat, ed., *History of Programming Languages* (New York: Academic Press, 1981): 661–685; K. E. Iverson, *A Programming Language* (New York: John Wiley, 1962).

APT. APT was the first programming language* developed to conduct computer-aided manufacturing. Developed in the late 1950s, APT launched an important application which today has led to computer-controlled manufacturing equipment, mechanical design using computers, computerized graphics, numerically controlled systems, and finally robots and robotic technologies. The use of APT, particularly to program numerically controlled tape systems (n/c), proved successful, encouraging the data processing community specifically and manufacturing companies in general to increase the use of computers for designing and manufacturing. The industrial leaders who first encouraged the use of APT were the large aircraft manufacturing firms supported by various research projects funded by the U.S. Air Force.

The initial steps involved the programming of machine tools whereby they were instructed how to cut out, for example, metal parts through the use of numerically controlled tape fed into a computational device which in turn guided the tool. One of the earliest examples involving this application took place at the Servomechanisms Laboratory at the Massachusetts Institute of Technology (MIT) in 1952 on a project supported by the U.S. Air Force's Material Command—long a heavy proponent of using computers. The objective was to develop an automatic programming language to control machine tools numerically. By 1955, using MIT's WHIRLWIND* computer, a system became operational. Interest in such programming went back further, however. In the early nineteenth century, programming machines had led to the development of various devices, most notably the Jacquard loom by 1805. More recently, in 1949, John T. Parsons and the Parsons Corporation proposed to the U.S. Air Force the development of a punched tape system to control milling machinery used in making templates, which in turn were employed in the production of rotor blades on helicopters. The contract was let in July 1949, and MIT subcontracted for the work the same year. Between 1952 and 1956, therefore, the Servomechanisms Laboratory gained additional experience in numerically controlled systems. That experience led the laboratory to propose the development of an automatic programming tool—what became known as APT— a suggestion accepted by the Air Force in 1956.

A team called the Automatically Programmed Tools (APT) Project was put together to work on the task. The following year a group of aircraft manufacturers joined with MIT to expand the project under the auspices of the Aerospace Industries Association. Although considerable portions of the language were developed by May 1957, a better working code was more widely available within the next year and the language was formally announced on February 25, 1959, at a press conference held at MIT. At that press conference, members of the press were given ash trays cut out by a numerically controlled machine programmed in APT. By then, APT II had been developed (in 1958) to run on an IBM 704,* with such additional functions as the ability to define successive cutter locations and to describe curves in programming, using English commands. APT III came out in December 1961. The Aerospace Industries Association adopted the language as its numerically controlled language and contracted with the Armour Research Foundation at the Illinois Institute of Technology to assume program maintenance of APT. Since then, the language has continued to evolve and be enhanced. It inspired such major pieces of software* as Lockheed Aircraft's CAD/CAM software tools which the International Business Machines Corporation (IBM) marketed during the 1970s. Software that could take the output of design drawings created with CAD/CAM packages today can operate with n/c tape-driven machines.

APT consists of three parts: (1) a series of instructions to perform the necessary geometric calculations to translate human commands while preparing machine-readable data in a standard format; (2) instructions for cutting, monitoring a machine's progress, and computing the coordinates for the machine in order to guide cutting tools exactly where to work; and (3) a postprocessor which converts control data into a format specific to a particular tool while ensuring that none of the device's peculiarities causes it to deviate from the instructions fed to it by APT in the first place.

By the mid-1960s hundreds of programmers had worked on this language in the United States. From the beginning, the team at MIT realized that many people would be involved; hence, they employed certain techniques to manage the creation of the language, techniques which, by the 1980s, had become standard methods for software development. The lab at MIT, with Douglas T. Ross as project leader, recognized that the definition of the language's semantics had to precede any work on syntax. In addition, APT's components had to be modular, which meant that coding conventions had to be organized and standardized. Each module had to work with all others during both the translation and execution phases. Finally, recognizing that the language would have to be enhanced in the years to come, the team wanted to design it in order to facilitate modifications. These requirements and controls made their work highly structured by the standards of the 1950s but were commonplace by the 1970s. This new method for managing the development of APT worked well. It is not clear whether or not other teams of programmers directly picked up the methodology. The only evidence today is that the next major software project in the industry— the development of operating systems for the IBM S/360* in the early 1960s— was managed similarly but that none of the developers were particularly knowledgeable about MIT's approach. Like so many developments within data processing, methodologies and technologies were frequently developed simultaneously by different groups unaware of each other's activities.

For further information, see: Douglas T. Ross, "Origin of the APT Language for Automatically Programmed Tools," in Richard L. Wexelblat, ed., *History of Programming Languages* (New York: Academic Press, 1981): 279–366; Jean E. Sammet, *Programming Languages: History and Fundamentals* (Englewood Cliffs, N.J.: Prentice-Hall, 1969).

ARITHMOMETERS. This word describes a class of machines which today we simply call calculators. The term was popular in the late nineteenth century and was used to name such adding machines as the Thomas Arithmometer* (1850) or the Baldwin*-Odhner* calculator (1875). Though adding machines, they could also perform multiplication and division, often relying on logarithmic tables. These were mechanical devices in the nineteenth century. As with many machines, by the early twentieth century electricity was incorporated into their design. Perhaps the earliest, and certainly the most famous, example of an electrified arithmometer was designed by Leonardo Torres y Quevedo** in 1913 and exhibited publicly in 1920. A problem was typed into the device, and after performing the necessary calculations, his machine typed out the results. Use of the term *arithmometer* declined during the 1920s and in the 1930s was replaced with the word *calculator*, which remains current.

For further information, see: Charles Eames and Ray Eames, *A Computer Perspective* (Cambridge, Mass.: Harvard University Press, 1973).

ARTIFICIAL INTELLIGENCE. This computer-related area duplicates functions associated with the intelligence of human beings in machines. These intelligent characteristics include reasoning, learning, and self-improvement. It is also the most controversial area of scientific research within the data processing community. Artificial intelligence (AI) is a very ill-defined field, for it encompasses work in philosophy, biology, psychology, electronics, and physics. Yet the subject of artificial intelligence is important because it has caused scientists to increase the capabilities of computers and has motivated them to do research that will lead to more sophisticated machines in future, some of which may be based on biological components and the behavior of cells (data processing's motivation for interest in DNA research).

Ironically, most workers in the field of AI never considered themselves computer scientists; rather, they all belonged to more traditional disciplines but were united in their desire to understand the functioning of the human brain and intelligence in general. The arrival of digital computers* spurred new lines of research and, in return, influenced the evolution of computerized technology, most recently, that of robots.

The history of AI is one of self-imitation, of understanding how the human mind operates. This curiosity led to three forms of motivation: replacement of human intelligence with machines; creation of relevant theories of human intelligence through simulations using computers; and assessment of existing technologies—both hardware and software—to suggest future areas of development. This history evolved through three phases. First, people imagined what intelligence was about and how their brains functioned. Second, the subject was discussed in philosophical terms for centuries. Finally, since the invention of electronic computers, more precise experimentation has been conducted to identify the characteristics of human intelligence scientifically.

Interest in AI may be traced back to ancient Greece. In *The Iliad* (850 B.C.) a bronze man named Talos guards Crete's beaches, while elsewhere in the story the reader is introduced to Pandora, created by Zeus to punish people for accepting a gift from Prometheus. Both creatures are presumed to have human-like intelligence and yet they are not humans. The same themes would appear in the nineteenth century in another humanoid made famous by numerous movies: in *Frankenstein*. In Egypt during the Hellenic era, gods were seen as automata who could talk, move, and predict the future. In about 200 B.C. Heron of Alexandria wrote descriptions of automata and, like others, envisioned them as having human characteristics. Yet, much earlier, around 1200 B.C., the Bible said that in the second of the Ten Commandments God cautioned humankind not to make up other gods which, in the context of the day when gods chatted with people, would have been anthropomorphic heresy and yet the creation of an artificially intelligent creature. These examples suggest that the early phase of AI was characterized by two traditions. The first or Hellenic view saw thinking devices, gods, and so on, as positive contributions that were both useful and attractive. Looking at the Ten Commandments and the Hebraic perspective, we are left with the impression that a second school of thought held such artificial creations to be repugnant, fake, and obviously condemned by Yahweh, the Hebrew god. (Ironically, many of the leading scientists in the twentieth century working in the field of AI have been of Jewish heritage.)

Interest in AI never waned, especially in Europe. Pope Sylvester II (999–1003), prior to becoming head of the Catholic Church, was known as Gerber. The notorious Gerber was said to have constructed a statue that talked, commenting on the truth of statements with "yes" or "no." Gerber's contribution to data processing, however, was quite important because he is credited with introducing the abacus* to Europe, borrowing it from the Arabs in Spain. Some historians claim that this avid student of the occult also introduced Arabic numerals to Western Europe. Whether or not he did, the use of such numerals, along with the use of Arabic mathematics, spread in Europe between A.D. 750 and 1400.

During this same period of the late Middle Ages, the use of complicated time pieces and other instrumentation became fashionable. Time pieces were particularly important because they not only regularized life and increased the idea of measuring events and things but also contributed to the manufacture of automata. These ranged from intricate human-like dolls that moved about when the hour hand struck, such as little figures going around in circles on disks, to birds popping out of clocks announcing the hour.

The Arabic influence on European science and philosophy was profound and is directly linked to the study of intelligence. In the first place, the Arabs carried on the Hellenic traditions concerning intelligence and automata. Second, in their writings, the Arabs argued that there was a distinction between that which was real (that is to say, in nature) and that which was artificial. They made no judgment as to whether one was more important than the other. These scientists experimented and pushed forward beyond the accomplishments of the Greeks. One team of researchers built the *zairja*, which they called a thinking machine. The astrologers who worked on this device assigned numbers to each of the twenty-eight classes of Arabic philosophy, and by manipulating these numbers one could gain new perspectives. Ramón Lull,** a Catalan living in the 1200s, saw the device and designed his own, called the *Ars Magna*. It has been described as a logical machine which its inventor hoped would exhibit reason and arrive at truth. The device reflects the logic of medieval Christian thought because he designed it with categories of wisdom in concentric circles which, when matched, answered important questions in all fields of knowledge. Lull actually built part of it, probably from metal discs. It inspired others to do research, and it continued to receive considerable publicity until the seventeenth and eighteenth centuries when a variety of calculating devices began to be constructed.

Following Lull, a separation between imagination and science took place. More rigorous tests were applied to validate statements and assumptions. Fantasy did not go away; recall the elaborate clocks that dominated many town squares in Europe in the fifteenth century. Wizards in the Middle Ages were identified with such fantasy as having automata working for them. Thus, for example, Paracelsus (1493–1541) claimed to have constructed a small intelligent man. This alchemist received considerable attention at the time. Then in 1580 another automaton, called Joseph Golem, was built by Rabbi Judah ben Loew in Prague. Golem supposedly was to be a spy, checking on the activities of the Gentiles. Golem was reportedly made out of mud and remained mute. He was also reputed to have spied and served as a "man" servant to the rabbi. This fantasy incorporated the use of supernatural power to give it life, special magical knowledge to operate it properly, and was a source of power for its user—themes that persisted in Western European mythology and literature for centuries. Similar traditions existed in Central and Eastern Europe but not in the Middle East. Other Europeans devised fantastic Golems. In the mind of Eleazar of Worms (1160–1230) were born instructions on how to make one. But by the sixteenth century, few people were dreaming about how to construct such devices since they were already making machines.

Mechanical statues, many obviously in huge clocks, became fashionable, the apex being the duck built by Jacques de Vaucanson in 1738. This animal flapped its wings, consumed water and grain, and even defecated. Although Vaucanson built other automata, his duck left the most important impression. He later commented that his objective had been to simulate digestive functions—a motive much like that of many twentieth-century researchers who sought to model another human activity: thought.

Automata, artificial intelligence, and thinking fake men survived longer in literature than in laboratories. By the early 1800s many stories of AI abounded in fiction, and no end was in sight even at the dawn of the Industrial Revolution. E.T.A. Hoffman's *The Sandman* (1815) appeared with Olympia, a female automaton; the ballet *Coppelia* (1870) featured mechanical beings that lived; and the opera *The Tales of Hoffman* (1880) also featured "living" mechanical beings. Mary Shelley, of course, introduced the most famous of them all, the humanoid in *Frankenstein* (1818). This particular automaton is relevant to the history of AI because it reflects much of the moral and philosophical concern and thinking about such devices. Here was science, driven by pure motives, creating something that was evil. Was it correct to attempt to create life? Morally, one could argue either side of the case, but the question has always been debated. Was the creature intelligent? Was it right to create intelligence if it could not be as humane as a person's? What role did spirit and soul play in the development of automata?

The dilemmas posed by *Frankenstein* were real and were not resolved. Yet fantasy continued. In the early 1920s robots came into Western thought again but more specifically as we think of them today—metal men—in Karel Capek's play *R.U.R.* Again the same issues were raised. In 1950 Isaac Asimov, a prolific writer on scientific themes, attempted to resolve the question with his "Three Rules of Robotics" in his book *I, Robot*. They were important because many have taken them to heart in dealing with the issue of building devices that were human-like. He said:

1. A robot may not injure a human being, or through inaction allow a human being to come to harm.
2. A robot must obey the orders given it by human beings except where such orders would conflict with the First Law.
3. A robot must protect its own existence as long as such protection does not conflict with the First or Second Law.

Work on automata continued after these laws were articulated. But earlier, during the nineteenth century, people turned to other types of machines that could handle the dull work of long and complex mathematical calculations. Charles Babbage** worked on his engines during the first half of the 1800s. Others also tried to build calculators to speed up mathematical calculations. The idea of engines helping human beings was consistent with the attitude toward technology in the age of industrialization.

In addition to the traditions of literature and mythology, AI derived a great deal from psychology, the study of how the human mind worked. Lull focused much of his work on an attempt to describe such workings of the mind. Like many who followed him, Lull assumed that the human mind could be replicated in the form of a machine. René Descartes (1596–1650) suggested that people did both rational and mechanical things. The body handled the mechanical and the mind the rational. Probably looking at all the clocks of his day, he concluded that animals were machines as were humans except that they also had a mind. Hence was born the idea of the mind-body distinction (the Cartesian dualism) which became a fixed part of Western thought for several centuries. Gottfried Wilhelm von Leibniz** (1646–1716), himself a builder of calculators, also agreed that mind and body were two separate issues but very well matched as if keeping time together. Leibniz sought to explain thought (reasoning) in a mathematical or algebraic manner. During the 1800s, George Boole** accomplished Leibniz's goal when he developed a method for explaining mathematical logic, giving a great theoretical and practical tool to twentieth-century designers of computers.

Thus, in the early modern period intellectuals were drawn to the idea that the mind could be explained rationally. After all, Sir Isaac Newton (1642–1727) had proven that mechanics could so be explained, leading the people of that age to assume that the same applied to the mind.

Others provided a philosophical base for the study of intelligence. Thomas Hobbes (1588–1679) argued that all human behavior was simply a reflection of internal motions generated by fear and self-interests. He thus introduced the idea of associative behavior. John Locke (1632–1704) said that the human was rational. David Hume (1711–1776) expanded on these ideas in his *A Treatise on Human Nature* (1739) in which he observed that complex ideas were collections of simple thoughts which ebb and flow in the brain. David Hartley (1708–1757) argued that nerves reacted to experiences, thus making it possible for the mind to have sensory experiences that could be accumulated. Julian Offray de la Mettrie, an obscure French doctor, argued in the 1740s that humans were simply machines. The French encyclopedists in general believed that humans were machine-like and thus technology had to be humanized.

The debate on the human mind-machine issue never abated, with all the important philosophers of Europe participating. Immanuel Kant (1724–1804), in *The Critique of Pure Reason*, argued that *a priori* guidelines in the mind caused it to force things outside of it to conform to these principles. Thus, one's view of the world determined the world's nature. Alfred Binet (1857–1911) devised tests to quantify intellectual capability and activity. Max Wertheimer (1880–1943), considered the father of Gestalt psychology, joined with the philosophers in the debate. He believed that perceptions were not elements in themselves but important structures in the mind. For him the mind took information and organized it while defining patterns.

In many ways these philosophers were dealing with concepts from Leibniz who argued that logical issues had to be expressed symbolically. Many twentieth-century historians consider him the father of cybernetics and possibly of artificial intelligence as an area of study. Beginning in the nineteenth century, scientists and philosophers have been struggling with the problem of how to symbolize logic. Boole together with the psychologists aided in this endeavor. Biologists tried to map out the brain and its functions. The debate continues. In 1958 a leading researcher in artificial intelligence, John McCarthy,** suggested that human knowledge had to be described in a formal representation, such as "first-order predicate calculus." He formulated theorems to define symbolic expressions. In the following decade, Alan Robinson introduced the Resolution Method by which a machine-oriented set of methods might be used to prove such theorems.

By 1940 scientists were rapidly formulating a theory of information, drawing less from mythology, more from literature and philosophy, and predominantly from psychology, mathematics, and physics. The effort first came together in

1948 in *Cybernetics,* a book written by Norbert Wiener** (he created the word for the occasion). Wiener dropped Newton's idea of energy, replacing it with the concept of information. He stated that the manner in which information was used to code, store, and so on, offered a better vehicle for explaining the mind's activities, not to mention those of electronic circuits and cellular reproduction. In this and in other writings, he often drew analogies between electronic machines and biological creatures. Joining with other colleagues, Wiener modeled the central nervous system as a means of describing many activities. He envisioned many actions originating in the nervous system, which pushed messages concentrically outward to muscles and organs. Information went back up through the nervous system when organs responded with senses. He was impressed with the close connections among communication, control, and even statistical mechanics, regardless of "whether in the machine or in living tissues." It was then only a small next step to combine the entire fields of communications and control for both machine and animal into one called cybernetics. The word *cybernetics* originates from the Greek, meaning steersman. That book gave birth to the modern age of information theory which subsequently played a profound role in the development of computers. Cybernetics gave focus and meaning to two generations of scientists working in the field of artificial intelligence.

Research in cybernetics in the 1940s and 1950s held out the promise that by modeling the behavior of neurons scientists would soon be able to mimic and thus reproduce human cognition. This idea originated just as digital computers* were becoming available and could be used extensively to simulate the human thinking process. Although it would not be done biologically, the manipulation of information in computers provided a halfway house for many scientists on their road to understanding the human mind. Meanwhile, the mind-machine motif survived into the 1950s and 1960s, joining a long heritage of such images. The difference this time, however, was the greater emphasis on the biological aspect and, more specifically, on nervous systems as transmission centers of information.

Work on intelligent machines as a method of modeling the mind had, of course, begun earlier. Leonardo Torres y Quevedo** (1852–1936) built chess-playing machines at the turn of the century in his attempt to understand cognitive behavior. He recognized, however, that the process was very complicated, and until then machines could only hope to replicate certain mechanistic human thought. But he recognized that the use of sensory data could dramatically alter human thinking. He was frustrated by the mechanical and engineering limitations of his day in Spain and France, limitations that had been overcome by the time the digital computer came into its own some fifty years later.

Others worked on machines to define and reproduce intelligence. Alan M. Turing** (1912–1954), a brilliant British scientist, presented a paper in 1937 in which he showed that certain types of mathematical problems could not be solved with specific processes. "Definite processes" were steps that could be repeated by a machine which he described theoretically. He developed and described this benchmark, called the Turing Machine, when he was only twenty-five years of age. The Turing Machine could theoretically do a large number of mathematical processes when they were expressed in the binary code of zeros and ones (originally, Boole's thought of stating logic, or algebra, as a series of zeros and ones). That is, Turing argued that if all the steps necessary to do something could be stated, they could be "programmed" and thus be executed within a machine. The concept is essential to our understanding of how a computer works. His idea emphatically suggested that algorithms—which are simply the series of steps necessary to solve a problem—could be described specifically within a set of criteria (that of the Turing Machine). If they could not, then they were unsolvable. His ideas profoundly affected the fields of mathematics and computers beginning in the early 1940s, focusing attention on the influence of thinking and cognitive behavior on the design of machines and on the study of the mind. His ideas, appearing as they did a decade before Wiener's book appeared, contributed directly to the concept of a theory of information.

One other important thought from Turing should not be lost: he linked the human being to the machine. Television, cameras, and microphones, for example, already imitated or extended some of humankind's functions in the 1940s. He saw that servomechanistic machines (robots) could replicate a few of the tasks done by a human's arms and legs. Turing went one step further, arguing that a baby's brain was an unorganized device waiting to be filled with experiences, knowledge, and sensations. He observed that if all problems could be phrased "find a number n such that . . . " then one had only to *search* within the brain for what was necessary to solve the intellectual problem—a perfect mode for a computer. Then in 1950 he published an important paper called "Computing Machinery and Intelligence" in which he introduced the scientific community to the Turing Test. Briefly stated, a person in one room, communicating with either another human or machine in another room via teletype, would be able to determine whether the other party was human or machine with a half-hour of conversation. If the interrogating person could not determine for certain whether the other party was a machine or person, then that machine could think. His idea was that, in order to carry out that conversation for such a long period of time, a machine would have to react to a large number of statements from the person—a thinking task. He suggested sample dialogues to illustrate his concept. For many scientists his test became the acid proof of a machine's intelligence for many years. It also provided motivation for the construction of game-playing devices for chess, checkers, and tic-tac-toe, which appeared in the 1950s and 1960s.

Turing's vision may have been limited by the lack of sophisticated development of computers (he died in 1954), but others were already linking the computer to the study of the mind. John von Neumann** (1903–1957), a U.S. scientist, encouraged the construction of digital computers in the 1940s, especially that of the ENIAC* during the late days of World War II. From their experience with ENIAC, the first electronic digital computer, many scientists, including von Neumann, gained an appreciation for the potential of such machines. In 1945 he defined the theoretical architecture of a computer much as we know it today; it had input/output, memory, and a processor. Von Neumann drew a connection between computer components and the nervous system. In describing his computer, the scientist used such words as memory, control, and control organs which came directly from language one would use in talking about the nervous system: associative neurons, sensory and motor neurons. In large part, these were the input/output functions he saw for a computer. However, he did stop short of declaring an ultimate connection between human thought and a computer's performance. In 1951 he stated that the two were in part different because nervous systems exhibited discrete and continuous behavior while computers could only do one or the other. Thus, discrete behavior represented the functions of digital computers and continuous behavior that of analog devices. To this day, computers are digital or analog.

All this activity failed to produce a coherent model of how the human mind worked, and during the 1980s that statement still held. But the machine-mind connection became stronger, and nowhere was this more graphically (or shockingly to some) obvious than when Marvin L. Minsky,** an AI researcher at Massachusetts Institute of Technology (MIT), boldly stated that "the brain happens to be a meat machine." The main lines of research in the 1900s had been done by physiologists and psychologists, and so Minsky's statement was clearly a product of the twentieth century's view of the subject. Yet it was this same set of scientists who contributed to AI largely because of their interest in the analogy with computers while defining the biological realities of the brain.

The first of these twentieth-century workers with a biological or human perspective was Warren McCulloch, a physician who eventually taught at MIT which, by the 1940s, had become a major center for AI research. McCulloch contributed a definition of computing machinery which allowed scientists a more precise construction of the idea of mind as machine than had been possible before. In 1943 he suggested that laws which described how the mind functioned could most profitably be sought by looking at the laws that governed information, not matter. He thus subscribed to Turing's ideas but added the role of neurons in more specific ways. McCulloch believed neurons were the key to understanding the functioning of the brain. Although today his views seem simplistic, in the late 1940s they were encouraging. His greatest impact was to give other scientists the confidence that they would eventually understand the functioning of the mind because he had shown them a way via biology.

During the 1950s biological connections to information theories thus became popular areas of research. Attempts were also made to propose how learning could be duplicated in machine-like environments. W. Ross Ashby, a psychiatrist, made the first major attempt in this area with the publication of his book, *Design for a Brain* (1952). He did considerable research which led him to articulate what would become two fundamental branches of research in AI. The first concerned the creation of intelligent behavior with little regard as to method.

The second involved the modeling of intelligent behavior as it was found naturally, particularly in humans. These two ideas or fields were more separate and distinctive in the 1950s than they would be later when cross-disciplinary efforts closed the gap between them.

In the 1950s Marvin Minsky began researching neurons and biology for answers about the brain, but by the middle of the decade he saw that understanding information processing would be a more fruitful line of study. While a young professor at Dartmouth College, he organized a conference in the summer of 1956 in order to bring together scientific researchers who were studying the functions of the mind. The conference is an important event in the history of AI and computers because it raised the collective consciousness of a group of scientists, giving them an identity with a unique field of study. Minsky developed the concept and term *artificial intelligence* to describe the work of this conference, giving a new field the focus and identity it had not had before. Along with Claude E. Shannon,** another researcher present at the conference, he published the proceedings of the conference, *Automata Studies* (1956)—a seminal work in the field of artificial intelligence.

Others who attended the conference were working in the field from different perspectives. Shannon, for example, had concluded that information theory represented the key to unlocking the mysteries of the brain. As early as 1937 he had articulated a method for using Boolean algebra to study switching systems for engineering. He became interested in chess-playing machines, specifically in how they could handle both numerical and symbolic information. Shannon was one of the first to argue that computers could be more than just large calculators, capable of using symbols to handle complex ideas, for example, symbols to represent words. Today that thought is a common descriptor of all computers, but this was not so in the 1940s. By the early 1950s Shannon had defined most of the major issues which AI researchers would struggle with in the 1960s and 1970s. He brought an engineering perspective to those working from a biological point of view. Although he never fully identified with AI, his ideas showed how to marry the ideas of a computer to those working with the functioning of the brain.

In the 1950s John McCarthy was also interested in what would become known as AI. In his search for formal ways to describe intelligent behavior, his perspective was highly mathematical. He eventually went on to create programming languages,* such as LISP,* in order to help identify the characteristics of intelligent behavior. He also constructed and commented on programmable robots in the 1960s and 1970s.

Thus, by the middle of the 1950s the hunt was on for an information-processing model. One who would contribute was a professor at Carnegie Technological Institute (later Carnegie-Mellon University) named Herbert A. Simon. In the early 1950s, when Simon was still at the RAND Corporation, home of much computer-based research in the 1950s, he and his colleagues worked on person-machine interactions. Assigned to a project for the U.S. Air Force, he simulated an air-defense system that involved person-machine relations. Allen Newell also worked with him and brought a heritage in logic to the project. In 1947 Simon published a book on organizational behavior, *Administrative Behavior*, in which he suggested that a person's conduct within an organization lay somewhere between the cause-and-effect action described by Sigmund Freud and that at the opposite end of the spectrum articulated by economists who saw the individual as "a preposterously omniscient rationality." He said that people made decisions through some reasoning process, with conclusions drawn from premises. In order to understand thinking, the premise as the smallest unit of analysis, and not the decision, should be studied. Simon later suggested that computers could handle the manipulation of symbols that were not expressed as numbers. His experience with the book combined with his later work at RAND led him to link the computer to the science of how the brain operated.

By the mid–1950s, therefore, a large percentage of the scientists working on the actions of the brain had linked their work to developments in computer science. By the same time, and as early as the late 1940s, the image of a computer as a giant brain had also gripped the public. Articles on electronic brains proliferated in the United States and to a lesser extent in Western Europe. The motif of the computer as mechanical intelligence remained intact down to the present. Yet some scientists warned colleagues and the public not to overuse this simplistic analogy (von Neumann, for example); others, such as Turing, Ashby, Christopher Stratchey, and Shannon, set the pace for the analogy by using it to describe how chess-playing machines could learn to play better. They and the public subscribed to the notion that computers could be designed to process information rather than simply do arithmetic. Chess-playing machines once again became fashionable, and contests were held between them and expert players. Moreover, the machines got better at the game as time passed.

One of the most famous was built by J. Cliff Shaw** who had been a programmer at the RAND Corporation in the 1950s and had worked with Newell and Simon on what became known as their Logic Theorist as well as their General Problem Solver. They built chess machines, designed programming languages, and articulated the characteristics of machines and mind in the 1950s and 1960s. They introduced the idea of list processing in programming languages to preserve expensive computer memory. This contribution proved far more valuable than chess-playing equipment. With list processing, a computer could keep a list of available memory, bringing into processing only that portion of a program that it needed next. It also offered a series of techniques for handling data structures. They introduced recursive computing as a means of establishing what subsequent instructions had to be executed, based on the sequences of steps taken earlier in a program. For many, man was beginning to create intelligent behavior in a machine. Thus, the Logic Theorist, developed in 1955–1956, and list processing showed that AI could produce human-like intelligent qualities artificially, though still in a primitive manner.

Yet all was not encouraging. By the early 1960s some were warning that AI and computers were not the end-all in our understanding of the brain's functions. The most dramatic arguments in favor of limiting computers in this arena were advanced by Hubert Dreyfus, a professor of philosophy at the University of California at Berkeley in the 1960s. He began with a paper entitled "Alchemy and Artificial Intelligence" published in 1967 and concluded with a book called *What Computers Can't Do; A Critique of Artificial Reason* (1972). He argued that human and artificial intelligence were not identical. Although he acknowledged that scientists might create some sort of artificial intelligence, that of the human was unique. He accused the AI community of overstating the importance of what they were about to accomplish, labeling them unrealistic. He predicted that AI would never work, happen, or provide the same intelligence as the human mind. He even accused scientists of not having made significant progress since the early to mid–1950s. During the late 1960s and early 1970s leaders in the AI community counterattacked, generating a large bibliography of publications critical of Dreyfus's views.

Although this argument brought public attention to the question of AI, particularly in the United States, it hardly dented the enthusiasm of a community of scientists. From a handful of some two dozen scientists in the late 1950s, over 250 were working on AI by the early 1980s. They were now heavily committed to the use of programming languages, computers for simulations and games, and, by the mid–1980s, to biology again (especially for the study of DNA) to define even more thoroughly information theory and the role of intelligence.

A new and specific twist that provided an opportunity to apply what was known about AI in the 1960s and to broaden understanding came with work done on robotics. The incentives were enormous. First, one could use robots where people could not go, such as in hostile environments with deadly gases, heat, or danger. Second, they could do repetitive tasks faster and more accurately than people. Third, they could be less expensive than humans and easier to manage. Fourth, they offered a fine opportunity to expand our knowledge of intelligence with considerable financial support from both businesses and governments. By the early 1980s the commercial market for robotic devices worldwide was projected to reach $4 to $10 billion by 1990. Furthermore, the concept of a mechanical being was still alive and well. The word *robot* was originally Czech meaning slavery and first came into English when Capek brought *R.U.R.* to the London theater in 1921. It received excellent reviews, and since then plays and movies have kept alive the image of robots. The use of an intelligent robot (one that can be programmed like a computer) attracted the attention of the AI community by the early 1960s.

McCarthy worked on such devices in the 1960s as an outgrowth of his interest in machines that could improve their own performance, such as chess-playing devices. In 1969, after nearly a decade of research, he published a paper outlining what he saw as the central concepts of the modern robot. One of the main points he stressed was that a robot could allow one to test intelligence. If an idea about intelligence could not be duplicated, perhaps in a robot, then that idea was probably incorrect. Thus, robots provided an excellent medium for testing empirically the validity of work done in AI. His work confirmed that the Turing Test (that a computer could fool a person into believing it intelligent) was passable, but he saw more. McCarthy argued that we should not fall into the trap of trying to make humans the benchmark against which to measure intelligence. He and his colleagues feared that a human's definition of people might be faulty and thus retard work on intelligence as exhibited by robots. He considered any robot or device intelligent if it had a model of its environment, could answer questions based on that knowledge, and did tasks requested of it which relied on its preprogrammed goals and the limitations posed by the machine's physical characteristics.

McCarthy noted that, in the construction of intelligent robots, the AI community would have to deal with four questions: (1) how could observations and conclusions based on observations be incorporated into the "brain" of a robot? (2) how could one represent information that was not gathered by the robot from its physical surroundings? (3) what would be the means for obtaining knowledge about a robot's world? and (4) how would such a machine assimilate and express that knowledge internally? These items remain on the agenda of researchers today.

Some progress on the resolution of these four issues has taken place. Beginning in the 1960s in Europe and in the United States, scientists built robots. Considerable work at Stanford University and at MIT, for example, led to devices that could see and had arms that could move. Televisions served as eyes and the computer as brains instructing hands, arms, and legs. Sensory devices in "hands" provided additional input to a computer within a robot to influence subsequent actions. One of the most famous of the early robots was Shakey at the Stanford Research Institute, where it was used to study vision data input in the 1970s. The first version of Shakey, which appeared in 1969, could see and do minimal problem resolution. The second had more memory and thus could be programmed to conduct additional analysis on what it saw. It was mobile and thus received considerable press coverage. By 1980 many industrial robotic devices had been made that were programmable. Most did not look anthropomorphic, that is, human-like with head, arms, and legs. Many appeared as boxes or pillars with an arm that performed some function, such as testing keys on a typewriter. In the early 1980s commercial models were available in many industries. Most major automobile manufacturing companies led the way with the use of thousands of such devices to paint vehicles, weld, and move parts.

There was one other area of research in artificial intelligence that had less to do with computers but that promised to influence their design in the years to come: language. Although scientists believed other animals, such as the porpoise, had language-like capabilities, language was seen as an exclusively human skill. The belief had long existed that, if one could communicate with a computer in a language natural to human beings, then that computer's intelligence could be established and use of it improved. Much work done on programming languages beginning in the 1960s involved the use of more natural-like programming languages intelligible to computers. In the 1980s research on the recognition of voice commands in human languages presented yet another potential step toward better person-machine communications.

The ability to program and to make computers understand languages offered another avenue of appreciation of how languages worked in the mind. Translation from one language to another via computer was an early dream (and still remains a dream) because the translations were too literal. Computer scientists and linguists quickly realized that a larger body of knowledge—the context of a piece to be translated—was necessary to the process. Most work on translators took place during the 1950s and 1960s. In AI, the study of linguistics led to projects concerning the manipulation of symbols and understanding (comprehension) by computers. Minsky's students, for example, did such work throughout the 1970s and into the 1980s. Joseph Weizenbaum** developed ELIZA at MIT in the 1960s to help. ELIZA was a question and answer machine that could simulate languages; more specifically, it was a system by which a psychoanalyst and a patient could talk.

By 1980 these avenues of research had produced a number of results. First, robotics as a practical method for studying intelligence was established as viable for the 1980s. Second, knowledge became a dynamic, nonstatic element in the mind proven by earlier experimentation. Symbols were created which could lead to information; they could also be easily reorganized, added, replaced, or deleted with the use of computers. This circumstance made any conception of the mind that much more complicated. Yet the idea of information theory as a vehicle useful for the study of intelligence had been proven viable. Moreover, observations about information management were influencing the design criteria of computers in the 1980s. Even so, the representation of knowledge remained one of the most controversial topics under discussion within the AI community.

Another concern of the 1970s and 1980s was with the issue of man-machine intelligence in moral terms, a problem that had plagued others for centuries. Even in the second half of the twentieth century, when computers had become an indelible part of Western civilization, the issues posed by philosophers and clerics about purpose and intent remained. Scientists also participated in the discussion. In 1976, for example, Weizenbaum published a book, *Computer Power and Human Reason*, in which he attacked his old field of artificial intelligence. He argued that AI should stay away from certain areas, such as where life itself might be attacked, where effects become irreversible or unpredictable, or where computers might replace humans in tasks requiring respect or love for people. Although Weizenbaum attacked a great deal of research done in AI as more mechanistic and less scientific, his moral concerns proved more important. He argued that people had adopted the metaphor of machine to describe themselves, suggesting a perspective about the human race that diverged sharply from the philosophical or religious vantage points of the past. He feared that in the process humankind would lose its humanity, its reliance on love, trust, and dignity, transferring them instead to machines which in turn might endanger our human qualities. His book was denigrated by the entire AI community, which understandably believed its own work was relevant, and his ideas remained minority views in the 1970s and 1980s despite the sales of the book.

Artificial intelligence is an ancestor to the modern computer, almost as important as mathematics, electronics, and engineering. No history of computing in the twentieth century would be complete without an appreciation for AI, which gave the data processing industry theories on information processing and numerous design criteria for the functions of computers and programming languages. Not the least of these was the idea that computers could do more than simple arithmetic. And as throughout history, AI held out the promise that humankind might yet understand the workings of the mind.

For further information, see: W. R. Ashby, *Design for a Brain* (New York: John Wiley and Sons, 1952); I. Asimov, *I, Robot* (New York: Gnome Press, 1950); C. Block, *The Golem: Legends of the Ghetto of Prague* (New York: Rudolf Steiner Press, 1925); A. Chapuis and E. Droz, *Les Automates* (Neuchatel: Editions du Griffon, 1949); H. Dreyfus, *What Computers Can't Do; A Critique of Artificial Reason* (New York: Harper and Row, 1972); Pamela McCorduck, *Machines Who Think: A Personal Inquiry into the History and Prospects of Artificial Intelligence* (New York: W. H. Freeman and Co., 1979); W. McCulloch, *Embodiments of Mind* (Cambridge, Mass.: MIT Press, 1965); Marvin Minsky, *Semantic Information Processing* (Cambridge, Mass.: MIT Press, 1968); David Ritchie, *The Binary Brain: Artificial Intelligence in the Age of Electronics* (Boston: Little, Brown and Co., 1984); Carl Sagan, *The Dragons of Eden* (New York: Random House, 1977); Claude Shannon and John McCarthy, eds., *Automata Studies. Annals of Mathematical Studies* 34 (Princeton, N.J.: Princeton University Press, 1956); Lytton Strachey, *Literary Essays* (London: Chatto and Windus, 1948); M. Taube, *Computers and Common Sense: The Myth of Thinking Machines* (New York: Columbia University Press, 1961); S. Turing, *Alan M. Turing* (Cambridge: W. Heffer and Sons, 1959); J. Weizenbaum, *Computer Power and Human Reason* (San Francisco: W. H. Freeman and Co.); N. Wiener, *Cybernetics* (Cambridge, Mass.: MIT Press, 1948; 2d ed., 1961).

ASCC. See HARVARD MARK I, II, III, IV

ASTROLABE. This device performed various calculations and observations related to astronomy. It was one of the earliest aids to calculation which emerged from the Middle East or Western Europe and was widely used by the Greeks and later in Northern Europe.

Although ancient documents claimed that the astrolabe could do over a thousand types of transactions, its early history remains shrouded in mystery and myth. The earliest documentary evidence of its existence, which comes from the Arab Empire and the school the Greeks ran at Alexandria, suggests that it was probably used to study astronomy. The earliest surviving copy of the instrument dates from 1062 A.D., yet Hipparchus of Bithynia (180–125 B.C.) was familiar with it. In the first century A.D. Ptolemy described how it worked, and, later, Theon of Alexandria (ca. 430 A.D.) and John Philoponus (ca. 625 A.D.) also wrote about the instrument, showing that it was used actively at the dawn of the Christian period.

An astrolabe looked like a number of plates made out of metal and was engraved with a spider web pattern, fitting into an outer shell called "the rete." This had a circle complete with a sequence of sharp points. The web pattern represented a stereographic projection of the heavens as it might seem from a particular latitude originating on the earth's surface. It included a map of the earth, while the larger circle on the rete was a zodiac representing the position of the sun in the sky on various days of the year. The sharp points indicated the location of important stars. The user rotated the rete over the engraved plates which were all connected together to determine the sun's position or that of stars for any hour of any day. It was used to establish the latitude of the user and to calculate the amount of daylight in hours for any day. Also known as a planetary equatorium, other applications included calculation of the hours of twilight at the start and end of a day and, most basic, establishment of what time it was either day or night.

This device permitted the casting of a horoscope. In addition, workers laboring in fields could be told how many hours to work or, conversely, it could determine how many workers would be needed to get a particular job done within a specified period of daylight. The study of the stars had long encouraged inventors to come up with aids, a source of inspiration still evident among computer scientists in the 1940s and again in the U.S. space program of the 1960s and 1970s. It remains unclear when people stopped using the device. It was still employed in Europe until nearly modern times and perhaps into the twentieth century in remote parts of the Middle East.

For further information, see: C. F. Jenkins, *The Astrolabe* (Oxford: Oxford University Press, 1925); J. D. North, "The Astrolabe," *Scientific American* (January 1974): 96–106.

ATLAS. This large British computer was constructed at the University of Manchester, home of much research on computer science in the 1940s and 1950s. It represented the third in a series of devices assembled by the Department of Electrical Engineering at Manchester in conjunction with Ferranti Ltd.† The first two had been the Ferranti Mark I and II. Initial work on ATLAS began in 1956. The first copy of the computer was installed at Manchester in 1962, a second ATLAS 1 at the University of London in 1963, and a third at the ATLAS Laboratory at Chilton during the same year. The three functioned until the early 1970s. The Chilton device was the last to be shut down (March 1973).

The development of ATLAS contributed to the design of integrated systems because it took advantage of recent enhancements in components and used an operating system* that was far advanced for its day. Perhaps before all other computers, ATLAS employed multiprogramming and one-level store, and introduced the concept of paging—all of which became common features of computers by the end of the 1960s. Its composite memory* of ferrite cores and magnetic drums operated as a one-level store managed by software. Paging contributed to this success because it swapped pages of data from memory in an organized fashion through the use of software. Paging did not become a common part of other computer systems until the late 1960s, making it possible quickly to have very large memory systems that were three, four, and more times larger than earlier capacities.

Heading the ATLAS project was Tom Kilburn who had worked with Frederick C. Williams** in the 1940s in developing the popular Williams tube memory system. Like the designers of the LARC* and STRETCH,* Kilburn sought to build a supercomputer beginning around 1956. No computer could realistically qualify as "super" unless its memory size was enormous. To avoid buildings full of components to achieve that goal, Kilburn's system had a real memory of only 16,000 words (still a good size for the period), but he then added 96,000 more words on a drum. Data from the drum would be paged into real memory when needed, thus making it appear that the computer's real memory was well over 100,000 words—huge for the late 1950s. Each page was made up of 512 words; all data in storage were associated with a page broken into groups of words never exceeding 512. An address register kept track of where all data were, so that the computer could tell which page to bring from the drum into core storage.

ATLAS also did time-sharing and hence could have various jobs going on at the same time, with numerous pages coming in and out of the computer associated with multiple jobs then running. Most remarkable of all was that it worked efficiently. It proved so effective that when job mixes were well balanced, the processor's utilization ranged as high as 60 to 80 percent. Machines that were advanced and widely used as late as five years in the future, such as the IBM S/360* Model 67, had difficulty reaching even 10 percent utilization. Advances in operating systems quickly solved that problem by the early 1970s, allowing computers to reach and exceed the levels originally set by ATLAS. Thus, the success of ATLAS, and the publicity it received within the computing community, encouraged computer designers to adopt many of its technical features in systems that appeared in the late 1960s and early 1970s.

A portion of memory, made up of 8,000 words of ferrite rods, housed an operating system to manage the computer. This "hairbrush" memory (so called because of the shape of its storage) sheltered the operating system in such a way that it was available on demand through software. It contained arithmetic routines, input/output control instructions, memory management code, and other features common to operating systems. The entire package was very fast, with operating system code living in the hairbrush memory functioning with a 0.4 microsecond cycle time. When combined with the overlapping of instruction times in all the units making up the ATLAS system, one had a very fast processor that spit out answers rapidly. It had a speedy arithmetic unit, arithmetic circuits to do calculations for addresses in the index registers, and all in an organized manner. As a result of these innovations, a floating-point addition occurred between 1.4 and 2 microseconds, multiplication in 4.7 microseconds, and division in 13.6 microseconds. Overall, when normal housekeeping transactions were also factored in, an instruction typically took 2 microseconds to execute. Some of the other statistical data on ATLAS included: 128 index registers, of which an application programmer could use 90, magnetic tape with one-inch tapes (later, magnetic disk storage was added to the second and third machines), paper tape and paper punch peripherals, line printers, and multiple channels from the computer to all peripheral gear.

The three original machines were never made into commercial versions because demand for such devices proved insufficient to justify the effort. Yet the three operated for nearly a decade performing calculations for scientists. Chilton's machine serviced a number of British research organizations and universities almost like a service bureau. A smaller device, also an ATLAS-class machine, called the TITAN or ATLAS 2, was built without magnetic drums but with a large core memory and thus did not employ paging. The TITAN was installed at Cambridge University where it operated until 1973. Two machines were eventually installed in the same community and a third at the Atomic Weapons Research Establishment located at Aldermaston.

ATLAS was a technical success that influenced the design of all large computers throughout the 1960s. Yet like other supercomputers of the 1950s and 1960s, such as the LARC and the STRETCH, it had its limits. It was best suited for large computational projects, which usually meant the military community or scientists at universities. The problem of who would buy a super large computer remains an ongoing one to the present time, even though computers grew in size and capacity over the decades. The problem with most supercomputers, and the issue that usually dampened their attractiveness to computer vendors, was their initial development costs and consequently the prices that would have to be charged possible customers. The cost of these machines could not readily be justified. As a consequence, they remained the purview of government agencies and a very few universities.

For further information, see: R. A. Buckingham, ''ATLAS,'' in Anthony Ralston and Edwin D. Reilly, Jr., *Encyclopedia of Computer Science and Engineering* (New York: Van Nostrand Reinhold, 1983): 142; Simon Lavington, *Early British Computers* (Bedford, Mass.; Digital Press, 1980).

B

BACAIC. This programming language for numerical scientific computing was also formally known as the Boeing Airplane Company Algebraic Interpretive Coding System (BACAIC). It was designed to run on an IBM 701 and became operational by 1955. Like many other programming languages* of the 1950s, BACAIC supported the computational requirements of scientists and engineers. It successfully allowed users to write mathematical expressions in notations commonly used by mathematicians. The one important exception to this description was that it could support constants. In addition, there were some other minor changes in syntax from what a mathematician might already know.

Little else is known about the language. That it came from Boeing was no surprise since that company employed many engineers. By the late 1940s airplane manufacturing companies were routinely relying on computers to manage the design of new products. Almost from the dawn of modern data processing software* tools were developed within the aircraft industry, a pattern that persists to the present. Almost all of the major CAD/CAM software packages within data processing came from that industry as logical outgrowths of the kind of scientific computing that packages like BACAIC supported.

For further information, see: M. Grems and R. E. Porter, *A Digest of the Boeing Airplane Company Algebraic Interpretive Coding System* (Seattle, Wash.: Boeing Airplane Co., July 1955).

BACKUS-NAUR FORM. Also known as the Backus-Normal form or BNF, this was the most important statement in defining formalized notation yet developed. It was produced in 1959 by John Backus** and Peter Naur,** with Backus generally considered the major author. Backus, the father of FORTRAN,* like other computer scientists of the 1950s, was concerned about standardizing functions of programming languages.* At this time a movement began to formalize syntax for such languages, much as they existed for human idioms. In order to describe any language, another has to be used (called a metalanguage), and the BNF was an early and important contribution to this effort. It was a statement describing IAL which, soon after, became known as ALGOL,* a proposed universal language which was eventually implemented. Backus described the elements essential to any programming language using symbols, connectives, and well-defined variables to help make up a language.

Backus presented these concepts in a paper delivered at the UNESCO International Conference on Information Processing in Paris during June 1959. In a discussion of the strengths and weaknesses of ALGOL, he offered a specific method for defining the syntax of a language and applied it to ALGOL. This paper opened a new era in the development of programming languages, marking the adoption of more rigorous methods in designing programming languages.

For further information, see: John Backus, "The Syntax and Semantics of the Proposed International Algebraic Language of the Zurich ACM-GAMM Conference," *Proceedings of the First International Conference on Information Processing, UNESCO, Paris, 1959* (London: Butterworths, 1960): 125–132; Jean E. Sammet, *Programming Languages: History and Fundamentals* (Englewood Cliffs, N.J.: Prentice-Hall, 1969).

BALDWIN-ODHNER MACHINES. Also known widely as the Baldwin Calculator, these devices represented a significant improvement over the Leibniz type of stepped-drum design. The Baldwin Calculator was a desktop machine that was lighter and smaller than earlier models and became available in the late nineteenth century. Like others of the period, it ushered in a new era in the use of calculators to aid mathematical calculations.

The difference in technology was important. In a stepped-drum gear, the result register could turn from 0 to 9 depending on how the setup mechanism that held the number wanted by the user was manipulated. The Leibniz approach involved gears, each of which had various numbers of teeth; the number of teeth depended on the position of the next gear along the drum being driven off the same shaft. Instead, the need was for a variable-toothed gear in which the number of teeth changed, thereby reducing the total number of gears needed. Such an improvement would help reduce the size of the entire calculator. The new machines solved this problem. Attempts at a redesign during the eighteenth and early nineteenth centuries had failed primarily because of poor machining techniques and, to some extent, lack of insight. The problem was solved simultaneously by Frank S. Baldwin in the United States and Willgodt T. Odhner** in Russia by the mid–1870s. Both machines operated essentially the same way and won quick adoption. The devices were usually called the Baldwin Machines in the United States and the Odhner Machines in Europe. Historians invented their own name: Baldwin-Odhner Machines, or simply the Baldwin Calculator.

Both the American and Russian versions involved a round disk with movable pins that stuck out from the edge of the flat dish. A lever attached to the middle of the disk controlled how many teeth protruded. Input depended on these levers.

Thus, in order to "input" the number 43 on a calculator, a lever on the first disk would be used to push out four pins and another disk protruded three more. The three would physically be to the right of the four. Addition would involve rotating these disks which shared the same shaft, and the answer appeared in the number of pins sticking out of the disks. The disks, being light and thin, took up little space, making the machines small and, of course, easy to use. Size was an important issue. Instead of taking up several square feet on a table, the new machines occupied the corner of many late nineteenth-century desks.

One of the most popular of the early devices came from the Brunsviga Company in 1885 and looked like an old-fashioned cash register. Between 1885 and 1912 this company manufactured more than 20,000 such machines. Brunsviga was only one of many companies selling them. By the early twentieth century, there may have been over 200,000 in existence.

These four-function calculators were used in accounting offices in both Europe and the United States. Large companies, such as railroads, kept track of tickets, salaries, rolling stock, and cash flow while insurance companies made extensive use of them to calculate actuarial tables along with normal accounting. Students in science laboratories in universities became another large group of users. As late as World War II, similar, though improved, versions were employed to prepare ballistics tables in the United States. Because of the large number employed, engineers were motivated to look for faster ways to computerize ever increasing amounts of data. The result of their search was the electronic digital computer* such as the ENIAC.*

For further information, see: J. A. V. Turck, *Origin of Modern Calculating Machines* (Chicago: Western Society of Engineers, 1921); Michael R. Williams, *A History of Computing Technology* (Englewood Cliffs, N.J.: Prentice-Hall, 1985).

BASIC. Beginner's All Purpose Symbolic Instruction Code (BASIC) became the most widely used language by owners of microcomputers such as those produced by the International Business Machines Corporation (IBM)† (PC), Altair, and Apple† during the late 1970s and the 1980s. It became the language of choice for the nonprofessionals interested in data processing, often using it with machines in their homes or as productivity aids at work. BASIC became one of the most widely used programming languages* in American schools that employed computers. In short, the language most frequently exposed individuals with little or no knowledge of data processing to computer science, primarily in the United States.

BASIC was first developed at Dartmouth College in the early 1960s as a teaching aid by John G. Kemeny and Thomas Kurtz who wanted an easier language than FORTRAN.* It has to be simple both to learn and to operate, particularly for those not interested in science. This effort was only one of many of Dartmouth's accomplishments in the field of data processing. It has sponsored various projects associated with computer science for years. Two of the more dramatic events in the history of modern data processing, for example, occurred at Dartmouth: (1) in 1940 George R. Stibitz** of Bell Laboratories† presented the first demonstration of the use of telecommunications between a terminal and a computer over telephone lines at Dartmouth; and (2) in 1956 John McCarthy** conducted a summer project on artificial intelligence,* which marked the first use of that term. That later conference also led to the establishment of a project that resulted in the programming language called LISP.*

With 75 percent of the student body majoring in the humanities and social sciences in 1961, exposing them to computer science by way of FORTRAN or ALGOL* was too much of a challenge. Kurtz, then a young professor and Dartmouth's first director of the Computation Center, and Kemeny, chairman of the Department of Mathematics, decided to develop their own system incorporating time-sharing, ease of use of equipment, and a new language that was simple to learn and use. Work began on the system in 1962 and the following year on what would become the first release of BASIC. In 1964 General Electric (GE)† installed a GE–225 computer to help in the project, and the first program in BASIC ran on May 1, 1964.

Refinements came quickly. A language was created in which the user never had to be concerned with object code, and access to it was always via a terminal, twenty of which existed on campus by the fall of 1964. Free access to terminals is a common sight on today's campus, but Dartmouth did it nearly twenty years earlier. The design of the language incorporated components from FORTRAN and ALGOL. During the 1960s students and professors at Dartmouth succeeded in refining the language. Years later it would run on Hewlett-Packard† equipment along with General Electric's.† In the 1970s the language ran on minicomputers made by Digital Equipment Corporation (DEC),† Data General Corporation,† and Wang. Just five months after the first edition of BASIC appeared, Dartmouth released an improved second version. By the time the third version appeared in 1966, the language had acquired many of the characteristics it has today. Yet other releases, produced jointly by GE and Dartmouth, appeared: the fourth edition in 1966 and the fifth in 1970. Dartmouth's final edition, the sixth in 1971, proved to be its cleanest and most fully documented version.

Copies of BASIC spread rapidly almost from the beginning. One reason why it was widely accepted was that GE used it on its Mark I time-sharing system. BASIC, in various forms, appeared at colleges and universities across the United States. Because of its grass-roots acceptance and modifications, by 1973 many variations of the language existed and there was no common set of standards such as was the case, for example, with COBOL.* Respectability for the language had come early in its history, thereby encouraging its wide use in educational systems. As early as 1966 the National Council of Teachers of Mathematics, contemplating whether to support BASIC or FORTRAN as the standard to use in education, elected to endorse BASIC. By then it was already evident that BASIC was a superior teaching aid when compared to FORTRAN. Years later, when microcomputers first appeared and most frequently at schools, it was only logical that BASIC would be modified to run on such machines.

Students have frequently been responsible for pushing back the technical frontiers in microcomputing, and the case of BASIC was no exception. For example, two students at Harvard, William Gates and Paul Allen, developed a version of BASIC that ran on an Altair microcomputer. That effort led to the establishment of the Microsoft Corporation,† one of the largest and most successful microcomputer software* manufacturing firms in the data processing industry. Microsoft was established in the 1970s while others were working on BASIC as well. Another gifted individual interested in microcomputing, Gordon Eubanks, also worked with BASIC in the late 1970s while in the U.S. Navy. His version, called BASIC-E, was different from Microsoft's. Gates's edition was an interpreted language, which meant that all instructions were directly translated into machine code. Eubanks's version was translated into what is called intermediate code, which in turn was converted into machine-readable code.

The advantage of Eubanks's version for purposes of selling compilers was that intermediate code could not be modified and its ideas could not be stolen by a user, even though the language itself could be used. Intermediate code could not be fully understood because it could not be read by humans. On the other hand, the full architecture and design of Microsoft's early versions were readable and thus subject to possible theft. Although Eubanks never copyrighted the first version of his code, a second, called CBASIC, was copyrighted and later marketed by his firm, Compiler Systems. By the early 1980s a number of software vendors were selling versions of BASIC, some of them licensed from the companies associated with Gates and Eubanks and others developed independently. By the time IBM's PC was beginning to make its impact on data processing (1982–1983), most micro vendors offered BASIC as a product. Peachtree Software, for example, sold versions for a variety of microcomputers, while other firms sold application programs written in BASIC. Table 2 illustrates a sample of BASIC.

In looking back on the explosive growth in the use of BASIC in 1977, Kurtz thought that even then over 5 million students had been exposed to the language. Yet the great period of growth was to come after 1977. Even at that time, he argued that it was readily accepted because "there are more people in the world than there are programmers." They did not understand data processing's technology but needed access to the power of computers; BASIC offered that capability. It eliminated the use of much technical jargon and many issues; it provided a large number of defaults so that users had to make fewer decisions when programming; and it operated rapidly in small systems. Commands were English-like, as were the language's descriptions of errors.

If one piece of software had to be chosen as the impetus for the growth of the microcomputer, it would be BASIC along with a host of application packages, some of which were written in this language. As of 1986, millions of copies of the language had been sold, and usage was no longer limited to children. It had

Table 2
Sample Program: BASIC

Problem: Do a curve fitting using the Least Squares Method.
Program:

```
1 REM CURVE FITTING PROGRAM
2 DEFINT I-L
3 DEFDBL A-H,M-Z
4 MX=99
5 EF=999
6 MD=9
7 DIM X(MX),Y(MX)
8 Q=M+1:DIM A(Q,Q),R(Q),V(Q)
9 Q=MD*2:DIM P(Q)
10 CLS:PRINT"   - LEAST SQUARES CURVE FITTING -":PRINT
11 PRINT"ENTER A DATA PAIR IN RESPONSE TO EACH"
12 PRINT"QUESTION MARK. EACH PAIR IS AN X VALUE"
13 PRINT"AND A Y VALUE SEPARATED BY A COMA.":PRINT
14 PRINT"AFTER ALL DATA IS ENTERED, TYPE"
15 PRINT EF;",";EF
16 PRINT"IN RESPONSE TO THE LAST QUESTION MARK.":PRINT
17 PRINT"THE PROGRAM IS CURRENTLY SET TO ACCEPT"
18 PRINT"A MAXIMUM OF";MX;"DATA PAIRS."
19 PRINT:J=0
20 J=J+1:INPUT"X,Y=";X(J),Y(J)
21 IF X(J)=EF AND Y(J)=EF THEN J=J-1:GOTO 24
22 IF J=MX THEN PRINT "** NO MORE DATA ALLOWED **":GOTO 24
23 GOTO 20
24 NP=J:PRINT
25 IF NP=0 THEN PRINT"** FATAL ERROR! ** -- NO DATA ENTERED":STOP
26 PRINT NP;"DATA PAIRS ENTERED"
27 PRINT:INPUT"DEGREE OF POLYNOMIAL TO BE FITTED";D:PRINT
28 IF D<0 THEN PRINT"** ERROR! ** -- DEGREE MUST BE>= 0":GOTO 27
29 D=INT(D):IF D<NP THEN 31
30 PRINT"** ERROR! ** -- NOT ENOUGH DATA":GOTO 27
31 D2=2*D
32 IF D>MD THEN PRINT"** ERROR! ** -- DEGREE TOO HIGH":GOTO 27
33 N=D+1
34 FOR J=1 TO D2:P(J)=0:FOR K=1 TO NP
35 P(J)=P(J)+X(K)^J:NEXT J:P(0)=NP
36 R(1)=0:FOR J=1 TO NP:R(1)+Y(J)
37 NEXT J:IF N=1 THEN 40
38 FOR J=2 TO N:R(J)=0:FOR K=1 TO NP
39 R(J)=R(J)+Y(K)*X(K) (J-1):NEXT K:NEXT J
40 FOR J=1 TO N:FOR K=1 TO N:A(J.K)=P(J+K-2):NEXT K:NEXT J
```

———
———
———

———

```
84 RETURN
85 IF L=K THEN 76
```

become a useful tool for those doing spreadsheet analysis, simple mathematics, and simulations in business. At home, BASIC was being used to develop budgetary and accounting systems, along with address files and other forms of recordkeeping on Apples, Altairs, PCs, and other micros from some 150 vendors.

In addition, such manufacturers as IBM, HP, and DEC made BASIC available in their minicomputers. Large computer vendors like IBM also offered compilers for use on large mainframes such as the 4300s and 308Xs. The versions were similar and ran in online systems. By 1986 BASIC was more widely used than any other language in general with the possible exception of COBOL. It clearly was the language of choice of the nonprofessional programmers.

. *For further information, see:* Paul Freiberger and Michael Swaine, *Fire in the Valley: The Making of the Personal Computer* (Berkeley, Calif.: Osborne/McGraw-Hill, 1984); J. G. Kemeny and T. E. Kurtz, *BASIC Instruction Manual* (Hanover, N.H.: Dartmouth College, 1964); Thomas E. Kurtz, "BASIC," in Richard L. Wexelblat, ed., *History of Programming Languages* (New York: Academic Press, 1981): 515–546; D. G. Mather and S. V. F. Waite, eds., *BASIC*, 6th ed. (Hanover, N.H.: University Press of New England, 1971); Jean E. Sammet, *Programming Languages: History and Fundamentals* (Englewood Cliffs, N.J.: Prentice-Hall, 1969).

BESM1. This Soviet computer was built in 1953 by S. A. Lebedev. This early Russian machine was constructed using Williams tube technology for memory,* and it operated at between 7,000 and 8,000 operations per second. The system employed paper tape* which read at the slow pace of twenty characters per second. Output emerged at a print speed of 1,200 lines per minute. Each line contained only ten figures, a small number when compared to an average of 120 characters per line found in Western Europe and in the United States in the same era. All data were numeric only. The computer used a word length of 39 bits, and its Williams tube memory was 1,024 characters in size, although in addition it had a read-only-memory of 376 positions. Access time for this memory was 6 microseconds, which made it one of the fastest of the early Soviet computers. A secondary memory drum was added which provided an additional 5,120 characters of storage—also very large by Soviet standards in the 1950s. The device employed 4,000 tubes as well.

BESM1 was built by the same engineer who had constructed MESM, one of the first electronic general-purpose computers in the Soviet Union. Technically, it was a calculator, and it was constructed in Kiev in 1947–1950. Most other Soviet machines of the 1950s emerged out of that design. The BESM1 was an obvious step forward from MESM, as was BESM2 which appeared in 1960. The most significant change in the second BESM machine was its memory; it now used ferrite-core memory of 1,023 words. Access time for this storage was 6 microseconds.

The names of these computers derived from Russian expressions. BESM are initials for the Russian phrase High-Speed Electronic Computing Machine, whereas MESM stands for Small Electronic Computing Machine.

For further information, see: Andrei P. Ershov and Mikhail R. Shuri-Bura, "The Early Development of Programming Languages in the USSR," in N. Metropolis et al., eds., *A History of Computing in the Twentieth Century* (New York: Academic Press, 1980): 137–196.

BINAC. BINAC was perhaps the first of the electronic digital computers* built in the United States to incorporate fully the functions of the stored program. Its construction began in 1947 and ended in August 1949 with its acceptance tests. The BINAC was a direct descendant of the ENIAC* and the EDVAC* and the predecessor of the first widely successful commercial computer, the UNIVAC.* It was constructed by John Presper Eckert, Jr.** and John W. Mauchly,** two of the most active computer builders of the early days of data processing. The BINAC was important technically because of its capability to store programs in a computer. Thus, users could dramatically simplify the use of such devices while increasing the speed and quantity of work assigned to them.

Although not the only computer under construction in the late 1940s, the BINAC was one of the earliest to be constructed by a company for a commercial customer. Eckert and Mauchly had formed a company bearing their names soon after World War II (1947) to build computers and were under contract with the U.S. government to do that. Government funding, however, proved insufficient to pay for the development of the federal computer. As a stop-gap measure to raise additional dollars, Eckert and Mauchly agreed to build a small machine for the Northrop† Aircraft Company named the Binary Automatic Computer, or simply BINAC. The company intended this machine to guide in-flight navigation of a missile code-named "Snark" for the U.S. Air Force. Under the terms of the contract for its construction, the computer would cost $100,000, with $80,000 to be paid up front to allow Eckert and Mauchly to work on their other computer project as well. This contract also made possible the design of a computer that could serve as the prototype for the UNIVAC, a general-purpose computer completed after the BINAC.

The BINAC used 31 binary digit words, 512 of which could be housed in the computer. Mercury delay lines provided storage. The machine had two processors, each of which was made up of 700 vacuum tubes. This number of tubes represented a major leap forward in miniaturization (to borrow a term used to describe what happened to electronics and computers nearly two decades later) and improved reliability over the ENIAC which used 18,000 tubes and took up more space and malfunctioned more frequently as a result of so many tubes failing. The improvements had come quickly. ENIAC's development took place in the early to mid-1940s and BINAC's hardly five years later. Already technology had progressed rapidly, as it was to do throughout the modern history of computing devices. The BINAC performed 3,500 additions and/or subtractions per second, or 1,000 divisions and multiplications in the same time. The binary numbering system was standard, although all output was in octal notation. The BINAC could not handle alpha characters; that function came with the UNIVAC

Table 3
Physical Characteristics of the BINAC

Architecture

Programming	Stored program
Data transmission	Serial
Number representation	Binary
Word length	31 bits
Other data types	None
Instruction length	14 bits
Instruction format	1-address
Instruction set size	25(32)[a]
Accumulators/programmable registers	2
Main memory size	512 words
Main memory type	Delay line
Secondary memory	Magnetic tape
Other I/O devices	Typewriter console
Error detection	Redundant CPUs

Performance

Clock rate	4 MHz (later 2.5 MHz)
Add time	0.285 ms
Multiply time	0.654 ms
Divide time	0.633 ms

Physical Characteristics (Approximate)

Vacuum tubes	1,400
Diode count	n/a
Power consumption	13 kW
Floor space (CPU only)	n/a
Two mercury tank memories, each	2'6" x 3' x 3'6"
Two power supplies, each	1'6" x 1'6" x 3'6"
One input console	3' x 2' x 3'

[a]Number of instructions used (number encoded).

and, as a direct consequence, the birth of commercial applications that required the use of words (e.g., inventory control, billing, and payroll).

Because Northrop originally intended to install this machine in an airplane for missile guidance, every attempt was made to keep the size of the device small. The use of fewer tubes obviously helped, as did the construction of small frames and other components. The end result was a much smaller machine than had previously been built. The two processors were each 5 feet by 4 feet by 1 foot. Other components consisted of two mercury tank memories, two power supply units, one input converter unit, and a console. Each of these units measured less than 3 feet by 3 feet by 4 feet (Table 3).

The machine, of course, was a stored-program computer which was probably the first built in the United States but not the only one in construction in the Western World. It was capable of solving a wide variety of mathematical problems as a general-purpose computer, despite the fact that Northrop wanted it for a highly specialized application and all input/output was numerical. Its operation included normal arithmetic functions, transferrence of data, and logical controls, many of which were typical of those available on other contemporary machines. Its reliability was dramatically enhanced in its design by the use of two processors which could monitor each other's transactions. In many ways, that approach presaged today's methodology of increasing reliability through redundancy of parts and processors, alternate data paths, and even the use of diadic systems all under the cover of one computer. In its simplest terms, the strategy used with the BINAC is the same in use today. Storage was also duplicated for the same reason. Thus, each machine would perform calculations, check the results with each other, and, if they concurred, would move on to the next set of calculations, if not, stop.

The console was a typewriter keyboard that became the method of providing to (talking with) the BINAC. Encoded magnetic tape could also be used.

Output came in the same two forms. Use of tape was an important step forward in the design of computers because with the BINAC came recognition that tape medium represented a practical and fast method of providing input/output capabilities for a computer. Although BINAC's tape system was awkward and unreliable, its use inspired Eckert and Mauchly to improve on the design of such a medium, going to metal-based tape for use in the UNIVAC later. After UNIVAC, tape drives became a standard part of a mainframe's configuration for input and output down to the present.

Construction of the BINAC progressed slower than specified in the original contract; it was completed in August 1949 instead of in May 1948. In its first demonstration, the BINAC solved Poisson equations. It prepared twenty-six solutions within three hours (time that included some hardware failures). This very fast machine (2.5 million pulses/second in its final form) performed some 500,000 additions, 200,000 multiplications, and nearly 300,000 data transfers (I/Os in today's parlance) in five minutes. This performance also satisfied the U.S. military in Eckert and Mauchly's other computer, allowing additional funding to flow to them because they had developed a prototype applicable to the government's needs, a prerequisite of the first contract. Northrop accepted the BINAC after the tests in Philadelphia and put the machine to use. The cost of construction—originally contracted for $100,000—was $278,000; Eckert and Mauchly had to absorb the difference.

How well did the machine work? Controversy over this question has continued to the present, with disagreement coming primarily from within the ranks of Northrop's engineers. In general, however, one may conclude that Northrop's management was disappointed with the BINAC. Besides being late in delivery and over budget, many components appeared to be either defective or not working as correctly or as efficiently as had been hoped. A great part of the problem was the dismantling and shipment of the machines making up this configuration from Pennsylvania to California, which the BINAC was not designed to experience. Its delicacy insured that the machine never worked the way its owner intended. Yet it was operational, even after reconstruction and modifications were made to it in California. Additional work on the machine might have improved its usefulness, but by then the lack of funds and Eckert and Mauchly's concern about getting on with the design and construction of the UNIVAC made any further efforts on behalf of the BINAC impractical.

The introduction of the UNIVAC line in the early 1950s and computers by other firms, in particular International Business Machines Corporation (IBM),† brought BINAC's history to an end. Born into a royal family of early computers important to the history of computing, the BINAC was one of many devices designed at the instigation of the U.S. government, especially to satisfy military requirements. Its greatest importance lay in BINAC's ability to prove that the concept of storing programs in a computer was possible and advantageous. BINAC thus represented a giant step into the modern world of computers.

For further information, see: A. A. Auerbach et al., "The BINAC," *Proceedings of the Institute of Radio Engineering* 40 (1952): 12–29; Nancy Stern, *From ENIAC to UNIVAC: An Appraisal of the Eckert-Mauchly Computers* (Bedford, Mass.: Digital Press, 1981) and her "The BINAC: A Case Study in the History of Technology," *Annals of the History of Computing* 1, no. 1 (July 1979): 9–20.

BIZMAC. This computer was under development at Radio Corporation of America (RCA)† during the 1950s. The company decided to begin construction of its own computer in 1952 and began to sell commercial versions of the machine in 1958.

Characters were made out of 6-bit words, with other bits used for detecting errors and correction. The machine had twenty-two instructions and used a three-address form, each of which occupied sixteen characters. Its memory* was very fast, using RCA's ferrite core technology to produce 4,096 characters of storage. Access speed approached 20 microseconds, and in addition, it had a drum that could house an added 4,096 characters. The drum operated at 10.24 revolutions per second. BIZMAC had an advanced tape system that wrote data to tape at 125 characters per inch and could move at the rate of 6.5 feet per second. It transferred data from cards at the rate of 150 cards per minute. Because the system was intended to be a very large one for use by commercial customers, it was so configured that up to 200 magnetic tape units could be attached to BIZMAC. In short, it would provide a crude online (or nonmanual) basis for accessing all information, for that size of configuration would obviate the need to mount and dismount tapes continuously. The idea of having machine-readable information always ready to be looked at was impressive and indeed radically new for its day.

BIZMAC also comprised a group of smaller computers which did specific tasks, such as managing the tape drives. Its concept was that a series of smaller computers, all interconnected by way of the mainframe, could be directed and coordinated in their efforts to retrieve and write out data on tape.

BIZMAC was technically an advanced system for its day. It was also poorly received by customers and equally badly marketed by RCA. Few were sold. The U.S. Army took possession of one for use in Detroit at a plant that built tanks. Several other copies were sold to other customers. Students of computer history have suggested that technical difficulties in the design and use of BIZMAC made it unattractive. Perhaps the strongest argument in this vein was the fact that the mainframe had to devote so much of its resources to managing the smaller computers that it was not cost-effective. Finally, by the time it appeared in the late 1950s it faced harsh competition from the IBM 305, which was the first computer to sport magnetic disks, signaling the end of tape drives as a direct-access device. Potential for such an approach, involving databases on tape, could not match direct access magnetic disks.

For further information, see: René Moreau, *The Computer Comes of Age: The People, the Hardware, and the Software* (Cambridge, Mass.: MIT Press, 1984).

BRUNSVIGA CALCULATING MACHINE. This device was one of the most widely used business calculators of the 1890s and early 1900s. Between 1892 and 1912, 20,000 were sold. Along with calculators made by the Burroughs Corporation,† Felt and Tarrant Company, and the National Cash Register Company (NCR),† these machines brought the management of large sets of numbers under more manageable control than had been seen since the development of calculus and algebra. These pieces of equipment were used primarily to calculate insurance tables, develop government statistics, and do normal day-to-day simple mathematics. Many navigational tables and a great deal of accounting were done with the Brunsviga.

The Brunsviga machine grew out of research done by Willgodt T. Odhner**, a Russian engineer. His design, based on what was then called the pin wheel and cam disk, became the basis of the calculator. Also known as the Odhner wheel or Odhner machine, it represented a series of patented products beginning in 1891. The Brunsviga calculating machine itself came about when Grimme, Natalis and Company of Braunschweig, Germany, acquired patent rights to the Odhner wheel effective March 21, 1892. This German firm thereafter manufactured what looked like a cash register to the casual observer, with a crank wheel protruding on the right. The firm modified the original design and improved on it, logging in over 220 patents in Germany alone related to the machine.

The basic design of the calculator rested on the use of pin wheels. Each wheel had up to nine pins, and a wheel would turn on a shaft driven by a crank. Each pin represented a number from 0 to 9; additional numbers would register when a wheel's capacity was achieved. Thus, for instance, the number ten was arrived at by turning one wheel through until all nine pins had been used, causing an additional turn on a second wheel which symbolized the new, larger number. Common mathematical tasks such as additions and multiplications were carried on by the turning of wheels. Multiple decades of numbers activated additional rows of wheels on shafts, all visible to the operator through slits in the machines' cover. In 1908 the Trinks-Arithmotype model appeared which printed results of calculations. Like computers decades later, the Brunsviga machine evolved as a product set of various devices, emerging as a more efficient and cost-effective system through generations of product introductions. They also became successively smaller and lighter—just as computers did later.

For further information, see: M. d'Ocagne, "Vue d'Ensemble sur les Machines à Calculer," Bulletin des Sciences Mathématique, 2ᵉ Série 46 (1922): 102–144; E. M. Horsburgh, ed., Napier Tercentenary Celebration: Handbook of the Exhibition (Edinburgh: Royal Society of Edinburgh, 1914, reprinted, Los Angeles: Tomash Publishers, 1982); E. Martin, Die Rechenmaschinen und ihre Entwicklungsgeschichte (Pappenheim: n.p., 1925); F. Trinks, Geshichtliche Daten aus der Entwicklung der Rechenmaschinen von Pascal bis zur "Nova-Brunsviga" (Braunschweig: Grimme, Natalis and Co., A.G., 1926).

BUG. This slang term is used by computer programmers and operators to describe problems and errors. Thus, a program that does not work correctly is said to have a "bug." The term may have originated during the early 1940s at Harvard University where the Mark I* computer was installed. On one occasion, when the computer malfunctioned, technicians found a dead moth in the system which they attributed as the cause of the system's failure. This famous carcass has been preserved for humankind taped to a page of notes; the notebook is kept at the Virginia Naval Museum. The term has become one of the most widely used and long-lasting in the history of data processing.

For further information, see: Valeri Illingworth, ed., Dictionary of Computing (Oxford: Oxford University Press, 1983).

BURROUGHS B5000. This series of computers was manufactured by the Burroughs Corporation† in the United States between the early 1960s and 1973. They represented the primary large computer offering from Burroughs, contributing to its revenues much like the S/360* and the S/370* processors did for International Business Machines Corporation (IBM).† The B5000 series was first shipped in 1963 and was last manufactured in 1973, although copies were still in use at the end of the decade. Burroughs subsequently brought out successors to the B5000 called the B6700 and the B7700 which it sold in the 1970s and 1980s, respectively.

The B5000 was made up of "cordwood" modules connected together. As Table 4 illustrates, it had all the usual components of a full configuration. The operator worked at a console at which he or she could perform two tasks: Halt and Load. That person could elect to load from disk drives,* drum memory,*

Table 4
Features of the Burroughs B5000

Cycle time	0.6 microseconds
memory	Ferrite core
Memory time	6 microseconds
Memory size	4096/stack, up to 8 stacks
Word size	6-bit character, 8 character and 1 parity/word
Auxiliary storage	Drums, disk, tape, 2 card readers
Size	9 bays, each 6ft. x 3ft. x 2½ft.
Output	Card I/O, 2 line printers

or cards. The computer used ALGOL 60* as its language and was equipped with an operating system.* In time, compilers were made available allowing one to program in FORTRAN,* COBOL,* and PL/I.* The machine could be configured with two central processing units (CPUs) and four input/output (I/O) channels. Because all storage components were accessible simultaneously and all channels could access this memory, multiprocessing became possible.

The computer represented a significant advance in the design of processors for its day. It was widely accepted by Burroughs' customers who used the configuration primarily for commercial applications; it had originally been intended for scientific users. It was a convenient computer because its operating system could accept unscheduled "hot" additional jobs to be processed immediately, easing the task of rescheduling work for an operator. Another feature involved the use of ALGOL; it was the most widely used adaptation of the language on a commercial system from any vendor. It was projected that it would be the great universal language in the early 1960s; while that expectation did not materialize, the Burroughs B5000 nonetheless proved it was a useful programming tool. The management of memory was masked in large part from the user and operator, making it "transparent" to each, and thus it eliminated much of the effort that it took in the 1950s and early 1960s to program its usage properly. The allocation of its memory was managed with some of the most advanced techniques of the period. Some even credited this computer with the first virtual storage function. With that, a computer would bring into the calculating portion of the machine (CPU) only those instructions and data needed for the next immediate execution of commands. The rest would be left in memory or in peripheral storage, thereby making it appear that the computer had more "real" memory than it did. That made it possible to run bigger programs with greater amounts of data. The machine was believed to operate slower than, for example, an IBM 7090 (a rival product), when processing arithmetic, but it was also less expensive.

For further information, see: R. S. Barton, "A New Approach to the Functional Design of a Digital Computer," *AFIPS Conference Proceedings* 19 (1961): 393; Franklin M. Fisher et al., *IBM and the U.S. Data Processing Industry: An Economic History* (New York: Praeger Publishers, 1983); E. Organick, *Computer System Organization: The B5700/B6700 Series* (New York: Academic Press, 1973).

C

"CAISSAC." This chess-playing machine was built by Claude E. Shannon**
of Massachusetts Institute of Technology (MIT), and it was named after Caissa,
the Muse of Chess. Chess machines have long interested computer scientists,
particularly those concerned with artificial intelligence,* inasmuch as they seem
an ideal way to test a machine's capacity to learn. Shannon, a scientist at MIT
during the 1930s and 1940s, wanted to build a machine that could play various
end games. Caissac could compute the advantages of various moves and notified
its rival of what move it wanted to make by lighting up the square it selected.
As a result of his work with this and other machines, Shannon wrote a classic
paper on programming computers to play chess in 1949 (see the following
bibliography).

For further information, see: Pamela McCorduck, *Machines Who Think* (New York:
W. H. Freeman and Co., 1979); Claude E. Shannon, *The Mathematical Theory of Communication* (Urbana: University of Illinois Press, 1949).

CARDS, COMPUTER. The concept of using cards with holes that can be
sensed by a machine and thereby cause it to perform functions first appeared in
Europe during the 1700s. In attempts to improve the speed and efficiency of silk
looms, various inventors developed devices that used cards with holes to
determine what patterns would be woven. The most successful of these efforts
were the Jacquard looms developed at the end of the eighteenth century. Within
a few years, tens of thousands of this type of loom were in use throughout
Europe, particulary in France, Spain, and the Netherlands. The positioning of
holes in cards determined the pattern to be woven by hooks that passed through
these to threads below. If a hole existed at a specific spot, metal hooks would
come down and through the card, and then pull thread up onto and through the
cloth. Thus, the positioning of a series of holes in a preestablished order of cards
provided a semi-automated machine that was programmable. Much like modern

computer devices, such machines required fewer personnel to operate than earlier devices, were efficient, and proved faster than hand-operated units.

The concept of punched cards developed further from the early 1820s through the 1860s when Charles Babbage** worked on his calculating devices. Envisioning the use of punched cards to direct the continuous actions of a calculating system of machines, Babbage designed a system by which instructions were fed to his engines using cards as input. Hence, by the time Herman Hollerith** began working with cards in the 1880s, the use of cards for automating machines and for providing them data had been a tradition for nearly 150 years.

Hollerith, however, turned the idea of using cards into a series of widely used and highly successful products that ushered in an age of data processing. It was to be an age characterized by the storage of information on a variety of data cards which could be analyzed by and read from card equipment (usually called card input/output or I/O devices). Hollerith's cards had holes to represent data that could be sorted to generate aggregates. Early cards made by vendors had round holes; later, square ones became common, although various-size cards remained common into the 1980s. The standard Hollerith card was the size of the U.S. dollar bill and thus could fit into cabinets originally designed to hold paper money. The standard eighty-column card of today appeared in its earliest form in 1928 from International Business Machines Corporation (IBM)† and measured 7 3/8 inches by 3 1/4 inches. IBM was the major supplier of these cards from the time Hollerith sold his company in 1911 until the late 1950s. Another widely used card appeared in 1969 and had ninety-six columns. Introduced by IBM, it was used on the System/360* computer. The card, 3 1/4 inches by 2 11/16 inches, could capture 20 percent more data than the eighty-column version. The larger card, however, remained the industry standard. Until the late 1960s, the majority of information was fed into a computer via the punched card.

The Hollerith card and its IBM follow-on were not the only ones available. Until about 1909 Hollerith enjoyed a virtual monopoly on his card. After that time others entered the marketplace, taking advantage of patent expirations on some of his card equipment. His largest competitor, the Powers Accounting Machine Company,† merged with the Remington Typewriter Company and the Rand Cardex Company, creating the Remington Rand Company.† This new company, like Powers before it, posed a formidable challenge to IBM's virtual card monopoly. Rand's cards were in some ways technically superior to the Hollerith variety. They had "die set" punching, which meant one could look at a card before the holes were completely punched out. This company's equipment was the first to have an automatic card feature and later a printing tabulator. Rand developed the first multiplying punch and later tied a typewriter to a card punch. During the 1920s the Powers Company also introduced an alphanumeric punch and tabulator which permitted twice as much data to be placed on a card. Between 1928 and 1931 IBM also introduced innovations that resulted in the familiar eighty-column card which became the industry standard.

Although we have no hard data on the number of cards across the entire data processing industry, it has been estimated that for the United States alone (the largest user of cards in the twentieth century) $800 million worth were sold in 1955, a figure that remained fairly constant through the 1960s. During the subsequent decade sales nearly doubled to about $1.3 billion. Even when we discount for the high inflation of that decade caused by financing the Vietnam War, we are still left with high shipments of cards as their costs fell. For the same market, actual shipments grew substantially during the 1950s and 1960s as the entire data processing industry grew during the 1970s. Total card shipments, as reflected in dollar values, thus grew from about 100 percent during the second half of the 1960s to nearly 300 percent in the next five years. In large part, particularly for the 1950s and first half of the 1960s, when cards were still a primary medium for input to a computer, these data help indicate the growth rate for the industry as a whole. Despite the increasing dependence on magnetic tape, disk, and later cathode ray terminals for data input and output in the late 1960s and throughout the 1970s and 1980s, the large and continued rise in the use of cards suggests the explosive growth in the use of data processing.

For further information, see: Geoffrey D. Austrian, *Herman Hollerith: Forgotten Giant of Information Processing* (New York: Columbia University Press, 1982); Montgomery Phister, Jr., *Data Processing Technology and Economics* (Santa Monica, Calif.: Santa Monica Publishing Co., 1975); Leon E. Truesdell, *The Development of Punch Card Tabulation in the Bureau of the Census, 1890–1940* (Washington, D.C.: U.S. Government Printing Office, 1965).

CDC CYBER SERIES. Control Data Corporation (CDC)† built the Cyber 170/700 series of computers beginning at the end of the 1960s and continuing through the 1970s. These machines, representing a family of widely used computers, ranged in size from a 170/720 of medium speed to the 170/760 and the 176 which were relatively powerful. The family replaced the earlier CDC 6000 which had also been a successful product line in the 1960s.

The 6000 series included the 6200, 6400, and the 6600. These were some of the most popular computers for scientific applications in the data processing industry. The 6600, introduced in 1964, may have been the most widely used scientific processor of the late 1960s. The 6500 and 6700 were dual processors and were also well received, particularly by the scientific and engineering communities in the United States.

The Cyber 70 series was first introduced in 1970 and replaced in 1974 by the Cyber 170 family. By the early 1980s the Cyber 170 group included the 720, 750, and 760. They were designed to be compatible with each other within the family. As the scientific community almost collectively required larger processors, CDC introduced them. Thus, for example, in 1969 CDC produced the 7600, followed by the Cyber 76 and the Cyber 176. Quantum jumps in capacity, typical of most new products introduced by various vendors, were evident. The 176, for example, operated at nearly twice the speed of a 760, done through faster memory* and more efficient circuitry. Vector processing also became available in later models.

A typical configuration consisted of a processor, central memory, satellite processors called peripheral processing units (PPs), and an assortment of typical peripheral (input/output) equipment. These included tape and disk drives,* terminals, and printers. The configuration was also enhanced with an Extended Core Storage (ECS), which allowed data to be swapped around the system. This system differed from many others in that it had a network of PPs which could be made up of as many as twenty-one processors, each with an arithmetic unit and memory. They could access either main memory or various peripherals. Multiple operations and jobs could operate concurrently in different ''functional units'' which were subsets of the entire configuration.

The size of these machines varied over the years. The 6600, for example, had a maximum memory of 131,072 words, each made up of 60 bits. In a 700 class machine memory rose to 262,144 words. ECS went by the boards in 1981 when Extended Semiconductor Memory (ESM) was introduced. That enhancement made memory faster and larger in capacity. The 700 series of the 1980s could hold up to 2,097,152 words or, in the parlance of the industry, 2 megabytes of core.

The various series of CDC's machines followed the industrywide practice of the 1960s and 1970s of evolving into more efficient technologies. It began with the 6600, the fastest and most advanced scientific processor at the time of its introduction. Subsequently, various CDC processors went from cycle times of 100 to 25 nanoseconds. Functional units rose by one, while they were redesigned to accept new instructions before completing a previous one which allowed work to pass through such units faster and more efficiently. Memory cycle times dropped by a factor of four, whereas available channels rose from 12 to 24. PPs grew from 10 to 20, and each PP's cycle time dropped by half. All of these developments occurred over a period of fifteen years.

The operating systems* for these machines were new. On the 6000 series it was called the Chippewa Operating System, named after the Chippewa Laboratory where it was developed in the early 1960s. In 1966 CDC replaced it with SCOPE 2, which included random disk accessing and support for ECS. SCOPE 3 appeared in 1967, and a time-sharing version called INTERCOM, two years later. The KRONOS operating system came out in 1970. It could handle much larger volumes of interactive users than its predecessors and had been developed out of an earlier joint project with McDonnell Automation in the late 1960s called MACE. With the 170, introduced in 1974, the company also brought out the Network Operating System (NOS) which was compatible with KRONOS, replacing it and SCOPE. An intermediate solution, compatible with SCOPE, was announced in 1975; it was called NOS/BE (/Batch Environment) for customers who would not convert to NOS directly or as quickly as the company wanted. CDC found most customers unwilling to make radical changes in their operating systems as a general rule. In addition to operating systems, these computers had compilers that enabled programming in APL,* COBOL,* ALGOL 60,* BASIC,* Pascal, LISP,* SNOBOL 4,* and FORTRAN.* In time DMS–170 was introduced as a database management system.*

For further information, see: J. E. Thoronton, *Design of a Computer: The Control Data 6000* (Glenview, Ill.: Scott, Foresman, 1970).

CHIP. This electronic component has been the basic building block of computers since the 1960s. It made possible the large decline in the cost for computers and magnetic storage devices for data while increasing the capacity of all processors to perform ever greater, more complex transactions. The importance of the chip to the second half of the twentieth century would be difficult to overstate because it clearly ranks alongside the development of the atomic bomb, exploitation of atomic energy, flight, and exploration of lower space. No computer built since the late 1960s has been without the chip. It is a small piece of semiconductive material through which electricity can pass from one section to another according to predetermined plans. Put another way, it is an integrated circuit of electronics without wires. A chip, which today is frequently smaller than a half-inch square, houses electrical circuits and instructions from humans on what to do with information (data). It also stores data used by the computer.

The chip represents a century of continued integration and management of electricity through more efficient circuits. By the early 1900s, in both Western Europe and the United States, a network of electrical power and associated cabling criss-crossed nations and dominated the skyline above many streets in towns and cities, a confusion of wires and poles. Mechanical equipment was being electrified while the use of electricity had created an enormous demand for new technologies, electrical engineers, large utility companies, and more products—in short, a new industry. Closely linked to electricity was the development of the light bulb by Thomas Edison. It was, to use other terms, a vacuum tube—a glass container with wires inside that could transmit electrical impulses. In the case of lights these filaments glowed, but for the purposes of electronics in general, their abilities to transmit electricity and later to amplify electrical currents were the key facts. Light was indeed important but not just as a precursor of technologies inherent to computing. By the late 1890s, and especially during the early twentieth century, scientists were attempting to amplify electrical impulses that passed through vacuum tubes in order to pick up radio signals effectively. The result was the three-element (wired) vacuum tube developed by Lee De Forest. This American mathematician turned engineer made possible improved electrical communications leading to powerful radio transmission and reception and, for the telephone industry, to long-distance telephone service. His single invention, modified and improved on through the late 1940s, expanded the electronics industry, making it a major component of the U.S. economy.

By the mid-1920s scientists at Bell Laboratories† were developing computer-like devices called relay computers, and, in the 1930s at the Massachusetts Institute of Technology (MIT), Vannevar Bush** was building analog computational machines. During World War II in order to meet the demands of the Allies for sophisticated calculating devices to break enemy transmission codes, to calculate firing tables for ballistics, and to create guidance systems for airplanes and missiles, a variety of digital computers* were built, from the British COLOSSUS* to the American ENIAC.* They relied on vacuum tubes to carry electrical impulses (instructions and data) through their systems. Some of these machines had over 15,000 vacuum tubes. Thus, the reliability of these tubes became a critical factor in the success of computers. Yet, too many tubes were blowing out or malfunctioning, making it difficult or impossible to operate these early computers as effectively as one would have liked, while posing a real threat to the construction of even bigger machines. Crude attempts to improve efficiencies involved pretesting tubes, using many electrical fans to blow heat off these tubes, and building vacuum devices to greater specifications of quality. But their limits were being reached. Another problem was size: 16,000 components the size of light bulbs made for very large computers, many of them room-sized. Yet following World War II, the demand for computers did not stop because of tubes. Research and development projects on computers rose sharply from less than a dozen to over twenty all over the United States and Western Europe by the end of 1947. That level of activity increased pressure to improve the quality of electronic components. With the demand for telephone service rising, American Telephone and Telegraph (AT&T)† in the United States was also feeling the pressure for better parts. Therefore, it had to manage more telephone calls at lower costs with greater efficiencies. It needed better switching gear—in other words, more computing technology.

By the late 1940s, therefore, the electronics industry had good reasons to shift its focus away from just transmitting electricity (for example, to operate a simple toaster at home or an engine in the factory) to the movement of information electronically. The use of symbolic logic (Boolean algebra's techniques) had been understood since the 1930s, and scientists wanted to apply this knowledge to manipulate electrical impulses. But the industry needed more efficient components that exceeded the speed and reliability of the vacuum tube and that would be smaller and less expensive. The answer was the transistor, clearly an important invention of this century. It represented a paradigmatic twist in viewing the problem of electrical circuits, and in the process it became the precursor of the chip. It would be difficult to imagine how the field of microelectronics could have been born without the transistor.

Already in the 1930s scientists had begun to explain in detail and more accurately the behavior of electricity, specifically the dynamics of electrical currents. They documented the behavior of electrons as they moved through solids (conductive material) and appreciated when and why they did not (such as with insulators). The list of electrical conductors was growing and so was that for semiconductors. These elements allowed electrical currents to flow through partially. Equally important to scientists, these were substances whose properties could be altered, giving one the opportunity to control the pattern of behavior of electrical currents. Today we call that, slightly inaccurately, programming a chip. The hope translated into a potential replacement for the inefficient vacuum tube. In 1945 Bell Labs led the way to a new effort by establishing a group to study the properties of germanium crystals, a known semiconductor. During World War II, while doing research on radar, scientists at Bell Labs had learned that this material represented a promising line of investigation. They were also under pressure to move quickly in finding better electronics to support AT&T's switching requirements. So the search was on for a solid-state electronic amplifier.

Two members of Bell Labs's staff, Walter Brattain and William B. Shockley,** went to work on the problem, studying silicon and germanium, both of whose electrical properties could be altered significantly. By the fall of 1947 Shockley's staff, including John Bardeen, had attached two small wires close together on one side of a sliver of germanium. When they ran electricity through it, Bardeen and Brattain discovered that this gadget could receive electrical impulses and amplify them. Then on December 23, 1947, they successfully demonstrated their device to the rest of the staff at Bell Labs—the first important instance of a solid-state electronic component with no moving parts. One of the witnesses turned the phrase "transfer resistance" (used to describe the device) into a shorter word: transistor. Although crude, and with a great deal of work yet necessary to perfect it, an amplifying device had been made that could theoretically replace vacuum tubes.

This breakthrough caught everyone by surprise both at Bell Labs and within AT&T. It came very quickly, within two years of initial work. AT&T patented the technology, and scientists went to work on improving it. On June 30, 1948, AT&T publicly announced the existence of the transistor. In the following decade, AT&T licensed many companies to make such devices. These in turn improved on quality and in the process destroyed the vacuum tube as the component of choice for essential electronics. The one notable exception was television where companies continued to use vacuum tubes until the late 1960s. But in the field of computers, the more reliable, smaller, less expensive transistor became the standard electronic component by the late 1950s. In 1956 three scientists at Bell Labs—Shockley, Bardeen, and Brattain—were recognized for their contribution with the Nobel Prize in Physics.

A great deal of attention was being paid to the study of solid-state electronics. The number of scientific papers published on the subject in the 1950s nearly rivaled the quantity produced on atomic physics. The U.S. government was also encouraging the use of transistor technology in products built for the military. It was also a time when computers were finally beginning to sell in quantity, creating the data processing industry and its insatiable demand for better electronics in general. Companies working on semiconductors and transistors grew and prospered. The traditional old-line electrical firms—Radio Corporation of America (RCA),† General Electric (GE),† Raytheon,† and so on—dabbled with computers or tried to constrain trade and progress and were found guilty by U.S. courts. Others, such as Texas Instruments (TI),† became involved in the new electronics, especially in solid-state technologies. TI, for example, dominated the manufacture of silicon-based products for a while in the late 1950s, while other firms grew up in the shadow of Stanford University in California, giving birth to what has been called Silicon Valley. By the mid-1970s this area was home to several hundred companies working with semiconductors.

In 1954 Shockley, of Bell Labs fame, left AT&T and moved back to his home state of California where he established his own firm to build transistors. Meanwhile, TI was working on its semiconductor projects. Both were motivated by potential business, particularly with the U.S. military establishment which was demanding yet further miniaturization of electronics (especially after the Soviet Union launched Sputnik in 1957). It should be noted, however, that prior to that event, the U.S. government had already concluded that rockets and miniaturized electronics would become critical elements in American defense systems in the years to come and thus had created a demand for newer electronic components. The companies working with semiconductors had also concluded that computer manufacturers represented a strong civilian market for new products.

Progress came quickly. In the mid-1950s, for example, new processes were developed for manufacturing transistors, driving their costs down so that they could even be put into portable radios by the early 1960s. During the mid-1950s research shifted away from germanium-based semiconductors to silicon which appeared to be more versatile. Shockley worked with germanium while another scientist, Robert Noyce,** became interested in silicon. Noyce headed up Fairchild Semiconductor Corporation,† which in the 1960s spawned a variety of semiconductor companies in Silicon Valley.

Thus, with scientists' interests in silicon on the one hand and with pressure for further miniaturization on the other, scientists formulated what today is called the concept of the integrated circuit (IC) or, more popularly, the microchip. The microchip was developed at Texas Instruments and at Fairchild Semiconductor at exactly the same time.

The idea scientists worked with in the late 1950s was simple. They wanted to take doped silicon, with a few wires, put these silicon pieces and wires together, and add resistors and other components, all to build a package or subsystem. This subsystem would next be clustered together on one "chip" of silicon. Noyce later commented that "the conceptual jump was from wiring together arrays of transistors to the realization that you could put all the other electronic functions on silicon as well." Jack Kilby,** at TI, was also working on the problem, which was difficult to execute. But both firms were successful. Their product, the chip, was thus a logical and evolutionary progression from the vacuum tube and its filaments to filaments on a sliver of semiconductor material called the transistor to multiple transistors on silicon which we now call the chip. Although the chip evolved from earlier technologies, its creation revolutionized the entire electronics industry.

The new technology created a battle between the two companies, which they finally resolved with cross-licensing. Various scientists next established their own firms to build chips and, then, in 1964, International Business Machines Corporation (IBM)† announced that its family of computers, the System/360,* would be built with chips. By 1969 approximately 15 percent of all the computers in the world were S/360s; the chip had become the basis of all computers from then to the present.

The wide acceptance of the chip required work in the beginning. By 1963, for example, only 10 percent of all electronic circuits were ICs. But demand grew. Fairchild, one of the dominant forces in the market during the first half of the 1960s, grew from several million dollars in annual sales into a $150 million business within a few years. Other variations of the technology appeared—bipolar and Metal-Oxide-Semiconductors (MOS), for example. MOS chips were slower than bipolar chips but also less expensive. Bipolar chips were used in many military projects because of their speed and in time were used in commercial computers. By 1965 twenty-five companies worldwide were making chips of various types. In the beginning they had a difficult time convincing many computer manufacturers to use chips, despite IBM's early commitment to them. Many managers felt threatened by the new technology, fearing the obsolescence of their careers a possibility along with major changes in their organizations. But by the end of the decade they had fallen in line or were forced out by this quickly developing competitive industry. By 1970 a new industry existed and had replaced much of the old traditional electronic components portion of the electrical manufacturing community. Today these new manufacturers are members of the semiconductor industry.

A major development in the history of chips came at the end of the 1960s when internal strife at Fairchild led many of its key engineers and scientists to leave, forming their own companies. One of these firms was Intel,† established by Noyce, Gordon E. Moore,** and Andrew Grove. Intel was the leader in the manufacture of chips throughout the 1970s and into the 1980s. The firm was created in 1968 with the idea of building semiconductor memories to replace core-type storages on older computers. Intel focused on MOS chips with large-scale integration (LSI). This company found ways of storing data cheaply on microchips. In 1970 it brought out the 1103 random access computer (RAM) chip, a product that stored just over 1,000 bits of binary data. As Figure 1 illustrates, such miniaturization meant that more powerful computer packages could be built which were smaller than those of the 1960s and yet much faster. As a result of such breakthroughs, companies such as Data General† and Digital

Figure 1
Trend of Input/Output Terminals versus Circuits on Modules

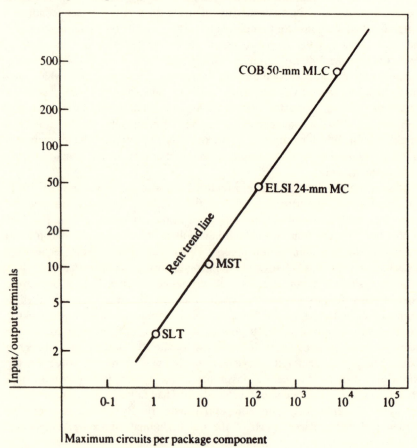

SLT: Solid Logic Technology (vintage 1964)
MST: Monolithic Systems Technology (vintage 1969)
ELSI: Early Large-Scale Integration (vintage 1973)
COB: Card-on-Board (vintage 1979)
MLC: Multilayer Ceramic (vintage 1979)

SOURCE: D. P. Seraphim and I. Feinberg, "Electronic Packaging Evolution in IBM," *IBM Journal of Research and Development* 25, no. 5 (September 1981): 618. Copyright 1981 by International Business Machines Corporation; reprinted with permission.

Equipment Corporation (DEC)† came into being with minicomputers. RAMs launched Intel in the 1970s, much as the integrated circuit had spawned the successes of Fairchild a decade earlier. After the 1103 came out, Intel announced the 4004, which was the first microprocessor usable in general-purpose applications. Intel called it "a microprogrammable computer on a chip," and indeed it was the closest to one available then. It consisted of five chips packaged together called the MCS-4. It could do logical operations on a set of binary instructions resident in its memory chips, which meant these could be embedded in computers or industrial equipment. During the 1970s and 1980s chips were imposed on motors and devices, making them "intelligent" (programmable), much as electricity had been implanted into machines to replace steam and belts in the late nineteenth and early twentieth centuries.

In 1974 Intel brought out the 8080, one of the most widely used chips in the history of data processing. In addition to improved speed and reliability, it was an 8-bit product, which meant it could accept a "word" of data made up of 8 bits each. The 4004 could only accept a 4-bit string at a time. The 8080 became the industry's standard and appeared in such diverse equipment as traffic lights, hand-held computers, and minicomputers. Yet within two years others were building similar chips, with twenty companies alone in the United States in 1975. The chip had advanced to the point where in 1975 it could, as part of a "card," execute more than 100,000 logical operations in a second. To do this it required only 5 volts of electricity. The portion of the card that served as the brains or microprocessor was slightly smaller than 1 square inch. From then to the present, the history of chips has been one of more miniaturization, shrinking costs, greater reliability, and more complex packaging of chips together in denser environments. They used less electricity and put out decreasing amounts of heat. Machines that required special air conditioning and specially equipped machine rooms were replaced with devices of greater power, the devices were the size of a desk and could be installed in an office. This was made possible by greater circuit densities (see Figures 2 and 3).

The growth in use of chips was staggering. In 1969, 750,000 microprocessors were sold for use primarily in computers and hand-held calculators. By 1984 the number ran into the tens of millions annually. Declining costs also increased demand for these products. By 1979 the data processing industry learned that costs for producing a chip fell approximately 30 percent (some said as much as 36 percent) each time a new generation of chip technology appeared. These new generations came about every six to seven years. Manufacturers learned what the yield could be on chips. Yield was measured as a percentage of a product always working when tested. Thus, making 10 to 20 good chips out of a 100 represented good yields, and as one learned more about a particular generation's characteristics, yields could go higher, in some cases to nearly 60 percent. The higher the yields, the more profit earned. The more chips built by a manufacturer, the greater the odds that yields could be improved. Packaging increased at a faster rate, with the number of circuits packed onto one chip doubling annually. The size of the chips themselves was growing very slowly and by the 1980s was becoming smaller again. All of these trends meant faster, less expensive computing (Table 5).

Capacities thus kept growing. One way capacity was measured was by the amount of data that could be put on a chip (Figure 4). In 1971 a RAM chip

Figure 2
Circuit Density Trend with Time

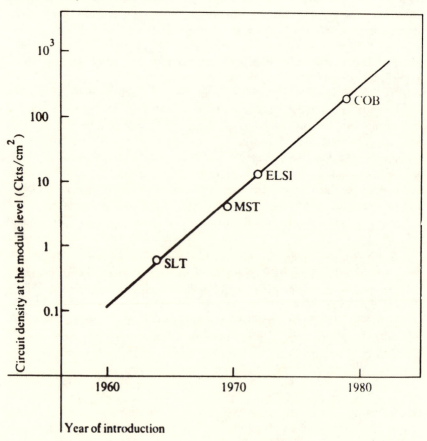

See Figure 1 for definition of terms. These were various packaging mediums for circuits and chips.

SOURCE: D. P. Seraphim and I. Feinberg, ''Electronic Packaging Evolution in IBM,'' *IBM Journal of Research and Development* 25, no. 5 (September 1981): 623. Copyright 1981 by International Business Machines Corporation; reprinted with permission.

typically held 256 bits of data. Each bit cost about two cents. In 1973 it was 1,000 bits at 0.5 cents per bit versus 4,000 bits at 0.24 cents each in 1975. In 1980, 16,000 bits fit on a chip at a cost of 0.03 cents per bit. Costs were halved again before the middle of the decade, while capacity went up twice as much as before. By 1986 1 million byte chips were available, and densities were triple what they had been in 1980. Put another way, the 8080 cost $360 in 1975 on the open market, and in 1980 better quality versions sold for less than $5 each.

The technological improvements of the 1960s and 1970s were largely the results of a small group of scientists, who would often share information at

Figure 3
Growth of Printed Circuit Density over Time

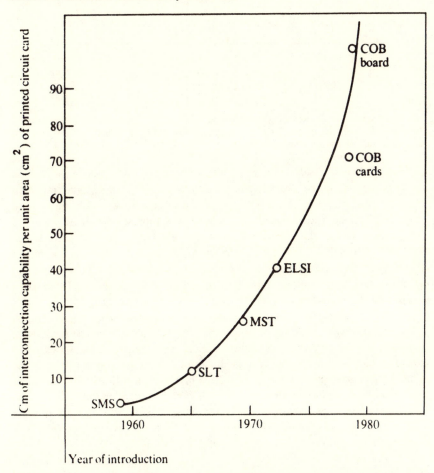

Only about 45 percent of the capability of the packaging medium of the time is used. The SMS is the earliest of this type of packaging. For defitions of the terms used, see Figure 1.

SOURCE: D. P. Seraphim and I. Feinberg, ''Electronic Packaging Evolution in IBM,'' *IBM Journal of Research and Development* 25, no. 5 (September 1981): 624. Copyright 1981 by International Business Machines Corporation; reprinted with permission.

conferences, although they quickly patented their successes and turned them into profitable commercial ventures. Companies within the computer manufacturing segment of the data processing and electronics communities were not as quick to make their own chips or to contribute to this particular evolution in electronics. One notable exception was IBM, the largest builder of computers from the 1960s to the present. Its early commitment to chips arose from its recognition of the value of semiconductor technologies.

Table 5

Speed versus Performance of IBM Computers, 1953–1979, Selected Years

Machine	Date	Relative Speed		Relative Price[a]
650	1953	1	1	1
360/30	1964	43	0.025	1,700
370/135	1971	214	0.011	19,000
4341	1979	1,143	0.001	1,143,000

[a]Relative price is the cost of executing an instruction in one second. See R. Moreau, *The Computer Comes of Age* (Cambridge, Mass.: MIT Press, 1984): 187–196, for a discussion of the impact of these developments on data processing in general.

When IBM decided that a new generation of computers was needed for success in data processing in the 1960s, it accepted the findings of a task force, reflected in the SPREAD Report, that a new generation of equipment would have to be built. One of the recommendations was that computers be based on chip technologies rather than on transistors. In 1961 the company made its commitment to solid-state logic and began to identify how to build computers and storage devices employing semiconductors. These products became part of the S/360 family of computers of the mid-1960s. Throughout the 1960s IBM was part of the industrywide race to reduce costs while increasing functions. It either built or purchased technology as economics dictated. IBM, a leader in the packaging and manufacturing of high-technology components since the 1940s, also made its own contributions, although these were comparable to those of the industry as a whole in the 1960s. Leadership in technological innovations came in the 1970s and early 1980s. By 1980 IBM had its own collection of chips and patents on semiconductors. For example, it found ways of putting ten circuits on a chip masterslice and, in 1980, 2,000. That is, an air-cooled chip module in the company's midsized 4300 computers of 1979 had as many circuits as a "fully populated" board of the S/370* of 1971 (Table 6).

Yet the greatest improvement came with the thermal conduction module (TCM) which was first announced with the 3081 computer. The TCM was made up of 100 masterslice chips, each with as many as seventy circuits. One TCM, about 6 inches square in size, thus had the same amount of logic gates as an IBM 370/145 computer of the mid-1970s. One could run a $400 million business in the mid-1970s with a 145, and now that power fit into a hand. This TCM afforded more capacity because many were packed into a computer and gave greater density at lower cost for IBM's larger computers of the mid-1980s. The 3090 series, announced in 1985, used an even denser TCM. The company had three chip manufacturing plants (in West Germany, Japan, and the United States) by the mid-1970s, and within a decade was experimenting with the 1 million bit chip in its laboratories. The 256K chip was IBM's standard for equipment used internally by the late 1970s. This kind of packaging made it possible for the entire industry to build smaller "boxes" that housed larger processors. The

Figure 4
Evolution of Logic Technology at IBM

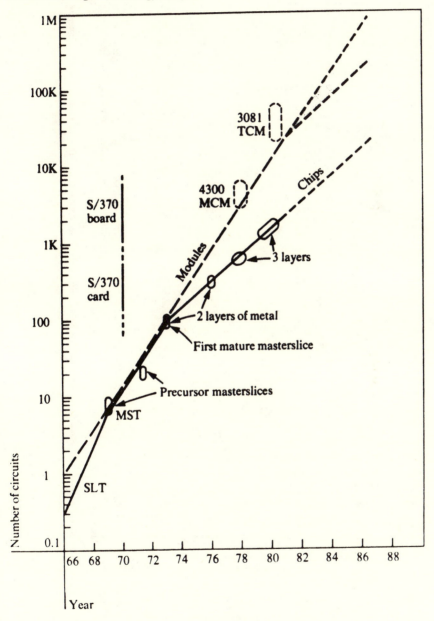

Table 6
Representative IBM Processor Memory Technologies, 1953–1979, Selected Years

Date[a]	Processor	Technology[b]	Cycle time (microseconds)	Volume/megabyte[c] (m³)	Power/megabyte (kW)
1953	701	CRT (1024)	12	240	980
1955	704	Core (50–80)	12	160	340
1959	7090	Core (30–50)	2.2	13	41
1965	S/360-50	Core (19–32)	2.0	1.5	5.5
1965	S/360-65	Core (13–21)	0.75	11	48
1968	S/360-95	Thin film (4096)	0.12	8.8	23
1971	S/370-145	Bipolar (128)	0.25	0.23	9.6
1973	S/370-145	Bipolar (1024)	0.25	0.028	2.4
1973	S/370-158,168	MOSFET (1024)	0.46	0.028	1.0
1974	S/370-158,168	MOSFET (2048)	0.48	0.014	0.36
1979	4331	SAMOS (65 536)	0.90	0.0009	0.05

[a]This is the date the first processor using this memory was delivered.

[b]The numbers in parentheses are the number of bits per CRT, per thin film substrate, or per semiconductor chip, or the core size in thousandths of an inch.

[c]This is the volume required for the memory array, support circuits, packaging, cooling space, and proportioned share of power supply, regulation, and distribution. The volume is extrapolated for memories having less than one megabyte capacity.

SOURCE: E. W. Pugh et al., "Solid State Memory Development in IBM," *IBM Journal of Research and Development* 25, no. 5 (September 1981): 586. Copyright 1981 by International Business Machines; reprinted with permission.

same story could be told for disk drives with greater levels of densities at lower costs for storing information in machine-readable form (Figure 5).

If the 1970s was the era of large-scale integration (LSI), then the 1980s belonged to VLSI or very large-scale integration of the type seen with IBM's components for computers. The 1980s began with the memory chip that could house 65,536 bits of data, which led microprocessors to equal the power of older mainframes (computers) using just three chips. In the world of VLSI, chips contained some 100,000 elements with paths between them as narrow as a few millionths of an inch. For some scientists in the early 1980s, there was concern that their knowledge of quantum mechanics and understanding of the physical limits of electrons might soon be reached. That problem would then call for new technologies in the 1990s to replace semiconductors. Yet with all this complexity in design and manufacture, the demand for such products actually rose, insuring that the business justification for continuing to work with semiconductor chips would remain strong. By 1984 there were well over 100 companies in the field, with some reports placing the number closer to 200.

Other data suggested the size of the demand for RAMs. The market had grown in the 1980s at about 36 percent annually, reaching $1.8 billion in 1983 on a worldwide basis. That demand only represented those chips used for computer memory, or nearly 20 percent of all manufactured components in the semiconductor industry. Forecasts in 1984 called for an approximately $10 billion market for RAMs alone by 1990, which was equal to three times the size of the entire semiconductor industry in 1983. Costs declined, suggesting that the sheer number of units shipped was enormous and growing. The 64K chip cost $500 for every million characters (bits), and the 256K chip dropped to $100 while the 1megabyte RAM was expected to drive the price for a million characters down to $60 in 1990 (Table 7).

The history of chips is one of dramatic innovation and yet of evolution from earlier technologies. Chips launched an industry and made computers and "intelligent" equipment pervasive in the industrialized world. As a building block of computers and related equipment, it helped make the data processing industry one of the more significant sectors of the world's economy. By freeing computers from constraints of capacity, poor reliability, prohibitive costs, and massive size, the chip made it possible for the data processing community to become the largest industrial sector in the world by the early decades of the twenty-first century.

For further information, see: Paul Freiberger and Michael Swaine, *Fire in the Valley* (Berkeley, Calif.: Osborne/McGraw-Hill, 1984); Dirk Hanson, *The New Alchemists: Silicon Valley and the Microelectronics Revolution* (Boston: Little, Brown and Co., 1982); René Moreau, *The Computer Comes of Age: the People, the Hardware, and the Software* (Cambridge, Mass.: MIT Press, 1984); E. W. Pugh et al., "Solid State Memory Development in IBM," *IBM Journal of Research and Development* 25, no. 5 (September 1981): 585–602; T. R. Reid, *The Chip: How Two Americans Invented the Microchip and*

Figure 5
Evolution of IBM Magnetic-Disk Technology

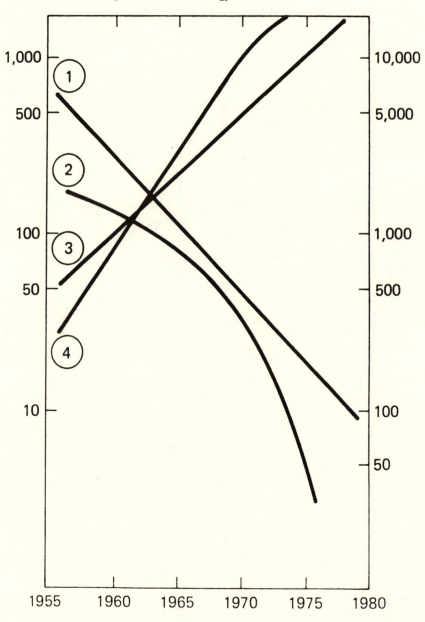

Table 7
Cost of Random Access Memory, 1973–1990, Selected Years

Chip Size	Price/million Characters	Year
1K	$50,000	1973
16K	5,000	1978
64K	500	1983
256K	100	1988
1Mb	60	1990

Demand for RAM Chips

Annual Volume (Billions bits)	Annual Sales (Billions $)	Year
˙ 8	$0.06	1973
650	0.30	1978
28,000	1.80	1983
800,000	8.00	1988
1,400,000	10.00	1990

SOURCE: Data were gathered and estimates made by Data Montgomery Securities, *Business Week*, April 2, 1984, p. 71.

Launched a Revolution (New York: Simon and Schuster, 1984); E. J. Rymaszewski et al., "Semiconductor Logic Technology in IBM," *IBM Journal of Research and Development* 25, no. 5 (September 1981): 603–616.

CLP. The Cornell List Processor (CLP) is a string and list processing language developed at Cornell University in the mid-1960s and represents an enhancement to another language called CORC. CLP added another type of data element known as an *entity*. Entities can vary in number at object time, and they can be subdivided into specific lists. Each of these lists has particular attributes which, according to historian Jean Sammett, are "simply variables with the entity actually being used as a subscript."

CLP was devised to make available to students at Cornell an easier-to-use string processing tool than had been available before. The development of this language reflected a pattern in many universities of creating languages for scientific research that were easy to use. It was used only at Cornell and thus never became widely available or important.

For further information, see: R. W. Conway et al., "CLP—The Cornell List Processor," *Communications, ACM* 8, no. 4 (April 1965): 215–216.

COBOL. COBOL was the most widely used programming language* in the United States during the late 1960s and throughout the 1970s. By the early 1970s most commercial applications were written in COBOL with an estimated 60 to 85 percent of all business applications using this language. Some industry watchers have even argued that well over half the lines of programming in existence in the United States in the mid-1970s were in COBOL. When compared to such earlier pre-COBOL tools as Assembler or FORTRAN,* it was a relatively easy language to learn and use. This relative ease of use contributed dramatically to the development of applications during the 1960s and 1970s. Its widespread acceptance made it easier for data centers to hire programmers who had experience with the language, which in turn encouraged further use of COBOL. During the 1980s; despite the availability and use of more powerful languages, COBOL remained one of the most accepted languages.

By the late 1950s requirements for programming and data processing had entered a period of significant change. Most computing in the early to mid-1950s centered on scientific and military applications, many of which were supported by the U.S. government. The largest applications and most extensive users of computers were still to be found in government agencies and in universities. Yet large corporations were increasingly using data processing for inventory control, accounting, finance, and payroll. They typically justified these uses through elimination of clerks or in increased response to market conditions (particularly with better inventory management). Corporations such as the International Business Machines Corporation (IBM),† Sperry Rand,† and Honeywell† marketed aggressively to businesses, driving up the demand for hardware and software* to support such applications. During this period U.S. military branches were also automating large and expensive functions that could be characterized as commercial applications. These included managing inventories, people, accounting, payroll, finance, budgets, modeling transportation issues, along with the more military uses associated with defense and fire control systems which were more similar to the traditional bread-and-butter applications of the 1940s and early to mid-1950s.

The growth in nonscientific applications was accompanied by an increase in computing power and a decline in the cost of computers. Hardware capacity was rising in the late 1950s, with more memory* possible and greater capabilities available for processing increasing numbers of transactions faster. By the early 1960s vacuum tubes had given way to transistors, thereby reducing computing costs and enhancing capacities, while dramatically improving hardware reliability. Many computer manufacturers recognized that they needed newer languages and better operating systems geared to business and nonscientific applications. By 1959 various projects were underway at Honeywell, IBM, Sylvania, and Burroughs† to develop English-like languages to support commercial applications. Languages commonly in use at the time were more scientific and mathematical in notation, such as FORTRAN, and, while effective, they failed to meet the need for new languages. This changing requirement for new languages that could operate in larger processors for different applications provided the backdrop for the development of COBOL.

The initial impetus to develop a specific new language came on April 8, 1959, when representatives from computer companies, users, and universities met at the Computing Center of the University of Pennsylvania to discuss business languages. At this meeting they decided to hold a follow-up, broader based gathering to begin the actual process of identifying characteristics for such a language. They asked the U.S. Department of Defense to sponsor the meeting. The Pentagon, recognizing the potential benefit of such a language, agreed.

The next meeting was held on May 28–29, 1959, with some forty attendees representing seven government agencies, eleven consultants and users, one university, and ten companies (Burroughs, General Electric [GE],† Honeywell, IBM, National Cash Register Company [NCR],† Philco†, Remington Rand,† Radio Corporation of America [RCA],† Sylvania Electric Products, and ICT). At this meeting it was decided that a new language would have to help reduce the cost of programming, be usable on a variety of computers, and be easier to use than existing systems. They agreed that, in order to make it easier to read and understand commands, it had to be more English-like than ever before. To implement such a language three committees were established: a short-range committee to determine what problems existed with current business-oriented compilers (compilers are used to translate a programmer's coding into terms a computer can understand); an intermediate-range committee to define specifications for a new language for use over the next two to three years; and then a long-range committee to complete the work. They projected that the first group would complete its work in three months and the last in several years.

The short-term committee performed the critical work, examining existing languages and defining the specifications for what become known as COBOL. Yet its initial mission was only to combine three existing business languages into one as a fast, interim solution to the problem of providing a usable common language. It was to be based on FLOW-MATIC from Remington Rand, AIMACO which had been developed by the Air Material Command in the U.S. Air Force, and COMTRAN, then under development at IBM. Remington Rand's language had first appeared in 1958, and the following year the Air Force finished its language and IBM published an initial release of its compiler and documentation.

The short-range committee was headed by Joseph Wegstein of the U.S. National Bureau of Standards.† The other eight members represented the U.S. Air Force, Honeywell, RCA, Burroughs, the Applied Mathematics Laboratory, Sylvania Electric Products, IBM, and Sperry-Rand. The most famous member of the committee was Jean E. Sammet,** who later became an outspoken advocate for COBOL and wrote extensively on its history and use. The committee held its first meeting on June 23–24, 1959, and completed its work when it submitted a final report in December entitled "COBOL—Specifications for a COmmon Business Oriented Language." This group developed a new language in three months, even though editing of the document and the language's details did not result in a published report until April 1960 when the U.S. Government Printing Office released *COBOL: Initial Specifications for a Common Business Oriented Language*.

The concepts underlying COBOL were developed during this first six months.

The establishment of this committee marked the first time a variety of competing manufacturers joined together to discuss specifications for a language. This was also the first time that the industry (manufacturers and users) agreed on such a major issue as the specifications for a language which had the potential for widespread use. Other industrywide meetings had been held in the past, and more would be scheduled in the years to come (especially through such organizations as SHARE† and GUIDE†). In 1959 participants in the development of COBOL were encouraged to cooperate with each other because the Defense Department had announced it would not accept proposals from manufacturers unless they supported such a common language on their equipment. Work was also rushed to prevent companies from completing development of their own business languages which, if allowed to continue, would have made it more difficult to adopt one standard for all. Remington Rand and IBM, for example, although members of the committee, were developing and releasing business-oriented compilers in the late 1950s and early 1960s.

After the short-range committee completed its work, both RCA and Remington Rand quickly produced COBOL compilers. In 1960 a Maintenance Committee was established to continue work on COBOL, cleaning up problems, keeping the language current with new technologies, supplying enhancements and guidelines requested by the data processing community, and working with such firms as RCA and Remington Rand as they released compilers. Pressure for more reforms grew. In 1961 new specifications were agreed on and incorporated in a major report called *COBOL-1961: Revised Specifications for a Common Business Oriented Language* (also published by the Government Printing Office for the Department of Defense which supported the revisions in June 1961). The importance of this particular document cannot be overstated because all subsequent revisions to COBOL by all manufacturers were to this set of specifications. These became the long-lasting reference to standards used by the data processing industry, despite other major updates issued throughout the 1960s.

Yet other changes appeared throughout the decade that were important and reflected the work of various committees. For example, in January 1964 the COBOL Maintenance Committee was divided into three subcommittees to deal with language, evaluation, and publication. The language group worried about COBOL's specifications, those on the evaluation subcommittee monitored user reactions and needs, and the third disseminated information, primarily using the *COBOL Information Bulletin* and government publications. Documents issued by vendors, such as those from IBM to describe their own COBOL compilers or as teaching aids, conformed to standards protected and defined by the Maintenance Committee.

COBOL gained broad appeal within the data processing world. From the beginning the short-range committee designed COBOL to have English-like terms such as OPEN, SORT, MOVE FIELD, and GO TO in order to make the text of any program readable by either the author of the program or, more important, by anyone else. The committee also avoided the more traditional use of algebraic notation in order to lend the language more to commercial than to scientific applications. It wanted inexperienced programmers to be able to use COBOL more quickly, and it also sought to enable persons other than the original author of the program to modify a piece of code if they so chose. By using English-like commands and redundant statements, the committee hoped to force a programmer to self-document what he or she was trying to do more frequently than before. This is in fact what happened, although years later many data processing managers would complain that standards for documentation were terrible. Even so, they were much better than those for languages used in the early to mid-1950s. COBOL was more verbose than other languages (such as FORTRAN) but only in order to impose discipline while increasing its ease of use.

COBOL was divided into four divisions: (1) the IDENTIFICATION Division which gave a program its name and always represented a minor part of the entire code; (2) the ENVIRONMENT Division, which caused the programmer to identify what hardware the code was to run on, such as the computer used; (3) the PROCEDURE Division, which was the critical one for defining what actions were to be taken with data—also known as the division with the executable operations defined; and (4) the DATA Division which was used to define information (files, length in characters, and so on) to be manipulated by the program. Until the present, these four divisions have remained intact, despite increases in the number and variety of possible commands, types of machines it could run on, and numerous variations of compilers. (See sample program, Table 8.)

There has been some controversy over what contributions earlier languages made to the fundamental architecture of COBOL. Jean Sammet, who participated in many of the early committees defining the characteristics of COBOL, argued that existing business languages of the late 1950s had varying influences. Her observations have been the most complete regarding the origins of COBOL. Sammet contends that FLOW-MATIC was quite influential because it was operational just before the original short-range committee went to work. The earlier language used full data names instead of FORTRAN-like symbolism. Commands in FLOW-MATIC were in English, such as ADD or COMPARE, proving that it could be done and thereby encouraging COBOL's designers to do the same. The committee also noticed that the designers of FLOW-MATIC had concluded that it would be more effective to separate descriptions of data and instructions from each other, a practice that is common today but was radically new in the late 1950s. The concept that data should be in a separate part of the program eventually led designers of programming languages* to create the capability of multiple programs using the same data. From that idea developed those concepts which today are embodied in such database management* software as IDMS and DL/1.

Those working on COBOL got very little from a second business language, called Commercial Translator, which at the time existed only as a list of

Table 8
Sample Program: COBOL

Problem: See REMARKS section of program below for description.
Program:

```
IDENTIFICATION DIVISION.
PROGRAM-ID.    'SORT360'.
REMARKS. THIS PROGRAM WAS WRITTEN TO DEMONSTRATE THE USE OF THE
         SORT FEATURE. THIS PROGRAM PERFORMS THE FOLLOWING TASKS -
         1. SELECTS, FROM A FILE OF 1000-CHARACTER RECORDS, THOSE
            RECORDS HAVING A FIELD-A NOT EQUAL TO FIELD-B.
         2. EXTRACTS INFORMATION FROM THE SELECTED RECORDS.
         3. SORTS THE SELECTED RECORDS INTO SEQUENCE, USING
            FILED-AA, FIELD-BB, AND FIELD-CC AS SORT KEYS.
         4. WRITES THOSE SORTED RECORDS HAVING FIELD-FF EQUAL TO
            FIELD-EE ON FILE-3 AND WRITES SELECTED DATA OF THE
            OTHER RECORDS ON FILE-2.
ENVIRONMENT DIVISION.
CONFIGURATION SECTION.
SOURCE-COMPUTER.    IBM-360 F50.
OBJECT-COMPUTER.    IBM-360 F50.
INPUT-OUTPUT SECTION.
FILE-CONTROL. SELECT INPUT-FILE-1 ASSIGN TO 'F401' UTILITY.
     SELECT SORT-FILE-1 ASSIGN 'SF1' UTILITY.
     SELECT FILE-2 ASSIGN 'F402' UTILITY. SELECT
     FILE-3 ASSIGN 'F403' UTILITY.

DATA DIVISION.
FILE SECTION.
FD INPUT-FILE-1 BLOCK CONTAINS 5 RECORDS
RECORDING MODE IS F
LABEL RECORDS ARE STANDARD
DATA RECORD IS INPUT-RECORD.

01 INPUT-RECORD.
    02   FIELD-A    PICTURE X (20).
    02   FIELD-C    PICTURE 9 (10).
    02   FIELD-D    PICTURE X (15).
    02   FILLER     PICTURE X (900).
    02   FIELD-B    PICTURE X (20).
    02   FIELD-E    PICTURE 9 (5).
    02   FIELD-G    PICTURE X (25).
    02   FIELD-F    PICTURE 9 (5).
SD SORT-FILE-1 DATA RECORD IS SORT-RECORD.
01 SORT-RECORD.
    02   FIELD-AA   PICTURE X (20).
    02   FIELD-CC   PICTURE 9 (10).
    02   FIELD-BB   PICTURE X (20).
    02   FIELD-DD   PICTURE X (15).
    02   FIELD-EE   PICTURE 9 (5).
    02   FIELD-FF   PICTURE 9 (5).
FD   FILE-2 BLOCK CONTAINS 10 RECORDS
RECORDING MODE IS F
LABEL RECORDS ARE STANDARD
DATA RECORD IS FILE-2-RECORD
```

Table 8 *(cont.)*

```
01    FILE-2-RECORD.
      02  FIELD-EEE    PICTURE $$$$$9.
      02  FILLER-A     PICTURE X (2).
      02  FIELD-FFF    PICTURE 9 (5).
      02  FILLER-B     PICTURE X (2).
      02  FIELD-AAA    PICTURE X (20).
      02  FIELD-BBB    PICTURE X (20).
FD    FILE-3 BLOCK CONTAINS 15 RECORDS
RECORDING MODE IS F
LABEL RECORDS ARE STANDARD
DATA RECORD IS FILE-3-RECORD.


01 FILE-3-RECORD PICTURE X (75).


PROCEDURE DIVISION.

      OPEN INPUT INPUT-FILE-1, OUTPUT FILE-2, FILE-3.
      SORT SORT-FILE-1 ASCENDING FIELD-AA DESCENDING FIELD-BB,
      ASCENDING FIELD-CC  INPUT PROCEDURE RECORD-SELECTION OUTPUT
      PROCEDURE PROCESS-SORTED-RECORDS. CLOSE INPUT-FILE-1, FILE-2,
      FILE-3. STOP RUN.

RECORD-SELECTION SECTION.
PARAGRAPH-1. READ INPUT-FILE-1 AT END GO TO PARAGRAPH-2.
      IF FIELD-A = FIELD-B GO TO PARAGRAPH-1 ELSE
      MOVE FIELD-A TO FIELD-AA MOVE FIELD-F TO FIELD-FF
      MOVE FIELD-C TO FIELD-CC MOVE FIELD-B TO FIELD-BB
      MOVE FIELD-D TO FIELD-DD MOVE FIELD-E TO FIELD-EE
      RELEASE SORT-RECORD TO PARAGRAPH-1.
PARAGRAPH-2 EXIT.
PROCESS-SORTED-RECORDS SECTION.
PARAGRAPH-3. RETURN SORT-FILE-1 AT END GO TO PARAGRAPH-4.
      IF FIELD-FF = FIELD-EE WRITE FILE-3-RECORD FROM
      SORT-RECORD GO TO PARAGRAPH-3 ELSE
      MOVE FIELD-EE TO FIELD-EEE MOVE FIELD-FF TO FIELD-FFF
      MOVE FIELD-AA TO FIELD-AAA MOVE FIELD-BB TO FIELD-BBB
      MOVE SPACES TO FILLER-A, FILLER-B WRITE FILE-2 RECORD.
      GO TO PARAGRAPH-3.
PARAGRAPH-4. EXIT.
```

This particular program, although simple, illustrates many of the commands as they appeared in an early release of COBOL as it might have been used on the IBM S/360 in the mid- to late-1960s when the language was initially used on a widespread basis.

SOURCE: Reprinted from *IBM Operating System/360 COBOL Language* (1965) GC28–6516. Courtesy of International Business Machines Corporation.

specifications. It was felt that many of the concepts inherent in Commercial Translator had been borrowed from FLOW-MATIC. Yet some specifics were used such as conditional situations in COBOL (where a program can authorize one to do such-and-such if X or Y condition exists). The design group also learned how to define multiple levels of data and how to use suffixes so that the same name could be used for different files simply by adding the name of a file to different files. A third business language, called FACT, had no impact on COBOL, nor apparently did the Air Force's AIMACO.

Initially, the committee intended COBOL to be used on what by 1959 standards were large computers. Manufacturers and designers wanted COBOL to run on such machines as the UNIVAC I* and II, IBM's 705* and 709, Honeywell's 800, and such others as the B-500 from Burroughs, Sylvania's MOBIDIC, and the RCA 501. COBOL eventually ran on many types of computers, satisfying yet another critical design criterion: that programs written and run on one brand of computers could, with little or no modification, operate on another. That characteristic has been preserved to the present time.

The major manufacturers of computers endorsed COBOL and moved quickly to develop compilers for the language. Ironically, the largest computer manufacturer, IBM, was nearly one of the last to introduce a compiler. Concerned that its own language would be compromised, IBM was initially wary about its customers' acceptance of COBOL. Yet with its fears allayed in 1961, IBM embraced COBOL and probably did more to foster its use in the 1960s and 1970s than any other organization. For the 7090 it had a COBOL compiler, and by the time it announced the S/360* in April 1964, it had made the language a critical piece of its software offerings. The S/360 announcement and IBM's hard push to sell this computer and new commercial uses for its products in general contributed significantly to the use of COBOL. Once available during the early 1960s, companies and many military data centers throughout the Untied States rapidly adopted COBOL. By the early 1970s it had become the preferred normal language for commercial applications. Foreign translations of COBOL led to new compilers in almost every major European language except French, and later it appeared in Japanese.

The fact that computer manufcturers supported the programming language, that the Department of Defense insisted on its use, and that data center managers in thousands of companies were eager for common languages all encouraged the rapid growth of COBOL. Acceptance came despite increasing criticisms, particularly by the early 1980s, that the language was too inflexible and that it reflected the work of a committee pushed to complete its work too quickly. Some critics argued that the design was too dependent on 1959's generation of hardware technology. Other languages, such as PL/1,* had easier coding conventions and better capability to support documentation. On the occasion of COBOL's twenty-fifth anniversary, *Computerworld* (the leading weekly publication of the U.S. data processing industry) published a harangue blasting COBOL as antiquated and ill designed. Yet such attacks defied the reality of

COBOL's experience. Throughout its life, there had been options such as FORTRAN, Assembler, and ALGOL,* not to mention more modern languages such as APL* and PL/1; yet COBOL remained the most widely used.

COBOL's language contributed to the evolution of other software tools. Perhaps its most important technical contribution was the concept that data, actions to be taken with facts, and the environment in which a program was to run were made discrete and separate parts of the code. Until the late 1950s the three were inseparable and difficult to work with, let alone modify. With COBOL the idea that each should be distinctive introduced a new era in the design of languages while improving the ease with which modifications could always be made. Perhaps as significant was the idea that data should be described so that it could be independent of any particular machine. The use of English-like mnemonics was another characteristic of this language.

Each of these features has appeared in subsequent languages and usually more efficiently. Other languages may have been easier to learn, especially by the late 1970s, taken less time to code to describe a command, and had more powerful data-handling capabilities. But these features appeared in large part because COBOL had proven them beneficial.

For further information, see: Robert Bemer, "A View of the History of COBOL," *Honeywell Computer Journal* 5, no. 3 (1971): 130–135; Richard L. Conner, "COBOL, Your Age Is Showing," *Computerworld*, May 14, 1984, pp. ID/7–12, 14–15, 18; "Time to Switch to COBOL?," *EDP Analyzer* 1, no. 11 (December 1963): 1–11; M. D. Fimple, "FORTRAN vs. COBOL," *Datamation* 10, no. 8 (August 1964): 34, 39–40; Jean E. Sammet, "The Early History of COBOL," in Richard L. Wexelblat, ed., *History of Programming Languages* (New York: Academic Press, 1981): 199–277 and her *Programming Languages: History and Fundamentals* (Englewood Cliffs, N.J.: Prentice-Hall, 1969); J. A. Saxon, *COBOL: A Self-Instructional Manual* (Englewood Cliffs, N.J.: Prentice-Hall, 1963); U.S. Department of Defense, *COBOL: Initial Specifications for a Common Business Oriented Language* (Washington, D.C.: U.S. Government Printing Office, 1960), *COBOL-1961: Revised Specifications for a Common Business Oriented Language* (Washington, D.C.: U.S. Government Printing Office, 1961); A. J. Whitmore, "COBOL at Westinghouse," *Datamation* 8, no. 4 (April 1962): 31–32.

COLASL. This programming language was developed at the Los Alamos Scientific Laboratory of the University of California to solve numerical scientific problems. The lab used this particular language primarily for production-oriented problems in the early 1960s. It also ran on one of the most famous computers in the history of data processing: STRETCH* (also known as an IBM 7030). In 1961, when it became operational, STRETCH was one of the technologically most advanced computers in the world.

COLASL was one of the two languages developed at the time within the same laboratory. The other was called MADCAP,* which was created by another group to solve numerical scientific problems as well. COLASL had good notational features for solving mathematical problems. While easy to read a program in COLASL, writing one was complicated because of strict rules. It was both procedural and problem-solving. It has been described as both a hardware and publication language. Although originally intended for use by nonprogrammers, it appeared that an understanding of computing was needed to use the language effectively.

The most obvious characteristic of this language and the one that insured it could be used for only a short period of time was its dependency on specific hardware. Once those pieces of equipment were no longer used, the language died. It relied on a specially built IBM 9210 Scientific Descriptive Printer which was used at Los Alamos beginning in May 1961; by the start of the 1970s these were gone. The four printers available to users of COLASL were attached to STRETCH. Their character set represented that of COLASL and in large part reflected classical mathematical notations. The language could be documented well by typing the English text descriptions of what was going on in red while typing the actual program in blue. Both would be loaded into the computer, but only the blue text (instructions) was actually executed. A user could also insert assembly language programs within those composed in COLASL. All data and constants were arithmetic. Because this language relied so heavily on the specific characteristics of STRETCH, it did not even have statements to handle input/output. Rather, it relied on assembly macros that used an existing package available with STRETCH. The language, however, did have its own set of routine mathematical functions catalogued in its own library.

Although it was used for only a brief period of time and quickly faded away with the introduction of newer computers by 1970, the language appeared when many others were being developed to solve numerical scientific problems. The greatest number of programming languages* developed in the late 1950s and early 1960s were precisely for scientific computing. For that reason, it reflected a common pattern of concern seen elsewhere in data processing.

For further information, see: K. G. Balke and G. L. Carter, "The COLASL Automatic Coding Language," *Symbolic Languages in Data Processing* (New York: Gordon and Breach, 1962): 501–537.

COLOSSUS. This British computer was built secretly during World War II and was made operational in December 1943. The machine was developed at Bletchley Park,† where a variety of computer-related British projects were undertaken, primarily associated with breaking enemy communication codes. COLOSSUS became one of the first electronic digital computers* built in Great Britain, and the experience gained from that project helped launch the British data processing industry's development of computers following World War II.

Little was known about the device until the late 1970s when the British government began declassifying some material relating to the activities at Bletchley Park. Although much remains to be learned, some technical details have become available. We know today that COLOSSUS was built in eleven months. It relied on a photoelectric reader capable of operating at 5,000 characters per second, used electronic counters and 1,500 valves, and performed in a parallel arithmetic mode (which meant it could perform more than one activity at the same time). It used punched paper tape* for input that operated at 5,000 characters per second—the same as the computer portion of the system. Like other machines built in the 1940s, it relied on electronic circuits to count, conducted binary arithmetic, and performed operations in Boolean logic. Its electronic storage registers (we would call this memory* today) were managed automatically by a sequence of operations. Other features that became common in future computers included conditional logic (otherwise called branching today), logically preset patch panels (switches), and telephone relays. For output it relied on a typewriter.

Professor M. H. A. Newman, a mathematician at Cambridge University since 1924, headed the development project. Also instrumental in the creation of COLOSSUS was T. H. Flowers, a British government expert on electro-mechanical digital devices employed in antiaircraft ranging. His experience with fire control systems reflected a similar background evident, for example, within the engineering department at the Massachusetts Institute of Technology (MIT) where computer development work for the military had been going on since the 1930s.

COLOSSUS was in fact several computers. The first was completed in 1943, and soon after the British government requested that others be built. Each one was an improvement over the previous machine. The second generation of COLOSSUS, called the MARK II (not to be confused with Mark* computers built in the United States and different from the British variants), was constructed in 1944. MARK II was built in several variations but was five times faster than the first COLOSSUS and was the result of work done by Dr. A. W. M. Coombs, another government engineer. MARK II had an effective speed approaching 25,000 characters per second. New enhancements over the older machine included an additional counter, heavier reliance on shift registers, and improved design of circuitry. The first MARK II became operational on June 1, 1944, just before D-day. Subsequently, other copies of the machine were made to do cryptoanalysis and other war-related work.

COLOSSUS generated experience with computers that engineers could use after World War II. With regard to the design of the machine itself, it was a special-purpose electronic digital computer that was program controlled, and it was similar in gender to such American devices as the ENIAC.* It is not clear how much dialogue took place between scientists and other engineers and professors since work at Bletchley Park was secret. Yet two observations are possible. First, because of the similarity of certain components in COLOSSUS and other computers, historians can see how dependent engineers were on existing technologies. Computer development illustrated how scientific or engineering progress evolved from existing pools of knowledge appreciated by all those concerned. Second, the experience gained in creating a stored-program machine (a computer that had its program within it instead of being keyed in by an operator command-by-command at time of execution) was a dramatic step forward over machines of the 1930s. Yet that dramatic evolutionary step was taken almost concurrently by computer designers elsewhere, particularly in the United States. That evolution grew out of existing technologies, learned experiences, and similar pressures to build more efficient machines.

The experience with COLOSSUS proved a positive one. Given the demand for additional copies of the computer, requests made at that time only by the British government, the performance of the machine must have been satisfactory. Several hours' worth of hand calculations could now be done in less than ten minutes. Reliability proved very pleasing. Fewer problems were experienced with this computer than with such electromechanical portions of the system such as the photoelectric reader or the typewriter. The actual amount of time between failures is not available, but those who worked on the machine have suggested that COLOSSUS malfunctioned less often than other contemporary computers.

For further information, see: B. Randell, "The COLOSSUS," in N. Metropolis et al., eds., *A History of Computing in the Twentieth Century* (New York: Academic Press, 1980): 47–92, and "Colossus: Godfather of the Computer," *New Scientist* 73 (February 10, 1977): 346–348; A. M. Turing, "Intelligent Machinery," a document dated September 1947, reprinted in B. Meltzer and D. Michie, eds., *Machine Intelligence* (Edinburgh: Edinburgh University Press, 1969): V:3–23.

COMIT. COMIT was one of the earliest list processing languages and, more important, showed how a language could search for particular string patterns and subsequently perform transformations on them. All subsequent string switching languages emerged from COMIT. Members of the Mechanical Translation Group within the Research Laboratory of Electronics and others working at the Computation Center, both at the Massachusetts Institute of Technology (MIT), jointly wrote the description of this language, completing it in December 1957, for a version to run on an IBM 704* computer. Dr. Victor Yngve, who headed the development of this language, created a unique one in that it was intended for use by professional linguists in research. COMIT was the first important programming tool designed to aid the study of linguistics.

Between 1957 and 1961 work progressed on the language. It was first run on an IBM 709/90 and later on an IBM 704. In September 1961 Yngve made the language available to whoever wanted it through SHARE† (an International Business Machines Corporation [IBM] users' organization), but with his people retaining responsibility for the maintenance and improvement of COMIT. Within two years, COMIT II had been designed within MIT and, by the summer of 1965, operated under MIT's Compatible Time Sharing System, having moved from a purely batch system to a quasi-interactive version. In 1965 Yngve left MIT for a position at the University of Chicago, taking with him the responsibility for further work on COMIT II. During the first half of 1968 he had another version of COMIT II operational on an IBM 7090/94, while yet further revisions were being readied to make the language run on an IBM System 360.* During the 1970s the language was used and enhanced but not to the extent it had been in the 1960s.

COMIT has been characterized as the string processor that introduced the idea of a rewrite rule in programming. In other words, it showed how a string could be changed in a particular way through specification of a pattern. This capability became attractive to others besides linguists, particularly to those who were trying to solve problems in manipulation of symbols. Others found it useful in retrieving information. COMIT was flexible and was both procedure-oriented and problem-oriented. It has also been called a problem-solver. Yet the most basic building block of COMIT—the *rule*—has been called nonprocedural. Although hardly ever equipped with sufficient features to call it an online system, this batch processor was relatively machine independent, despite the fact that it more frequently ran on IBM's computers than on those of other vendors. An exception to this was a version called LECOM which ran on a GE 225 computer at a number of universities in the early 1960s. At Washington State University a jump was made from IBM's second-generation computer to its third-generation System 360s. Subsequently, others followed suit.

Although the definition of the language was always written in ordinary English, no serious effort appeared to have been made to structure a formal system of notation. Consequently, not all of the commands and functions were fully understood. The language lacked arithmetic functions of any consequence, which probably represented the greatest difference between it and almost all other programming languages.* COMIT was not, of course, intended to be a general-purpose programming language; thus, the omission of arithmetic features was not considered a serious problem. Some users in the 1960s, however, thought that it had no convenient manner to assign their own names to strings—a feature that would become available on many other string processing languages.

Yet COMIT was an example emulated by designers of string processing languages. Its world was frequently that of the university or the commercial designer of programming languages. A commercially available version of this program apparently was never developed. It was influential, however, lending many functions to other languages. The CONVERT commands in LISP* came from COMIT as did others in Formula ALGOL.* The most important of the string and list processing languages, SNOBOL,* borrowed from COMIT. The use of such languages became somewhat easier after COMIT. This was the case, for instance, with LISP's notation, even if it was more difficult to learn than COMIT.

For further information, see: V. H. Yngve, "A Framework for Syntactic Translation," *Mechanical Translation* 4, no. 3 (December 1957): 59–65, "COMIT as an IR Language," *Communications, ACM* 5, no. 1 (January 1962): 19–28, and his "COMIT," ibid., 6, no. 3 (March 1963): 83–84.

COMMERCIAL TRANSLATOR. Commercial Translator was one of the first computer languages developed at the International Business Machines Corporation (IBM)† for business data processing. By January 1958 specifications for a business-oriented language had been written for a product called COMTRAN. The intent was to create a language with English-like syntax. This language was renamed Commercial Translator, and it made three important contributions. First, it applied mathematical formulas to a business language which in the past had been done only for scientific software packages. Second, it introduced the use of IF . . . THEN . . . functions in a programming language. Third, the language contained the idea of using various levels in describing data. The use of the *Picture Clause* in COBOL* came directly from the Commercial Translator.

The language first appeared in 1959 when the short-range committee was working on what would become COBOL.* As one of the few business-oriented programming languages* available, it influenced some of the characteristics of COBOL. The 1960's version of the Commercial Translator contained features which the committee had defined for COBOL. Between 1960 and the end of 1961 IBM supported its Commercial Translator, arguing that COBOL was not yet sufficiently defined to motivate the company to abandon its own product. Furthermore, some customers were already using it and could not be denied support. Hence, IBM made it run on its 709, 7070, 7080, and 7090 computers. At the same time, however, the U.S. government threw its support behind COBOL, pushing IBM into offering compilers for both languages. Finally, with the introduction of the System 360* in 1964, IBM withdrew support for the Commercial Translator. By 1970 there were only a few users of this early language; COBOL had supplanted it as the most widely used programming language for business applications.

IBM's software* engineers had cut their eye teeth on the Commercial Translator. Their later programming languages reflected the experience gained with this early software, one that had introduced new and interesting technical features. Although hardly remembered by historians, it was important for IBM's development. Company designers learned how to introduce floating-point numbers in a business language. In Commercial Translator they created a truth operator, included the capability to specify functions, added a parametric substitution (the *DO* command), gave it an absolute value operator, and allowed programmers to assign values to a condition name. These functions increasingly appeared in one form or another in subsequent languages.

For further information, see: Jean E. Sammet, *Programming Languages: History and Fundamentals* (Englewood Cliffs, N.J.: Prentice-Hall, 1969).

COMPTOMETER. Comptometer was the first key-operated adding machine and thus one of the earliest such devices to resemble the modern adding machine. At the turn of the century the term *comptometer* was used to describe other adding machines that were similar in appearance and function. The original comptometer was manufactured by the Felt and Tarrant Manufacturing Company of Chicago. This company introduced the original product in 1887 at a time when a number of such devices were on the market along with other mechanical machines, for example, cash registers. Like the typewriter, it was a new and additional aid to what the 1970s would call office automation. Prior to this machine, commercially available devices could rarely add and subtract. The comptometer performed all four mathematical functions in one device. The operator simply pushed buttons to perform calculations, much as one would today without any additional steps. The one big difference was that the comptometer was a mechanical machine, whereas today's either are electrical/mechanical or, like the hand calculator, rely on logic technology as do computers. The advent of the comptometer greatly encouraged the use of mechanical devices because it was an easier and faster method than older and often manual methods. The calculator was accurate and could carry transactions up to eight digits. Its principal marketplace was commercial, although copies appeared in universities, laboratories, and government agencies as well.

During the 1890s various models appeared with enhanced features. These were designed to reduce the size of the machine while improving the capacity or convenience of mathematical functions. Thus, for instance, an interface guard was introduced into some of the later models that prevented more than one key from being hit from a misdirected punch by an operator's finger. Therefore, if an operator's finger landed on one key hard and another lightly, the latter would not be depressed. Another enhancement involved giving the user the capability of deliberately depressing more than one key simultaneously. That function allowed more rapid depressing of keys and hence faster data input into the machine or permitted multiple mathematical transactions. It was most useful in trying to create multiple columns of data, each with more than one digit of data per row.

By 1900 the success of this device had made the comptometer an industry standard. By World War I it was one of the top four machines in use within the United States, competing with devices marketed by Burroughs,† the National Cash Register Company (NCR),† and Felt and Tarrant. The comptometer and variants of its technology disappeared by World War II as newer devices became available. After the war digital computers* and electronic desk calculators dominated.

For further information, see: D. E. Felt, *Mechanical Arithmetic or the History of the Counting Machine* (Chicago: Washington Institute, 1916); E. M. Horsburgh, ed., *Napier Tercentenary Celebration: Handbook of the Exhibition* (Edinburgh: Royal Society of Edinburgh, 1914), reprinted, Los Angeles: Tomash Publishers, 1983.

COMPUTER. A computer is a functioning device that performs computations without the intervention of people while it is in progress. More specifically, a computer has also been defined as a machine that can perform calculations, sort data, file away information, edit, and otherwise manipulate (process) facts. There are two basic types of computers: digital* and analog.* Digital computers perform their tasks on discrete pieces of information such as numbers (digits) and letters (characters), whereas analog machines perform their operations on variable quantities of information. These variable quantities can be such data as weights, voltages, or lengths.

Considerable controversy remains over the origins of the word *computer*. One early use of the term occurred in 1398, when a scribe named Trevisa, in discussing the maintenance of a calendar, wrote the phrase "compotystes . . . departed by twlue mones, in sixe euen and sixe odde." This English source of the word was apparently not the first use of the word, for earlier it had been employed to describe those who did day-reckoning. Another Englishman, Sir Thomas Browne, wrote in 1646 about "Calendars of these computers." By his day some Europeans had already begun to make mechanical aids to calculations. One of these inventors, Blaise Pascal,** had built a calculator in 1652 which he called the *Pascaline*.

Yet other terms evolved over the years which served as possible bases for the word used to describe calculation machines. Thus, the word *abacus** (the oriental frame with rows of balls for counting) originated in Greece and later became a Latin term for slab, when the term changed from *abax* to *abak*. These early words described a device that consisted of a board covered with sand on which one would trace figures and, after completion of an exercise, erase them. By the seventh century slabs could also be erased while calculations or calculating was done using strings of stones (*counters*). The Latin word for such stones or pebbles was *calculus*, the diminutive of the earlier words *calx* or *calc*. Those words were also the origins for lime and limestone yielding *calcium* and other variations of *calci* later into English. The word *calculus* in English, therefore, was as much a medical term for "stony formations" as it was a phrase for describing arithmetic figuring. *Callus* refers to bony substances and *callous* to hard skin.

The origins of the word *computer* have also been linked directly to words associated with mathematics. For example, the word *count* has various forms, as does *counten* in Old French along with *cunter*, *conter* (to tell) and, more recently, *conte* or tale. Latin contributed *computare* and *compute*. The Latin origin is of particular importance because it merges two words: *com*, which means together, and *putare*, which means to think or reckon. When put together to form *computare*, we come close to the word used today to describe the function performed by modern computers. By the fourteenth century, Europe had returned to the more Latin version of the word—to *compte*, from which, by approximately 1500, came the new spelling of *comptroller*. Today that word, and its variant *controller*, describes someone who is responsible for counting money or is in charge of managing it, or for auditing accounting practices. The word *comptroller* originally also evolved from the marriage of two Latin words: *contra*, meaning against, and *rotulus*, meaning register. The second half of the word may well have come from *roll* or *rote*, the words used to describe a person who checked rolls.

A number of tangents and compounds evolved from the words relating to computer. These include the more obvious words used to describe related functions: account and accountable, recount and discount, not to mention such distant cousins as count (a sixteenth-century title in English that originated from the French *comte* and Italian *conte*). From Latin emerged *comes* and *comit* to imply a coming together, hence *companion*, later a friend of a king, in short, a *count*. The second half of the word *companion* may have come from the Latin word *panis*, meaning bread. Thus, companion may have meant the person who broke bread with a specific individual. This kind of action could quickly have led to *comrade*, to the French term *camarade* or the Spanish phrase *camarada*. An interesting variant was *camera* from Latin meaning box or room. The device for taking pictures today comes from the term *camera obscura* (dark box) which originally meant roommate.

Another source of meaning for the modern word is from *counter*, a word that originally described someone who counted and today refers to a function in a computer as well. Yet there was a distinction between the person who counted and a machine that did calculations or calculating. The distinction had to be made because many words came from counter, originally perhaps *countre* in old English, and from the French *contre* and, earliest of all, *contra* meaning against from Latin. Thus, such words as counteract, countermand, and counterpart came from the first word counter and its variants. Despite all this confusion over the origins of words, it appears that the specific phrase *computer* came more directly from calculate than from counter.

But regardless of the word's origins, its use derived from an action statement, calculator. In the seventeenth century European machines developed for the purpose of aiding arithmetic were described as calculators and calculating devices, using the appropriate variant of these words, depending on whether the inventor spoke French, English, German, or Italian. By the mid-nineteenth century, calculators were commercial products, and the notion of machines processing numbers had been well established. The linkage with mathematical terms remained hard and fast throughout the century, with commercial products even being called arithmometers* and compu-this or that. With the evolution of such devices into computational machines of some sophistication in the late nineteenth century and during the early decades of the twentieth century, computational devices evolved into computers. By the 1930s the term was commonly used in research facilities, such as at the Massachusetts Institute of Technology (MIT). These electronic devices could only process electrical impulses, which in various combinations represented numbers (digital) or continuous streams of feedback (analog). The analog contraptions made at MIT by Vannevar Bush** or the digital machines (such as the ENIAC*) of the 1940s were conveniently called computers. In fact, they performed computations of numbers. Even analog devices were thought of in this manner. By the time such equipment could handle data based on letters and not just numbers, the term *computer* had come into common use. Since the late 1940s, therefore, computer has come to mean a distinct set of devices that can process numbers and perform logical sequences of functions.

For further information, see: International Business Machines Corporation, *Vocabulary for Data Processing, Telecommunications, and Office Systems*, 7th ed. (New York: IBM, 1981); Jack B. Rochester and John Gantz, *The Naked Computer* (William Morrow and Co., 1983); Joseph T. Shipley, *Dictionary of Word Origins* (New York: Philosophical Library, 1945).

COMPUTER, ANALOG. There are two types of computers, digital* and analog. The analog has a much longer history than the digital, although these older devices played a less significant role in the history of data processing in the twentieth century. Analog devices were easier to construct and so appeared first. By the 1920s they were the subject of considerable research and development. In the years just before World War II the increasing scientific knowledge about computational equipment formed part of the technical basis for many of the configurations of the 1940s which today we call digital computers. Despite their limited impact on the history of data processing, analog devices preserved their distinctive niche in data-handling equipment throughout the twentieth century.

Analog is an adjective describing devices that measure or process information in a continuous manner. The most common example always cited of an analog device is the wristwatch which shows time when looked at. A clock that shows time in specific numbers, for example, 9:41, would be called a digital device. It would be an analog unit if time were told by the position of the little and large hand or pointer aimed at a particular number on the clock's face. In the example of the wristwatch, time is expressed in a continuous manner. Another illustration is the slide rule* where the answers are given as a series of numbers on a continuous scale. On the other hand, a pocket calculator, because it shows an exact answer in precise language (e.g., a number), is a digital device. Formal definitions of analog computers always state that these instruments process analog data. Analog data are items of information typically representing physical quantities that are continuously changing or variable. Such information consists of data "whose magnitude is made directly proportional to the data or to a suitable function of the data" (IBM *Data Processing Glossary*). That is, answers are given relative to other answers.

Put into more intelligible terms, such data are representations of something going on right now, such as the amount of gasoline available in a car as measured by the fuel gauge (another analog device). During the past forty years electronic analog computers have been used to measure more than to calculate. They have monitored voltage, pressure, and temperature, and some simple analog devices have included barometers, thermometers, and the simple voltmeter. The word *analog* (also spelled analogue) suggests precisely the kind of answer obtained:

analogous to something else. That is, data appear in continuous form, an approximation of the true answer. Digital, on the other hand, gives an exact answer. Thus, 2 plus 2 in digital form is four, while in analog the answer lies between 3 and 5.

Analog devices have been used to measure physical phenomena and for applications in engineering, physical sciences, and process control. An oil refinery, for example, relies heavily on these devices to manage the flow of raw material throughout the plant. Many shop floor computers that control either an assembly line or a particular process in that line (e.g., the painting of a truck) were frequently analog computers, and they were programmable. Analog computers were either general-purpose devices that operated on a wide variety of data, such as in process control, or special-purpose machines to do very limited tasks (e.g., only paint trucks). It is important to understand the concept of the analog because it has existed throughout the history of computing technologies and its fundamental distinction from digital has not varied dramatically. At worst, digital devices have created some confusion because they can perform many of the tasks done by earlier analog equipment and often cheaper and better. Their distinctions were most evident in the 1930s and 1940s, diminishing in the 1950s and 1960s. Knowledge of their history makes it possible to understand why they attracted so much attention within the field of computing in the early days of computers.

The applications that first and most dramatically provided incentives to develop analog devices were astronomical observations and then navigation. The first problem was how to measure and predict future movements of stars and planets. In the beginning this concern grew out of astrology and religion. Later, when men sailed, they needed assistance in using astronomical knowledge to guide their paths across the seas, particularly when they sailed out of sight of land by the Middle Ages. It was easier to build a model of the heavens based on thousands of years of observations and thus to measure the movement of stars and planets against that model than to calculate mathematically their precise behavior. That last step would await the development of appropriate mathematical aids (such as calculus) and a better knowledge of astronomy in the nineteenth and twentieth centuries. Analog devices, therefore, were an obvious first step. Analog devices have been found that date back to 400 B.C. A common early device was the astrolabe* with its variant for astrologers called Orreries and later other variants used just for navigation.

The astrolabe has been traced back to the Greeks but may have existed earlier. The earliest documented use of this device is about 180–125 B.C., during the life of Hipparchus of Bithynia. The astrolabe measured the movement of the heavens. Known also as a planetary equatorium, it calculated future movements. It looked like a metal tray with dials and numbers. By rotating these retia over the plates, the positions of various heavenly bodies could be determined. Such devices were used in Europe into the early modern period.

A step toward sophistication was taken with the development of a device that had a drive mechanism. The earliest of these devices was Greek and was made up of gears and plates called the Antikythera device.* Dials represented the positions of the sun and moon throughout the period of a lunar calendar and on a daily basis. Such gadgets first appeared in Europe during the Greek period. Knowledge of these was apparently lost (perhaps after the fall of the Roman Empire) and did not apparently reappear until 1575, remaining for some 300 years thereafter.

Although watches and clocks represented another major step in the evolution of analog devices and made their appearance in the late Middle Ages, usually as large clocks planted in towers, their history did not contribute directly to that of data processing.

Applications or needs always drove technology in new directions, and that was certainly the case with other analog devices. Tide predictors* fall into this category. Concern for the ebb and flow of tides existed from the earliest days of sailing because a miscalculation could cause a sea captain to rip the bottom off his ship as he entered or left port. It was never sufficient to observe that the tide was rising or falling through normal eye observation. Because sailors had noticed a correlation between the movement of the heavens and the behavior of tides, a series of attempts were made to correlate the two in predicting the future behavior of the seas and oceans. By the 1400s a device had been invented that calculated tidal patterns for various European cities. In the 1800s Lord Kelvin developed a mathematical formula for predicting tidal behavior using harmonic analysis. Thus, by the middle of the nineteenth century, a device existed to predict tides using mathematics as well as pulleys and gears. Others were built in the early twentieth century based on his formula, the most famous of which was that of the U.S. Coast and Geodetic Survey. This agency began construction of a device in 1905 and put it into use in 1911, keeping it in operation until the 1960s. It computed data in mechanical form and may have been the longest operating computational device in existence.

The slide rule represented yet another development in the world of the analog. During the nineteenth century it came into its own as one of many aids to computation. In its modern form, it could be found as a ruler with a movable slider in the middle providing estimated answers and, like the astrolabe of old, as a round dish with numbers and arms. It remained the single most popular and portable calculating device for mathematicians and especially engineers until the advent of the hand-held digital calculator in the 1970s. Generations of students in science, mathematics, and engineering grew up with this device beginning in the 1880s. Earlier versions had existed for some 200 years but were used by only a small number of individuals.

As mathematics came into its own with greater complexity in Europe during the seventeenth and eighteenth centuries, the need for help grew. One area of concern that led to the development of analog devices involved what became known as the differential analyzer.* Scientists and mathematicians faced the problem of how to calculate areas under a given curve. This is known today as the integral of the given function, expressed correctly as:

$$\int f(x)dx$$

which became the symbol of a problem to solve with the use of analog equipment. If the function $f(x)$ was not known in some analytical form, mathematicians ran into difficulty in calculating elementary functions and, even if known, they did not do very well. Their answers were iterative. Since the early nineteenth century, attempts had been made to develop mechanical devices to help; at the end of the century William Thomson (Lord Kelvin's brother) built a machine made up of a rotating disk, a sphere, and a cylinder that could do the calculations involved in an elementary integration.

Yet the devices that proved most useful to the development of computers in general came from the Massachusetts Institute of Technology, home of many important computational projects in the twentieth century. There, Vannevar Bush** built machines in the 1930s that were practical enhancements to earlier inventions. His machines were sophisticated by the standards of the day, used electricity and mechanical parts, and became available to scientists. His idea was to have a string wound around a spinning shaft being pulled. As it tightened the shaft tugged at the other end. The end of the string being controlled was in turn connected to an output cylinder, while at the other end it was used to drive gears. These gears would pass the results to other equipment that computed answers. Bush wanted to use this equipment, which became known as a differential analyzer, to solve significant problems using differential equations. Such calculations needed to be performed when analyzing a network of electricity, such as that of an electrical company. Bush constructed three different machines in the 1930s to accomplish his mission and published articles on their technology which were widely read by those who, in the 1940s, built digital machines.

Other scientists made Bush analyzers, as they were nicknamed; five were known to have been constructed. Perhaps the most famous of these were built at Manchester University by Douglas R. Hartree** and A. Porter and a second at the Moore School of Electrical Engineering† at the University of Pennsylvania where, in the 1940s, engineers built the ENIAC*—the first digital electronic computer. Although these devices had some electrical parts, they were primarily mechanical. Not until the late 1940s could they be characterized as electrical analog computers in the full sense. The same comment applied to digital computers.

The years between the two world wars were fruitful ones for analog devices. In addition to the construction of harmonic analyzers, enhancements to tide predictors, and the advent of mechanical differential analyzers, work commenced in other areas. For example, producers of electricity in the United States sought ways to analyze the flow of their product across complex networks of wires and for that purpose encouraged and sponsored the development of network analyzers. Those used by General Electric,† for instance, measured the flow of electricity and performed calculations to predict the demand and behavior of this substance within a network. Electrical networks were simulated as early as 1925. A similar requirement existed for telephone communications, and thus Bell Laboratories† also focused on such technologies in the 1920s and 1930s. In the case of the network analyzers of the 1920s and 1930s, they could only examine steady-state problems involving drops in voltage along a line, measure possible flows of currents along particular lines, and model how electricity might be delivered to different points using various paths. By the 1960s such devices could also measure and calculate transient conditions which developed, for instance, during faults or while networks were being switched on or off. Digital computers could also model networks using software* designed to do just that. One early illustration of that task was the use of GPSS,* a simulation package, in the 1960s which could be used to analyze loads on telecommunication lines. In the 1970s students with little knowledge of computers were using that software along with other programs on high-speed computers.

George A. Philbrick produced some advances in analog devices during the 1930s. This engineer pioneered the use of what is called high-gain d-c amplifiers in computational devices. In simpler words, he was one of the first to use operational amplifiers in a computational machine. He sought to produce a computer that could model a system in an electronic computational device. However, his work did not receive the publicity necessary to influence other research.

The work that did receive adequate attention was done at Bell Labs where the use of amplifiers in computers was emphasized. This project grew out of the work done by C. A. Lovell and D. B. Parkinson. They used such devices to build the M-9, an antiaircraft gun director during World War II. Designed by them and built at Western Electric, the device was intended to set the sights of a gun on a moving airplane as the gunner turned knobs to guide the sight of the weapon toward enemy aircraft. The M-9 Gun Director represented a combination of a mechanical differential analyzer (yet much smaller than Bush's) and electronic analog calculations. The device soon came to the attention of other researchers.

The most important of these early scientists was J. B. Russell at Columbia University who, armed with a contract from the National Defense Research Committee, built a general-purpose electronic analog computer, the first such device. In May 1947 he and his colleagues published the first article describing their work, probably the earliest document to detail the use of an operational amplifier as a component of a general purpose computer. Their machine was based on the M-9, but unlike the more specialized device, it could work with broader mathematical problems.

Developments in early electronic analog computers came quickly and at the same time as digital devices. The 1940s were years during which many computer-based technologies that had been in the labs were converted into war-related devices and others for commercial use. In the late 1940s in particular, these machines were converted into broader, general-purpose pieces of equipment that became more electronic and less mechanical in form. Demand for such devices, both analog and digital, increased, leading to the birth of the first computer companies and to the conversion of others from being vendors of office and card punch equipment into data processing giants as we know them today.

In the case of the analog computer, this pattern of behavior was best illustrated by the Reeves Instrument Corporation which, in 1947, developed an analog device for the U.S. Navy. Other companies soon joined the action, developing the forerunner of the more modern analog computer (1947–1948). The exact date remains debatable since the Reeves device was built in 1947 and was used extensively in 1948. From that machine emerged a more sophisticated one used by the military called the REAC. Between 1947 and 1949 several companies were established in the United States to build analog devices. Their machines were used to do network analysis, perform military applications such as directing guns and bomb sighting, track events in wind tunnels, simulate flight, compute polynomials, and solve problems in differential analysis.

The class of machines that had been in use the longest and most effectively through the late 1940s was the network analyzer. Why did these devices not dominate the world of analog computers? Why were other analog devices invented and sold in the late 1940s and early 1950s? The answer might be economic. Using data from 1955 illustrates the problem. A large analyzer cost the same as a large digital computer or mechanical differential analyzer, ranging from $100,000 to $500,000. An analog computer of the same period cost far less. A second problem involved components. The parts used in network analyzers allowed errors to creep into the system. These caused problems in accuracy and resulted from resistance in inductors, leaking capacitors, and inductance of resistors. Yet such devices remained very useful for measuring electrical power and its distribution. Although no evidence has yet surfaced to prove it, since the market for such devices was quite limited, all that were needed were already in the hands of electrical companies and were loaded up into their accounting books with very long depreciation schedules. It was that very accounting practice which in the late 1970s and early 1980s caused enemies of American Telephone and Telegraph (AT&T)† to call for a breakup of the Bell empire in order to foster appropriate market conditions to introduce more current technologies in the field of telecommunications.

Network analyzers notwithstanding, by the late 1940s and early 1950s a new class of analog devices called the mathematical analog computer was emerging. This type of machine initially included the old mechanical differential analyzers and even digital integrating differential analyzers. These mechanical machines were accurate up to five digits (figures) but by the mid-1950s were being replaced with digital equipment. The digital equipment was less expensive and even more precise. Electronic differential analyzers were also easier to use than mechanical ones.

Analog technology in the 1950s appeared in many applications. Typically, these involved having data (answers) in figures of four or less and were used to solve problems involving collections of simultaneous differential equations that had either constant or nonconstant coefficients. Output was presented to an end user graphically, such as on a plot, similar to what a cardiologist obtains when measuring the behavior of the human heart. Thus, such equipment became popular tools for an engineer measuring a particular phenomenon.

The components of an analog computer of the 1950s and 1960s were similar. D-c amplifiers, called operational amplifiers, served as integrators or collectors of data fed to the system. Potentiometers were used to set coefficients in a problem. A third collection of parts allowed an end user to start and stop computations and data gathering. With time sophisticated hardware and then software* appeared that would allow a user, for example, to hold the solution of a problem at any point in the process of calculating the solution. The first efforts at such controls involved the use of relays, but by the 1960s software came to dominate. Relays disappeared as transistors, and later chips* became the heart of all computers.

By the end of the 1960s analog computers were appearing in industries, first in laboratories and engineering departments and then on the shop floor tracking the flow of events, materials, temperatures, and other environmental conditions. Thus, for example, a paper-making machine of the late 1960s had analog computers attached to monitor the percentage of pulp and water being mixed together to make a certain class of product. In laboratories and hospitals they monitored bodily functions, for example, heart rates and temperatures by the early 1970s.

Analog devices were inherently more restrictive in applications and in the specificity of output and so never reached the level of use seen with digital computers. During 1972, for example, U.S. manufacturers of analog computers had shipped only $47 million of such devices as opposed to $1.790 billion in digital products coming off the assembly line. To put this into further perspective, in the same year peripheral equipment for digital computers amounted to $2.148 billion, and additional memory for digital equipment sold for $1.518 billion. The sale of data entry equipment, usually a low figure in any data processing department's budget for equipment, and such telecommunication equipment as modems (to connect telephone lines to terminals) outsold analog devices. Data from the 1960s and 1980s reconfirmed the fact that analog devices remained in a minority.

Yet this technology remained as current as that of digital machines. Analog computers sold in the 1970s by Electronic Associates, Inc., Systron-Donner, Inc., Applied Dynamics Corporation, and Telefunken reflected the same hardware components available in reliable digital equipment. One of the leaders in analog devices for shop floor applications in the early 1980s, Hitachi, had programmable machines using high-capacity chip technology. These machines interfaced with digital computers to drive automated equipment on the one hand and to produce a variety of management reports on the other.

The analog device underwent several clear lines of evolution. The first involved mechanical devices from the ancient era to the age of the Industrial Revolution which were designed to aid astronomy initially, then navigation, and finally mathematics. A second phase began in the 1800s and led to more complex hardware, all mechanical, culminating with electronic machines in the 1930s and 1940s. The next era, which continues to the present, saw very sophisticated devices that could handle a large band of specialized and general-purpose applications, both large and small. But as with digital devices, technologies bred new advances on old ones. The demands of the marketplace and competition by the 1950s caused greater advances over those of the 1930s. Analog computers did not prove to be as versatile as digital computers and have therefore always been relegated to a lower status. Yet in the 1930s at least, research on analog devices contributed handsomely to the rapid development of digital computers in the 1940s, a factor often overlooked by those who have taken greater note of the advances in electrical engineering alone.

For further information, see: Vannevar Bush, *Pieces of the Action* (New York: William Morrow and Co., 1971); P. A. Holst, "A Note of History," *Simulation* 17, no. 3 (September 1971): 131–135; C. F. Jenkin, *The Astrolabe—Its Construction and Use* (Oxford: Oxford University Press, 1925); Clarence L. Johnson, *Analog Computer Techniques* (New York: McGraw-Hill Book Co., 1956); D. H. Pickens, "Electronic Analog Computer Fundamentals," *Proceedings of the IRE* 25, no. 3 (August 1952): 144–147; J. R. Ragazzini et al., "Analysis of Problems in Dynamics by Electronic Circuits," *Proceedings of the IRE* 19, no. 2 (May 1947): 444–452; Michael R. Williams, *A History of Computing Technology* (Englewood Cliffs, N.J.: Prentice-Hall, 1985).

COMPUTER, DIGITAL. This class of computers represents one of two essential types; the analog* is the other. After the 1940s the digital became the more widely used of the two. It was also the most frequently used machine for scientific and commercial applications. In addition, digital computers served as the technological basis for the rapid growth of the data processing industry. These computers ranged in size from the very large processors used to support entire organizations with thousands of users down to the microcomputers of the 1980s found on desks in offices and in homes.

Digital computers became more widely used as general- and special-purpose equipment than the analog because the digital operated on information in some discrete form whereas the analog performed operations in a continuous manner. Analog devices proved particularly useful for process-oriented applications of some limited scope, such as in regulating the flow of materials in production. On the other hand, a digital computer could perform a wider range of functions and hence was used for specific and varied applications in both business and science.

A digital machine measures and represents all quantities or information as discrete digits. Thus, a clock that shows the time as 12:10 is a digital device, whereas the analog has a big arm and a little arm, with the shorter one pointed in the general direction of the 12 and the other at the 2. Digital devices have been used for many centuries; however, electronic digital computers have only been in existence since the 1940s. Some research on their function did take place in the 1930s. On the other hand, analog devices, also endowed with a rich heritage, were already available in electronic or electromechanical form in the 1930s. Yet they never became as useful as digital computers did by the end of the next decade.

Most historians of the electronic computer think primarily in terms of the digital. They also refer to specific machines as belonging to a particular "generation." Each of these generations, marked by specific years, also reflects different technologies from which computers were built. The history below is structured around four generations. The first generation, which covers the years 1946 to 1953 or to 1959, witnessed the birth of the electronic digital computer, particularly the ENIAC,* BINAC,* and EDSAC.* Those who believe the first generation lasted from 1946 to 1959 include the UNIVAC* series in this categorization. The second generation commonly refers to those machines designed and built between 1959 and 1964. The third generation marked the arrival of the IBM System 360* in 1964 and ran until 1969/1970, when the fourth generation, the one we are still in, followed. Many observers within the industry argue that we are about to enter the fifth generation, a proposition that caused much controversy by the mid-1980s.

The categorization of computers by generation, although an historical convenience, is not completely tidy. Not all vendors migrated to a new generation at the same time; therefore lags occurred, particularly at the end of an era. For example, when International Business Machines Corporation (IBM)† began to ship S/360s in 1965—a third generation computer—at a time when everyone else was still selling second-generation products; yet historians refer to the end of the second generation almost arbitrarily and cleanly as 1964. To make these chronological breaks relevant, historians generally link a computer to a particular generation based on the type of technology it used rather than on when it was sold and installed. The word *generations* is used in this sense below. In this manner a generation served as a bellwether of new or different technologies. Unlike the case with other technologies (e.g., steam and electricity), historians of data processing have not yet characterized eras or generations by applications, that is, by periods of change in the use or impact of a particular invention.

The first generation (1946–1959) was characterized by the use of vacuum tube technology. Main memory* in a computer involved the use of delay lines, electrostatic tubes, and magnetic drums. These machines were quite slow, ranging from 40 to 40,000 microseconds for cycle time of memory. These computers also used magnetic tapes for peripheral storage along with magnetic drums. Such computers were typically controlled by subroutine libraries and some interpreters. By the early 1950s assemblers and compilers began to make their appearance along with rudimentary operating systems. There were no multiprogramming capabilities, time-sharing, or telecommunications in general.

Common examples of first-generation equipment included the ENIAC, EDVAC,* SEAC,* SWAC, Harvard Mark III and IV,* the UNIVAC I and later 1103,* WHIRLWIND* and the IBM 701* and 702 for the period 1946–1953. The years from 1953 to 1959 witnessed the arrival of the widely used IBM 650,* the 700 series (704*, 705*, 709), the UNIVAC II, and entrants by Burroughs† (205, 220), the National Cash Register Company (NCR)† (120, 200), Datamatic (1000), and Radio Corporation of America's (RCA's)† BIZMAC*.

The earliest work on the electronic digital computer was conducted in the late 1930s at Iowa State University by John V. Atanasoff** who worked out some of the design characteristics later evidenced in computers. However, he did not make a fully functioning system. The first electronic digital computer was in fact the result of work done by John Mauchly** and J. Presper Eckert** at the Moore School of Electrical Engineering† at the University of Pennsylvania. Their machine, called the ENIAC, became operational at the end of World War II and inspired a whole series of devices invented by the same developers or others aware of Mauchly and Eckert's work in the late 1940s. This work, as a whole, advanced the collective knowledge of computer science in the United States and in Europe. The ultimate progeny of this line of computers were the UNIVACs of the 1950s.

Other origins of the first generation included the work of Howard H. Aiken** at Harvard University and his Mark I* (completed in 1944) with the assistance and support of IBM. Although based on mechanical technology, using rotating shafts, electromagnetic clutches, and counter wheels, the device nonetheless worked. At Bell Telephone Laboratories† work had been in progress, also since about 1937/1938, on relay computers called the Model I through VI (the last finished in 1949). In Germany Konrad Zuse** worked on the Z* series in the late 1930s and again in the 1940s after World War II, based on the technology of relays. At various British government agencies and universities, war-related projects led to the construction of the COLOSSUS* class machine and, after World War II, to the EDSAC,* built by Maurice V. Wilkes** and his students at Cambridge University. Other important developmental work went on at the University of Manchester. Thus, the period of the 1940s could be characterized as one of rapid and positive developments involving the use of existing electronic technologies lashed together to create the first digital computers. Every major effort was motivated by war-related needs, with the exception of Bell's relays which were needed to support research on telephonic projects. Harvard's may have been another exception; otherwise war was the first motivator, and afterwards military support continued to help.

Associated with these events was the development of the stored-program concept, articulated most effectively by John von Neumann** in the mid-1940s. This concept required that computers be able to take a program (set of instructions), store them within itself (on storage devices), and be able to call into the computer's processor appropriate programs and data to be executed. That concept has governed the architecture of all computers down to the present day.

When the 1950s began, there was a growing awareness that computers could be used for more than just scientific or military application, that, in fact, there were commercial possibilities. Such a belief, although limited, was sufficiently widespread to spur electronic manufacturers and inventors to develop more devices. Yet is was a small world. In 1950, for example, the United States had some twenty computers and automatic calculators worth an estimated $1 million.

The vacuum tubes used in machines of the 1940s were also improved on and used in the 1950s. Improvements included miniaturization, the use of less electricity, and increased reliability. Between 1950 and either 1954 or 1955 scientists searched for very fast memories with enough capacity that were also reliable. The result was the development of ferrite-core memories that allowed larger programs to run, working on more data. Between 1955 and 1959 scientists sought to use software* (programming languages,* aids, and operating systems*) to enhance the capabilities of computers while making them easier to use. The pressure to reduce the cost of memories and of computing also influenced the evolution of machines.

The problem of memory resulted in the use of better vacuum tubes. But with the introduction of radar in the 1940s a new form of memory, called delay lines, became possible. They could store up to 100 times the amount of information as a vacuum tube, and so after the war, scientists explored their possible use. The first important implementation of delay lines came in the University of Manchester's MADM (Manchester Mark I*) computer. Other implementations included the EDVAC, EDSAC, BINAC,* WHIRLWIND, and UNIVAC I. The UNIVAC I was a particularly important milestone, for it represented the first widely known commercial computer. During the 1950s, the word *UNIVAC* was synonomous with the term *computer*. Made available for the first time in 1951, it still used 5,000 vacuum tubes (as opposed to 18,000 in the ENIAC) and in 1953 had the first high-speed printer in the data processing industry. Electrostatic tubes were also used for memory for some computers, known as the Williams electrostatic tubes, which in effect looked like large television sets. These tubes were appealing because they were very fast, with an access time of 25 microseconds. They were also inexpensive when compared to other options. The Williams tube appeared on the IBM 701, 603,* 604,* and 605. The IBM Card Programmed Calculator (CPC)* also reflected this technological heritage. The CPC was a popular machine, and some 700 copies were sold. Von Neumann, at the Institute for Advanced Study at Princeton since 1946, constructed a machine based on the Williams tube in 1951. He called it the IAS computer* and used it in 1952.

Despite advances in capacity and speed due to the Williams tube, it soon became obvious that it was too sensitive to electromagnetic disturbances. Scientists also turned to magnetic drums to provide large capacity at low cost and at reasonable speeds and thus began to consider this approach as a viable option. Magnetic drums had a coating that allowed surfaces to be magnetized. They became available in speeds ranging from 5 to 25 milliseconds. The idea, developed for the IAS in 1946, first became widely available on the ERA 1101* in 1950. Meanwhile, the sale of computers reflected the result of even better devices. Thus, for instance, IBM sold more than 1,000 copies of the 650, while other firms introduced their own products. Drums, however, could not compete with electronic devices in general for speed and failed to satisfy a rapidly growing demand for capacity in excess of what contemporary technology could provide.

The final hardware answer for first-generation equipment was the ferrite-core memories used in the IBM 704. Ferrite cores were rings made out of ferrite. Wires with electrical currents passing through the center of the donut-shaped device could cause the surface of the cores to magnetize to represent information. RCA, for example, did considerable work with such technology which the firm introduced in the BIZMAC computer in 1952, although the machine did not become a product until 1958. The UNIVAC 1103A was the first computer to offer such technology (1954). Its memory ran fifty times faster than that on the UNIVAC 1103. IBM's 704, introduced in 1954 and shipped the following year, had memory that ran at 12 microseconds and initially had a storage capacity of 4,096 words, each made up of 36 bits. It was developed by Gene M. Amdahl,** also one of the fathers of the S/360* and later the designer of his own computers in the 1970s. Ferrites remained a popular technological base for memory devices until 1970 even though transistors and chips* had become dominant by then (Table 9).

Early peripheral equipment used existing technologies. Thus, punched card equipment, which had been in various stages of evolution since the 1880s, provided an early set of input/output devices. Because of the layout of data on cards which ran in such equipment, tape and disk files later used a similar format. It took until the end of the 1970s to break away completely from these early formats. Punched paper tape* had been in use as a form of auxiliary memory since at least 1939 when George R. Stibitz** had employed it at Bell Labs. By the early 1950s, and first on the UNIVAC I as a business device, magnetic tapes provided additional storage. With the introduction of the Hurricane computer in the same era, organizing data on tape in blocks of fixed length (the standard way of handling data down through the late 1970s) first became available. Some of the more commonly used tape drives* included UNIVAC's and the IBM 726.

Tapes were convenient. They packed data very densely on their medium, making it possible to store several hundred thousand characters on one reel. That density continued to increase over the decades. Tapes were portable: they could be taken from one building to another or across the nation. However, all information was in sequential order, one record after another, which meant that to get at something in the middle of the tape, the user had to read all of the first half to get to the desired information.

During the first generation, work began on yet another form of storage which

Table 9
Some Computational Devices, 1939–1954, Selected Years

Memory type	Date	Machine	Fast-memory Capacity	Access time (seconds)	Word Length	Addition time (seconds)
Electromechanical						
Relay	1939	Bell Labs[a]	400 bits	0.5	7 decimal digits	0.3
	1944	Harvard Mark I	60 words	0.5	23 decimal digits	0.6
Vacuum Tube						
Flip-flop	1946	ENIAC[a]	20 words	0.001	10 decimal digits	0.0002
	1948	SSEC[b]	8 words	0.001	14 decimal digits	0.0002
Williams tube	1948	MADM[b]	32 words	0.0001	32 bits	
With Miniaturized Vacuum Tubes						
Delay lines	1951	UNIVAC I	1,000 words	0.0003	12 decimal digits	0.0005
Williams tubes	1953	IBM 701	2,048 words	0.00003	32 bits	0.00006
Ferrite cores	1953	UNIVAC 1103	1,024 words	0.00001	36 bits	0.00003
	1954	IBM 704	8,192 words	0.00001	36 bits	0.000024
	1954	CAB 2000	128 words	0.00001	22 bits	0.00046[c]

[a]Calculator.
[b]Computer.
[c]Discrepancy between access time and addition time owing to the fact that what was called secondary memory was actually main memory.

SOURCE: R. Moreau, *The Computer Comes of Age: The People, the Hardware, and the Software* (Cambridge, Mass.: MIT Press, 1984): 68. Copyright 1984 by MIT Press, all rights reserved; reprinted with permission.

became known as magnetic disks. Looking like a stack of phonograph records with surfaces that could be magnetized and hence have data stored on them, they first attracted scientists in the late 1940s. As part of the effort to develop the EDVAC, work was done on magnetic disks as early as 1948. In the following decade two disk technologies emerged. The first, called rigid disks, was used for computational equipment in the Minuteman ballistic missile project and was built by the Autonetics Corporation. Flexible disks, first introduced as a plastic product, came out in 1952. The first widely used disk was the RAMAC which attached to the IBM 704 computer in 1956. The Random Access Method of Accounting and Control (RAMAC) could store 5 million characters. It was considered a fixed disk, that is, the phonograph record-like units were built into the disk drives,* which then read from and wrote to this technology. At the end of the 1950s work began on the disk pack which could be removed from the disk drive and thus be carried out. These first appeared on the IBM 1311 in 1962. Each of these early disk packs could store 3 million characters; by the mid-1970s capacity exceeded 70 million. Disk drives were more attractive than tape because they enabled the user to get to specific information directly and not sequentially, much as one might move the needle of a record player over to the specific piece of music desired and then lower it to the record. This sped up one's ability to get to machine-readable information. Second- and third-generation devices were expensive but became less so as the technology improved, particularly during the fourth generation.

The period of the first generation was also a Golden Age of programming languages.* Many of the high-level languages used today first appeared during this era. The most important included FORTRAN* (the most widely used language for scientific and engineering applications) and COBOL* (the most widely used language for commercial applications by the end of the 1960s and during this period under development). These were also the years when operating systems first came into existence to handle basic household functions in computer systems.

To appreciate the importance of these developments, we must understand how programming and software were handled in the earliest first-generation computers. Instructions for applications, movement of data and records, use of peripheral equipment, the execution of instructions, and so on, within the central processor had to be detailed by programmers in low-level language code, that is, in terms intelligible directly by the computer. Beginning in the 1950s, languages were developed that allowed coding in idioms that were closer to human languages, first using algebraic notation (or something that looked like it) and later more English-like commands. Then using a translator or compiler, these instructions were converted into terms which hardware understood. The result of these developments, along with the creation of operating systems, made it possible for programmers and data processing people to know less about the inner workings of computer technology in order to make it work effectively.

This trend toward ease of use continued down through each generation of computers.

The first generation established the agenda for digital computers for decades to come. Out of the first generation emerged the standard configuration, also known as the von Neumann machine, which was made up of a memory and a central processor with a variety of input/output equipment. The latter included card readers and punches, printers, tape and disk drives, and a variety of other memory and storage units. Terminals also made their appearance. Operating systems* and high-level languages were defined in theory and practice, leaving to future scientists the task of expanding on these theoretical and early practical plateaus. The accomplishments of these years, therefore, have drawn considerable attention from historians.

The second generation (1959–1964), although shorter in duration than the first, saw significant changes. In the area of computer equipment, it was the period during which vacuum tubes finally became a thing of the past, while transistors (precursor to the chip) emerged as the standard component technology for digital computers. Magnetic core memory came into its own as a standard feature, whereas main memory cycle times improved again to 2–10 microseconds (down from 10–20 microseconds at the end of the first generation). Additional enhancements to the speed and reliability of magnetic tape and disk storage were evident with a continuing drop in the cost of such technologies. Software* continued to develop, while FORTRAN came into full bloom with COBOL right behind it. Operating systems were now standard fare on all major systems, even if limited in function, whereas telecommunications first emerged in a normal computer system's configuration. Such common features of the 1980s as interrupt systems and virtual memory made their earliest, if tentative, appearances along with multiprogramming. Common examples of second-generation computers included the Philco 2000, CDC 1604 and 3600, IBM 7000, the very popular IBM 1401, Ferranti Atlas, RCA 301 and 501, Honeywell 800, and the UNIVAC III and 1107.

Computers of the second generation illustrated the impact of the "technological imperative" at work. Greater reliability of equipment, particularly of memories (using ferrite core), and the increased speed produced by the use of transistors in turn drove prices down and usage up. Machines became smaller in size relative to the power they had and used less energy. The second generation, through the transistor, came into contact with semiconductors for the first time. Either germanium or silicon was used as the building block of the transistor which, by the end of the 1950s, had been converted into what today is called the chip. Not until the mid-1960s, however, would chips play an important role in the architecture and construction of computers.

The transistor, which so revolutionized electronics in the 1950s and 1960s and which has been discussed elsewhere, had a similar influence on computers. Because they were so small and led to further research and to the use of semiconducting materials, it became possible during the second generation to begin the manufacture of integrated circuits on the surface of silicon. Consequently, since 1948, the actual size or volume of circuits began to shrink by a factor of 10 approximately every five years throughout the period of the second, third, and fourth generations of computers. By the 1980s the five-year window had shrunk to four, suggesting even faster miniaturization.

The earliest known use of transistor technology in a computer was in the SEAC,* a machine of the 1950s used to study meteorology. But it was a machine in transition since it also had 750 vacuum tubes and nearly 10,000 diodes. Philco Corporation† developed its own device in 1954 called the TRANSAC S-1000 and two years later the CXPX for the U.S. Navy. The first commercial machine was a civilian version called the Philco 2000. As Table 10 shows, many others followed. Another of the early machines was the Atlas Guidance Computer Model 1 constructed in 1956–1957. Scientists at Cape Canaveral used it between 1958 and 1961 to help launch rockets and space ships. Control Data Corporations† 1604 (also known as the CDC 1604) launched the new company, which remained an important vendor as of 1986. Some of the other totally transistorized machines which were second generation in their technology included the GE 210, the IBM 1401* (the most popular of the series), IBM 1620,* IBM 7090, NCR's 304, and RCA's 501.

All of these machines encouraged a new era in processing. Many were so employed that multiple programs or jobs could run in them, instead of one at a time as had been the case with first-generation equipment. The new batch processing systems could handle multiple jobs because their operating systems could effectively control the flow of work through a system. As a consequence, users reduced the amount of time wasted between jobs while expanding the number of applications available. As Table 11 suggests, the number and variety of applications (at least in this survey) grew during the period of the second generation and by 1969 (third generation) had shown no slowdown. These data also suggest the kinds of applications that first went on computers. Interestingly, most historians of computing thought that in the main only accounting and scientific applications dominated second-generation machines. Although that remains true, the table also provides evidence of other applications, such as in manufacturing, suggesting a use for computers earlier than otherwise thought.

The year 1959 is important for data processing. New languages appeared, standards for others were set, and many computers were introduced, thereby establishing the existence of the second generation. Univac Division brought out the LARC,* and the following year IBM made available the 7030, also called STRETCH.* In Europe Machines Bull† (a French computer manufacturer) introduced the Gamma 60. Such devices were designed to meet the needs of scientific users and so could handle large quantities of data and complex calculations. STRETCH introduced the use of disk drives on a large computer; all major systems after that had the same data storage capability. The business community also took to computers aggressively and made fewer distinctions between those designed for scientific users or for commercial applications. Thus,

Table 10
Characteristics of Some Second-Generation Computers, 1959–1963

Computer	Type	Word Length (bits)	Number of Words	Memory Type	Access Time (microseconds)
Gamma 60	O	24	8–32K	Ferrite	10
Burroughs 5000	B/D	13	4–32K	Ferrite	6
			32–65K	Drum	8,500
CDC 1604	B	48	8–32K	Ferrite	4.8
CDC 3600	B	48	32–262K	Ferrite	1
Honeywell 800	D	12	4–32K	Ferrite	6
IBM 1401	D	Variable	1.4–16K	Ferrite	11.5
IBM 1620	D	Variable	20–60K	Ferrite	20
IBM 7030 (Stretch)	B	64	16–262K	Ferrite	2.2
IBM 7040	B	36	4–32K	Ferrite	8
IBM 7070	D	10	5–10K	Ferrite	6
IBM 7080	D	Variable	1K	Ferrite	1
			80–160K	Ferrite	2
IBM 7090	B	36	32K	Ferrite	2.18
IBM 7094	B	36	32K	Ferrite	2
PDP 1	B	18	4–65K	Ferrite	5
Philco 2000–210	D	8	8–32K	Ferrite	11.5
			32K	Drum	25,000
LARC	D	12	100	Ferrite	1
			10–97K	Ferrite	4
			6,000K	Drum	68

Table 10 *(cont.)*

Computer	Times for Basic Operations (microseconds)			Price in U.S. Dollars (millions)	Number of Instructions in Code	Number of Addresses per Instruction
	+	x	÷			
Gamma 60	200			1.5	115	1-3
Burroughs 5000	10	37	63	0.5-2	62	Variable
CDC 1604	7.2	25.2	65.2	0.750	92	1
CDC 3600	2	6	14	1.4	51	1-2
Honeywell 800	24	150	312	0.980	43	3
IBM 1401	230	2,100	2,600	0.125	32	1-2
IBM 1620	56	496	1,686			2
IBM 7030 (Stretch)	1.5			5.7	73	1
IBM 7040	16	33.5	18.5	0.625	200	1
IBM 7070	72	924	792	1	106	1
IBM 7080	11	100	253	2.1-3.8	227	1
IBM 7090	4.36	4.36-30	6.54-30	2.9	268	1
IBM 7094	4	4-10	6-18	3.1	28	1
PDP 1	10	20	30	0.120	225	1
Philco 2000-210	14.8	69.9	73.8	1-2	76	1
LARC	8	8	28	7		1

Table 10 *(cont.)*

Computer	Number of Index Registers	Indirect Addressing?	Floating Point?
Gamma 60	0	Yes	Yes
Burroughs 5000	6	Yes	Yes
CDC 1604	6	Yes	Yes
CDC 3600	64	Yes	Yes
Honeywell 800	3	Yes	Yes
IBM 1401	0	No	No
IBM 1620	16	Yes	Yes
IBM 7030 (Stretch)	3	Yes	Yes
IBM 7040	99	Yes	Yes
IBM 7070	0	Yes	Yes
IBM 7080	3	Yes	Yes
IBM 7090	7	Yes	Yes
IBM 7094	0	Yes	Yes
PDP 1	8	No	Yes
Philco 2000-210	99	Yes	Yes
LARC		Yes	Yes

B = binary
D = decimal
O = octal

SOURCE: R. Moreau, *The Computer Comes of Age: The People, the Hardware, and the Software* (Cambridge, Mass.: MIT Press, 1984): 132–133. Copyright 1984 by MIT Press, all rights reserved; reprinted with permission.

Table 11

Computer Applications per Hundred Installations, 1956–1969, Selected Years

Sample	81 IBM 650 Sites	British Survey 1589 Sites		Research Institute of America Survey						IDC Report	
				Small Computers in Mfg Companies		Medium Computers		Large Computers		720 Companies	2000 Companies
	1956	1964	1969	1965	1968	1962	1968	1960	1968	1969	1969
Payroll	65	59	69	70	85	56	73	59	82	67	56
General bookkeeping				31	57	39	58	37	64	15	14
General ledger Accounting	25									65	52
Financial Accounting		40	82								
Accounts receivable				43	64	39	51	41	64		
Accounts Payable				35	57	25	58	33	55		
Order processing & billing				43	64	53	58	41	64		
Order analysis										30	12
Billing		36	67							31	21
Sales analysis & control				55	85	42	66			56	31
Sales forecasts							37				
Inventory control		40	64	39	71	28	58			74	41
Production schedule (includes control)	33	19	32							33	14
Cost accounting	35									23	13
Labor distribution										13	6
Mailing lists				16	50	44			64	2	4
Engineering & scientific	53	28	61			37		22	36	10	8
Information storage & retrieval						37			46		
Savings and demand deposit											
Student records											15
Other applications	34	54	81	58	177	68	153	74	207	58	143
"Other"/"miscellaneous"		12	45				36			5	5
Grand total	245	288	501	390	710	350	730	370	910	482	440

SOURCE: Montgomery Phister, Jr., *Data Processing Technology and Economics* (Bedford, Mass.: Digital Press, 1979): 137.

vendors had to drop their distinctions as well and design equipment that could handle both types of applications. This was accomplished by the end of the second generation.

Users increasingly wanted more multiprocessing with their systems. Although work on such features had been ongoing since 1954, primarily at the U.S. National Bureau of Standards† in the beginning and later at universities and within data processing companies, and had first become commercially available in 1956 with the IBM 305, it became one of the most attractive qualities of the LARC and the Gamma 60. In addition, these machines made consoles for operators and engineers common, along with channels to attach peripheral equipment easily, the use of bytes, and, of course, multiprocessing.

Smaller computers were also innovative. These included the IBM 1401 which was introduced for commercial applications by those who had relied on card punch technology in the past. All the 1401 models cost less than their predecessors and collectively became one of the most widely used computers of the period. A total of 10,000 1401s were sold, a huge success considering that other computers sold only several hundred copies or less. The other noteworthy event of the period involved the introduction of the PDP 1, the first of a long string of PDP* computers for intermediate-sized computing requirements. It had the best price/performance ratio of the early computers and cost about $120,000. The PDP 4, first delivered in July 1962, cost half as much. Fifty of the first were built, and sixty-five of the second.

Second-generation machines in general offered greater independence between processing and handling of input/output using operating systems. Hence, operators had to know less about programming than before, and in turn programmers and users needed to know less about a computer's operation. Multiprogramming came into its own for the first time as did time-sharing systems, such as that developed at the Massachusetts Institute of Technology (MIT) in the late 1950s and early 1960s. The first interactive computing capability appeared on the IBM 305. SABRE,* a project shared by IBM and American Airlines, was a passenger reservation system that allowed many of the functions and technologies of the second generation to come together. These included the use of terminals, large magnetic disks, better operating systems, and more reliable computers. The fundamental concepts behind SABRE served as technological agendas for years to come; in modified form the same application operated in 1986, relying on more modern hardware and software, but in essentially the same manner.

The net result of all these changes, introductions of new products, and usage of computers was the growth in the number of systems installed. As Table 12 indicates, the raw number of computers installed rose sharply during the second generation when compared to the first (through 1959). In terms of economic trends, Figure 6 suggests the reason. The cost to compute was dropping, and, although the data show the trend only for IBM's equipment, the same can be said for the industry as a whole. More small than large computers were being sold, accounting for a great portion of the rapid growth. In 1964 the value of

Table 12
Annual Shipments of U.S. Firms and Year-End Installed Base, 1955–1965
(Worldwide Computer Activity)

Year	Shipments $ (million)	Install Base	
		Number (thousand)	$ (million)
1955	65	0.25	180
1956	170	0.80	340
1957	280	1.7	600
1958	260	2.9	1,030
1959	600	4.5	1,600
1960	720	6.5	2,270
1961	1,100	9.3	3,260
1962	1,400	12.6	4,510
1963	1,710	16.8	5,970
1964	2,200	24.0	7,920
1965	2,400	31.0	11,100

Data are for general-purpose computers. The U.S. portion of these shipments represents the lion's share.

SOURCE: Based on "EDP Industry Report," March 1971, International Data Corporation, as reproduced in modified form in Bruce Gilchrist and Richard E. Weber, eds., *The State of the Computer Industry in the United States* (Montvale, N.J.: American Federation of Information Processing Societies, 1973): 11.

all installed computers was six times that of equipment at the end of the first generation.

Operating systems also told the tale. In 1955 the operating system used to control an IBM 650 only had about 5,000 instructions. In 1959, on the IBM 7090, it approached 1 million. The first third-generation computer from IBM, the S/360, had an operating system with several million instructions. In 1970 the IBM S/370* exceeded 10 million instructions. By then small distributed processors also had operating systems with over 700,000 lines of instructions (such as the 8100's operating system DPPX from IBM in 1978), while earlier devices in the PDP family were heavily loaded with instructions. The other trend involved the cost of programming. The percentage of all sums for a system expended on software in 1955 was approximately 15; by 1964–1965 this figure had grown to 50 percent, and by the end of the 1970s to 80 percent.

The second generation did not end with the introduction of the IBM S/360 in April 1964. Although that collection of computers signaled the start of the third generation (1964–1970), it would be several years before other vendors could replace their old product lines entirely. In the case of IBM, the first shipments in quantity of the new product did not take place until 1965. Components in these machines were made from monolithic integrated circuits and reflected the first of the medium and large-scale integration that characterized the technology

Figure 6
Improvement in Performance/Price Ratio for Some IBM Scientific Machines

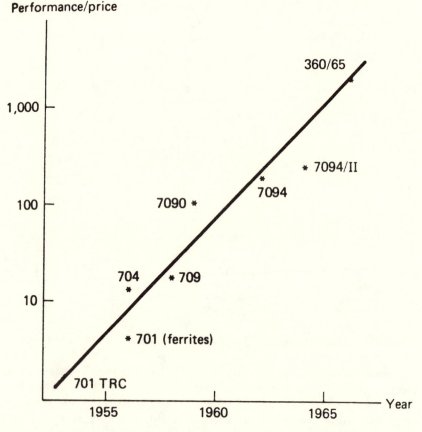

Ratio is expressed as the number of operations performed at a price of $1,000.

SOURCE: R. Moreau, *The Computer Comes of Age: The People, the Hardware, and the Software* (Cambridge, Mass.. MIT Press, 1984): 190. Copyright 1984 by MIT Press, all rights reserved; reprinted with permission.

of fourth-generation computers. Memories were built out of magnetic core and by the end of the 1960s reflected large-scale integration. The main memory's cycle time went first from 0.5–2 microseconds down to 0.020–1 microseconds, representing significant increases in the speed of operation over second-generation equipment (2–10 ms). Magnetic tape and disk drives handled greater amounts of data, faster, less expensively, and more reliably. Operating systems obviously became more sophisticated, and a greater number of interactive computing languages, such as BASIC,* PL/I,* and APL,* came into their own. For the first time in any serious manner, database management* software also appeared.

All major computer vendors introduced third-generation equipment between 1964 and the end of 1969. The best known devices included the Burroughs 5500, the CDC 6000 series and 3300, the UNIVAC 1108, the Honeywell 200 series, RCA's Spectra 70, the NCR Century, the GE 400 and 600, and the centerpiece of that generation, the IBM S/360.

This family of computers was the most successful product in the history of American business. It exploded onto the scene in April 1964 in the form of over 150 separate products, ranging from computers to a complete set of peripheral equipment, operating systems, programming languages, migration aids to assist customers moving from older systems to this one, and terminals. These computers were compatible, that is, programs that operated on one model could, with little or no modifications, operate on larger models. Peripheral equipment generally could be used with computers of various sizes. Operating systems functioned compatibly on all computers. The compatibility of machines was a first and became one of the most distinguishing characteristics of the third generation. Prior to that, computers of different sizes were not generally compatible, nor were their operating systems or peripherals. That all changed with S/360. From then on all vendors provided compatibility of systems on an increasing basis down to the present.

The impact of compatibility and other features was enormous. Prior to S/360, if a company wanted to move up to a bigger computer, it had to rewrite all of its programs, learn to operate new equipment and operating systems, and establish new managerial procedures. That process, in addition to being traumatic, cost at least as much as the new equipment, gear that might have to be replaced again in just a few short years. S/360 signaled the end of ''conversions,'' one of the most dreaded experiences a data processing manager could have. Furthermore, compatibility represented the recognition of an unquantified fact known across the industry: the investment in application programs in many ''shops'' exceeded the value of installed equipment. This fact was not lost on vendors seeking ways to move their customers to more and bigger equipment.

S/360 relied on what came to be known as Solid Logic Technology (SLT), which in English meant that IBM had decided to base its computers on chip* technology, departing from transistors which had been the rockbed of second-generation products for the entire industry. It was a bold decision, at least as startling as the industry's decision to standardize on transistors just a few short years earlier. Some have argued that the decision was even bolder than the transister decision since chips had been tested less and were a more unstructured technology. Yet they offered the argument that, in order to make the quantum leap forward as requested by IBM's management, they had to take the technological gamble that silicon-based chips would offer the increases in performance and capacity needed while reducing costs, which would break open data processing. It was the ultimate in the technological imperative at work. No example since then has been as dramatic or as important to the history of the industrialized world's economy.

In addition to relying on chip technology, the S/360 swept into its net an accumulation of many innovations that had been floating around the industry in the previous several years. This included, for example, expanded character sets available on computers all the way down to the IBM 029 keypunch, the standard for the decade and deep into the next. More and different applications thus became possible, always the driving force which ultimately drove demand higher for more computing. The new computers came in models 30, 40, 50, 60, 62, and 70; others were added throughout the 1960s. The bigger the number of a model, the more memory and power a computer had. A model 20 was introduced later along with a model 44 (the latter just for scientific computing). The product was incredibly successful despite some technical shortcomings (for example, lack of virtual storage). By the end of 1969 approximately 15 percent of all computers in the world were S/360s, and IBM had doubled in size.

Others sought to take advantage of the concept of a family of computers. Of those already mentioned, RCA's Spectra series was one of the more important because it was built to be compatible with the S/360. RCA intended the Spectra to compete head on with the S/360; it failed, although sales were not insignificant. Various models of Spectra conformed to available models of the S/360 (30, 40, and 50). The Spectra did reasonably well until RCA decided to abandon sale and support of computers because the cost of competing in the future would have been too high. The research and development costs for anyone wishing to compete in computers was rising. For example, IBM spent a minimum of $500 million to design and develop the S/360 and had to bet the very existence of the company on success. Investments of that size and greater became common in the 1970s and 1980s. Yet from that time when the S/360 was announced, the stakes in the data processing game had escalated, not just for computers but across the board in software, hardware, and marketing. For that reason, it was understandable why so few companies introduced new computers in the 1970s (less than 100 firms usually), a decade when the data processing industry continued to expand at between 10 and 28 percent annually. It also explained one reason why IBM's technology became the industry's standard throughout the late third generation and completely during the fourth.

Just as the cost and performance of second-generation equipment fostered even greater use of computers, third-generation technologies had the same effect. As Table 13 indicates, in 1969, at the height of the third generation, a sampling of major users of equipment in the United States had spent over $3 billion on computers. For the first time, certain businesses could not operate without the use of computers. Airplane manufacturing companies listed in the table, for example, had made the greatest commitment to such technology and continued to lead the way in the early 1970s. By the fourth generation, banks were also investing heavily in computers, to the tune of nearly 5 percent of their operating budgets by the end of the 1970s.

Table 13
Computer Concentration in the United States, 1969

Company	Total (million $)	Per $1 K Sales ($)	Per Employee	Per $1 K Assets ($)
The Top Ten				
General Motors	228.9	10.06	302	16.3
General Electric	220.0	26.30	550	38.3
Boeing Aircraft	140.0	42.80	986	64.0
Ford Motor	139.0	9.80	334	15.5
McDonald-Douglas	98.5	27.30	788	74.1
Lockheed Aircraft	93.9	42.40	988	101.0
Westinghouse	89.2	27.10	646	39.3
Sperry-Rand	85.0	54.40	841	77.6
Shell Oil	84.0	25.30	2,153	19.9
North American Rockwell	82.6	31.3	724	60.2
Other Companies				
Standard Oil, N.J.	58.24	4.13	385	3.47
McGraw-Hill	13.78	37.45	1,060	44.89
Gulf & Western	10.32	7.86	138	5.02
National Dairy Products	9.36	3.86	199	9.87
Top 100 Firms	3,306.50	13.95	408	15.05

The data measure computer concentration by computer values in U.S. dollars. Included are non-computer companies in which over 50 percent of all revenues are derived from manufacturing or mining.

SOURCE: Based on data gathered by *EDP/IR*, October 9, 1969.

The fourth generation (1970 to the present) of digital computers was less distinguishable from the third than were the first two periods. The third and fourth used chips made of silicon for components in processors and memories. Both continued to improve the speed of computing and to increase the capacities of all equipment. Telecommunications gear became more prevalent in fourth-generation configurations, as did the use of online computing, distributed processing, and minicomputers. But beginning with late third-generation equipment, vendors, including IBM, sought to introduce products in an evolutionary manner rather than in some dramatic revolutionary form. Thus, the S/360 family changed and improved as new models and peripherals appeared after 1964. The same applied to its software.

Throughout the 1970s and into the 1980s—the period of fourth-generation equipment—product introductions came in a continuous stream. In 1970 IBM introduced the S/370 which also relied on chips and the already available virtual storage capability. Throughout the decade other S/370s appeared with denser chips, larger memories, and declining costs of computing. In 1979 IBM brought out the 4300 series which, by the mid-1980s, had evolved into over a dozen models. At the high end, the 3000 series developed into the 308X series and, in 1985, into the 3090 series. Packaging became more dense. Chips of 16K in the early 1970s were already past 256K by the early 1980s.

Because of the fuzziness between what constituted late third-generation and early fourth-generation computers, historians quibbled on how best to categorize technologies so close to their time. It was generally accepted that the IBM S/370 seemed to provide some sort of delineation between the two, if not as dramatic as between the second and third. What other computers should go into either late third-generation or early fourth-generation remained questionable. Yet in that no-man's land that constituted one or the other, one would have to add the CDC Cyber 70 series, Digital Equipment Corporation's PDP 10 and 11, the Honeywell 6000, Burroughs 6700 and 1700, not to mention numerous mini-computers, all of which were fourth generation with few connections to earlier systems.

The important features of fourth-generation equipment were not necessarily technological, although their improved efficiencies and capacities could not be denied. Rather, it was the increased use of such equipment by people who had decreasing requirements to understand their inner workings that represented a distinguishing characteristic from the past. As Figure 7 shows, the number of workers in the United States relying on computers to do their jobs or to help them do their jobs (perhaps more important) continued to rise while the requirement to understand them rose at a much slower pace. The amount of online information grew at about 40 to 45 percent in many years, particularly as access speed for such devices jumped, along with capacities to store more, and all at declining costs. The growth of terminals during the fourth generation typically ranged from year to year between 10 and 30 percent. In the early 1970s vendors talked about having one terminal for every 100 office workers. By the end of the decade the ratio approached 1 for every ten and by the mid-1980s closer to 1 for every four. In some companies in the United States by 1986, one terminal on every desk was almost a reality.

The economics of fourth-generation equipment also tells a story. As Table 14 illustrates, using just IBM's computers as a bellwether, costs continued a steady downward spiral as measured by the expense of executing instructions within a processor. Measured another way, using one megabyte of memory on a computer also helps. In 1970 the purchase price for that amount of storage would have been about $600,000, yet at the end of the decade it hovered at $15,000. The trend continued in late fourth-generation years. Thus, the shipment of memory in the United States in 1979 approximated 41 gigabytes and grew to 872 in 1984. Disk storage in the same two years went from 156 million units to almost 1.1 billion.

All of the machines discussed in third- and fourth-generation computers were digital. However, many of the component technologies available in third- and fourth-generation computers appeared in analog devices. But because analog

Figure 7
Growth of Dependence on Computers, 1955–1990

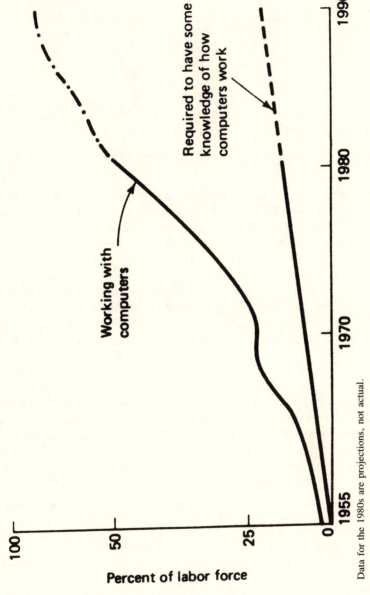

Data for the 1980s are projections, not actual.

SOURCE: James W. Cortada, *Managing DP Hardware: Capacity Planning, Cost Justification, Availability, and Energy Management,* © 1983, p. 12. Reprinted by permission of Prentice-Hall, Englewood Cliffs, New Jersey.

Table 14

Cost of Processing Using IBM Computers, 1955–1985, Selected Years

Year New Technology Shipped	Cost/million Instructions
1955	$40.00
1961	2.00
1965	0.40
1971	0.11
1977	0.08
1979	0.04
1981	0.02
1983	0.01
1985	0.005

SOURCE: IBM Corporation; see also James W. Cortada, *Managing DP Hardware* (Englewood Cliffs, N.J.: Prentice-Hall, 1983): 13–26.

units changed so slowly in function or demand, the evolution in digital computing was convenient for cataloging shifts in technology, usage, and demand over the past forty years. Indeed, data processing pundits in the mid-1980s, looking over the horizon for new and dramatic improvements in technology, talked of a fifth generation. Because such a new generation did not yet exist, however, historians referred conveniently to four.

Each generation drove upward the demand for more computers (Figure 8). The growth of the data processing industry could be largely tied to such improvements in technology neatly and effectively. In fact, the lesson was that improvements in technology did encourage the use of new devices. It was a lesson learned from earlier tools, such as counting tables and the abacus,* later from steam and electricity, but one that remains valid today. The relative convenience and cost of computing in the mid-1980s made data processing as accessible as electricity. The process that led to such availability and use remains linked closely to the arrival and evolution of digital computers. No other circumstance had as dramatic an influence on data processing in general than digital computers and their associated peripheral equipment, operating systems, and programming languages. Its history allows us to appreciate how the modern data processing industry came into being.

For further information, see: Stan Augarten, *Bit By Bit: An Illustrated History of Computers* (New York: Ticknor and Fields, 1984); Paul E. Ceruzzi, *Reckoners: The Prehistory of the Digital Computer, From Relays to the Stored Program Concept, 1935–1945* (Westport, Conn.: Greenwood Press, 1983); James W. Cortada, *An Annotated Bibliography On the History of Data Processing* (Westport, Conn.: Greenwood Press, 1983); Dirk Hanson, *The New Alchemists: Silicon Valley and the Microelectronics Revolution* (Boston: Little, Brown and Co., 1982); Simon Lavington, *Early British Computers* (Bedford, Mass.: Digital Press, 1980); N. Metropolis et al., eds., *A History of Computing in the Twentieth Century* (New York: Academic Press, 1980); R. Moreau, *The Computer*

Figure 8
The "Multi" in Historical Perspective

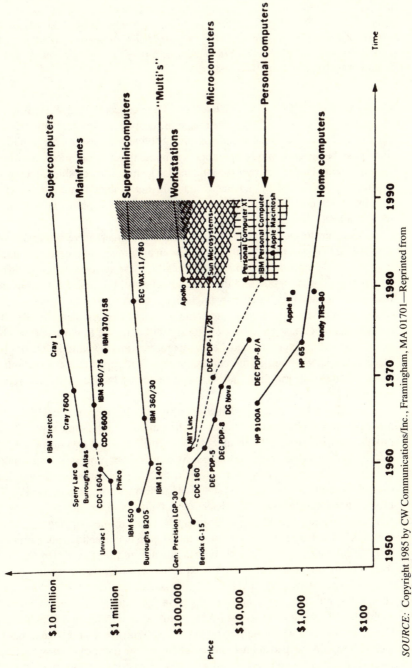

SOURCE: Copyright 1985 by CW Communications/Inc., Framingham, MA 01701—Reprinted from *Computerworld*, October 14, 1985, p. ID/23.

Comes of Age: The People, the Hardware, and the Software (Cambridge, Mass.: MIT Press, 1984); Emerson W. Pugh, *Memories That Shaped an Industry: Decisions Leading to IBM System/360* (Cambridge, Mass.: MIT Press, 1984); Joel Shurkin, *Engines of the Mind: A History of the Computer* (New York: W. W. Norton, 1984); Nancy Stern, *From ENIAC to UNIVAC: An Appraisal of the Eckert-Mauchly Computers* (Bedford, Mass.: Digital Press, 1981); Michael R. Williams, *A History of Computing Technology* (Englewood Cliffs, N.J.: Prentice-Hall, 1985).

COMPUTER, FAULT-TOLERANT. The concept behind fault-tolerant computers is that they will continue to operate despite failures that might develop, usually in hardware. Such machines are absolutely necessary for applications that must be in continuous operation. Among these applications are defense systems, large banking networks, guidance systems on space ships, and medical applications. Many of the advances made in fault-tolerant systems found their way into more conventional computers by the late 1960s, particularly as the cost of components dropped.

With regard to the design of such systems, computers have been built which automatically recover from faults, that is, with no human intervention, continuing to perform all calculations correctly. Typically, these systems comprise different components that do tasks, and there are always multiple copies of components. Known as redundancy, a system might have two or more processors, each of which simultaneously executes the same job and then compares results. If one module malfunctions, the others obviously will continue to work and be able to signal the overall system, and hence operators, that a malfunction has taken place at a specific point in the configuration. That approach has been reflected in such features as masking faults from users or by building dynamic recovery procedures.

Masking faults is accomplished simply by having, for example, three processors for each module, each doing the same job at the same time. The three processors will vote on the results of their work. Thus, if two or more answers match, that result is presented to the next sequence of processors to act upon or is presented by the system as a whole in the form of output to people. In the case of dynamic recovery, the system will perform self-repair. In the 1980s that usually involved the computer detecting a faulty component and cutting it off from use, while initiating a replacement to pick up the crippled unit's work, meanwhile tracking the problem for eventual human correction. Such procedures have been used effectively with both hardware and software.* In the case of software, the tactic has frequently been called error-correcting code.

The desire to have fault-tolerant systems existed from the earliest days of computers, but only in the early 1950s were computers designed specifically as such. One of the earliest was the SAPO, built between 1950 and 1954 in Prague, Czechoslovakia, by Antonin Svoboda.** Relying on conventional hardware technology, SAPO had relays and magnetic drum memory,* as well as three processors that voted on results. Error-correcting retry functions were incorporated into the management of memory. Later, Svoboda built EPOS, a system that represented an improvement over the first. Svoboda reportedly built these machines because parts were so unreliable and because he and his associates feared that they might be punished if their computers did not function well. He was one of the first to work with that kind of computer.

In the United States the incentives to build such a system came from defense, the requirements of the space program, and the need for reliable telephone networks. The National Aeronautics and Space Agency (NASA), for example, supported many efforts beginning in the late 1950s and extending through the 1970s. The first device built for NASA was the Orbiting Astronomical Observatory (OAO), an on-board computer constructed out of transistors which employed a fault-masking design. The second device developed for NASA was used on the Saturn V as its guidance computer. It had triple processors and duplicated available memory. Next came the JPL Self-Testing-and-Repairing (STAR) computer. The intent was to have a computer that could survive for ten years on long flights to other planets. By the late 1960s it was possible to have such a device with dynamic recovery features which, in the layperson's terms, allowed a computer to fix itself as it broke down, on the fly, so to speak. Although never used since flights to other planets with astronauts on board were not launched, it nonetheless pointed the way to more advanced fault-tolerant systems. The U.S. Air Force supported the construction of an on-board machine also called the Fault-Tolerant Spaceborne Computer (FTSC), a more advanced STAR with greater capacity. Another early system, particularly important to the history of such computers in the 1960s, was the Electronic Switching Systems (ESS) which were installed at telephone switching offices all over the United States.

During the 1970s research on fault-tolerant systems accelerated. At NASA, for instance, systems were constructed for use on various spaceships. More precisely, various companies built these for NASA, thereby spreading experience with fault-tolerant technology across both data processing and defense suppliers. Software Implemented Fault Tolerance (SIFT) employed programs for NASA to improve the reliability of hardware, whereas the second, called the Fault-Tolerant Multiprocessor (FTMP), a hardware solution, complemented earlier work. Commercially available machines began to appear in the 1970s and 1980s. One of the earliest was the Pluribus, a second-generation device. Tandem Corporation built a series of machines in the late 1970s and early 1980s. By the mid-1980s it was even supplying such machinery for International Business Machines Corporation (IBM)† to sell to its customers. Tandem was the leading manufacturer of such equipment by 1984/1985.

For further information, see: A. Avizienis et al., "The STAR (*Self-Testing And Repairing*) Computer: An Investigation of the Theory and Practice of Fault-Tolerant Computer Design," *IEEE Transactions, Computers* C–20, no. 11 (November 1971): 1312–1321; *Proceedings of the IEEE (Special Issue on Fault-Tolerant Computing)* 66, no. 10 (October 1978).

CORAL. Class Oriented Ring Associative Language (CORAL) was a minor string and list processing program of the 1960s. Like many other software tools of the 1950s and 1960s, it came out of the Lincoln Laboratory† at the Massachusetts Institute of Technology (MIT), the brainchild of L. G. Roberts who used the language for online display of graphics and to do systems work. Actually created in 1964, CORAL ran on a TX-2 computer. It has been described as a language that employed list structures called *rings*. Its key concept was that all elements, with the exception of the first, would point to that initial element that represented the beginning of the list or to the element immediately behind the last element.

The language was machine dependent; that is, it ran on only the TX-2, and many of its characters were a reflection of the character set of the Lincoln writer keyboards.

For further information, see: L. G. Roberts, "Graphical Communication and Control Languages," *Information System Sciences: Proceedings of the Second Congress* (Washington, D.C.: Spartan Books, 1965): 211–217.

CORC. This programming language* was designed to solve numerical scientific problems at Cornell University, using the Burroughs 220 and the CDC 1604 computers. It was designed to allow students in other than data processing courses to learn a language quickly in order to solve problems. Its most unique feature for that historical period was its ability to correct errors while compiling programs. Little else is known about the use of this language.

For further information, see: R. W. Conway and W. L. Maxwell, "CORC—The Cornell Computing Language," *Communications, ACM* 6, no. 6 (June 1963): 317–321.

CPS. This small, PL/I*-like programming language, officially known as the Conversational Programming System, was one of the first online tools developed in the 1960s to solve small scientific problems. It was the result of a joint effort by Allen-Babcock Corporation and International Business Machines Corporation (IBM)† to develop a small subset of PL/I. Work began on the language in early 1965, and the initial release became operational in late 1966. The designers sought simplicity of use by having programmers sitting at terminals while preserving a great deal of the useful syntax of PL/I.

This project allowed Allen-Babcock and IBM to experiment with the effects of microprogramming on computers. It was originally designed to operate on an IBM S/360* Model 50, and various machine instructions (called microprograms) were developed in a successful attempt to make the language more efficient. Today, the use of microprogramming within hardware is a common feature because it makes languages and hardware more efficient and faster. This was not a common characteristic of computers in the early 1960s and did not become a dramatically new and widely used feature until the advent of the S/360.

CPS was frequently confused with another language, RUSH (Remote Use of Shared Hardware), which was a variation of CPS and served as the programming base for time-sharing sold by the Allen-Babcock Corporation. IBM's version was CPS. During the late 1960s and throughout the 1970s, CPS was the most widely used version. IBM marketed PL/I very effectively in the late 1970s, expanding the demand for CPS and other related programming tools.

CPS was one of the earliest successful online programming languages* designed to do small numerical scientific problems. It grew up along with other languages serving the same purpose and at a time when computing capabilities with larger computers, more complex operating systems,* and online programming were becoming available. CPS was also one of the first such tools that allowed a novice programmer to learn a powerful language such as PL/I. Experienced programmers also found it very useful for quick, short problem-solving. It has been widely used to perform financial and accounting analysis, and to solve other analytical problems involving the use of numbers (e.g., sales analysis).

For further information, see: Jean E. Sammet, *Programming Languages: History and Fundamentals* (Englewood Cliffs, N.J.: Prentice-Hall, 1969).

CSIRAC. The Computer of the Australian Council for Scientific and Industrial Research (CSIRAC) was built between 1947 and 1951 as Australia's first stored-program computer. It was the creation of Maston Beard and Trevor Pearcey—two scientists who worked on the development of radar systems during World War II and who concluded that further research without the aid of computational devices to handle the heavy mathematical workload would be unproductive. These two scientists went to work at CSIR after the war and were able to obtain funding necessary to begin work on a computer.

Their motivation and the characteristics of their computer reflected concurrent developments in computer science in other countries, particularly in Great Britain and the United States. The machine finally built was small and slow but functioned. It initially had an execution rate of 500 instructions/thousand, a rate that was soon doubled. It used 20-bit words which, although short and thus generating less precise mathematical answers, made the design of the machine easier to perform. It was made out of 2,000 vacuum tubes; later, germanium diodes were used in its circuitry. CSIRAC relied on magnetic drum memory,* which was then common on other systems, with capacity to store 1,024 words. Disk storage was added in 1954. It used eighty-column card input/output equipment, a teletypewriter, and photoelectric tape—again all of which were quite typical of computers from the late 1940s and early 1950s.

CSIRAC became operational by June 1951, although its seven-man team had performed initial tests on portions of the system as early as 1949. It continued to be changed and improved until mid-1956 when it was given to the University of Melbourne. While still at the Division of Radiophysics, it was employed in doing calculations on cloud-physics and on radio astronomy. After being moved to the University of Melbourne, CISRAC became the centerpiece of the Computing Laboratory where it was used until November 24, 1964, when it was replaced with an IBM 7044. Today the CSIRAC is on exhibit at the Chisholm Institute of Technology in Australia. The Institute claims that this is the only complete first-generation stored-program computer in existence.

For further information, see: Maston Beard and Trevor Pearcey, "The Genesis of an Early Stored-Program Computer: CSIRAC," *Annals of the History of Computing* 6, no. 2 (April 1984): 106–115; Trevor Pearcey, "An Automatic Computer in Australia," *Mathematical Tables and Other Aids to Computation* 6 (1952): 167–172.

CULLER-FRIED SYSTEM. This online programming language for solving numerical scientific problems was developed in the 1960s. It was the merger of two separate programming systems. The first, developed by Glen Culler in 1961, was called the On-Line Computer System (OLC) and was used for scientific computing at the University of California at Santa Barbara. Remote terminals at the University of California at Los Angeles and at Harvard University made it possible for others to use the language for teaching and research. The other programming system was created at TRW† when it was called the Space Technology Laboratories at Redondo Beach, California, in 1964. That language was also intended for use by engineers and scientists. The two merged, operated on highly specialized hardware (called an RW-400), and were interactive.

The Culler-Fried system was one of many languages that emerged in the early 1960s to do online programming to resolve scientific problems. It was of the same generation as such programming languages* for numerical scientific computing as AMTRAN,* Lincoln Reckoner,* and MAP.*

For further information, see: A. Ruyle et al., "The Status of Systems for On-Line Mathematical Assistance," in *Proceedings, ACM 22nd National Conference* (1967): 151–167; Jean E. Sammet, *Programming Languages: History and Fundamentals* (Englewood Cliffs, N.J.: Prentice-Hall, 1969).

D   _____

DATABASE MANAGEMENT SYSTEMS. The principles of managing information and the software* associated with them represent a critical part of any computer system. The issue of how best to control data within a computer has always drawn the attention of scientists who design systems and of users who continually seek more efficient ways to seek specific data, load it back into computers, and protect this vital asset. By the 1960s software emerged to help manage information in machine-readable form. With that software was born what we today call database management systems. No appreciation of the software in a computer is possible without understanding the evolution of database systems. Such software complements the use of high-level programming languages* and operating systems.*

The term *database* means a collection of information (data) in a computer system. The data may or may not be interrelated. With database systems, an attempt is frequently made to reduce redundant copies of information by providing a mechanism for sharing a copy of the data with multiple applications. Thus, for example, someone's home address (known as a data file) could reside in the computer's storage and be used when necessary by a billing application and later by a program that generates a paycheck. Prior to the use of database management systems, the same address file would have been attached to both the billing and payroll programs. The risk of errors existed (since two files now had to be kept up to date) while requiring twice as much storage on disk drives.* The promise of database management systems was reduced redundancy while allowing programmers to write software packages without having to worry about managing files as before. An equally exciting and ultimately more important benefit was the ability to create a reservoir of information which end users could get to through the use of such tools as programs generating reports and queries, all with little or no knowledge of the data processing way of programming. By the 1970s this last prospect was very important because there were not enough programmers in the United States to satisfy everyone's requirements. The average data center had a backlog of unwritten programs ranging from two to four years' worth of work and a latent demand that exceeded that. Furthermore, some 60 percent of all programs written in the 1970s were to query existing data files, all of which could be offloaded onto nonprogramming people if databases were properly set up and managed and if appropriate software tools became available to get to the information—both conditions were satisfied during the same decade. For the end user and the data processing professional, databases and their managing software therefore offered the opportunity to link related information together in a logical fashion, using indexing systems, so that accessing such data quickly became easier.

The term *database* has been used in various ways. It first appeared in 1964 in data processing literature and apparently was initially used by U.S. military personnel to describe collections of information that their end users wanted to get to through time-sharing systems. Meanwhile, in businesses the term in vogue was *integrated data processing*. But by the late 1960s the phrase *database* had been adopted by the entire data processing community, particularly in the United States. It was employed as the expression of a collection of data, primarily a consolidation of information from various applications, within a single computer. The next logical step involved the creation of software to do this which were called database management systems (DBMS). These in turn gave rise to languages and other software tools that relied on DBMS to control and manage information. The development of database systems in the 1970s and 1980s led to the use of the term *database technology* to describe a raft of developments related to DBMS.

The earliest effort to manage data rationally within a computer system was made in the 1950s. Known as "master files," these were collections of information attached to specific application programs and could not be shared with other software. With such files two or more copies of the address file would have been needed. This manner of maintaining information continued to be the most widely used technique until the late 1970s. By then the requirement for others (besides a data center's personnel) to get to information led to the wide usage of database management systems. Thus was born the concept of data independence which was so commonly discussed in the 1980s. Much of the history of database management systems involves the creation of tools to structure data in usable forms, also called data structuring methods; the development of high-level data languages to manipulate information; and the creation of data-protective facilities to manage who is to use information while insuring its privacy and accuracy.

By the late 1950s researchers were beginning to report the successful formulation of file management routines that could sort files regardless of content and that led to the availability of sort/merge software tools which became a staple in every data center during the 1960s and 1970s. From the earliest days, then, the focus was on developing techniques for manipulating and sharing information. The need to reduce redundancy and hence increase the requirement for sharing was as much motivated by economics as by function or technology. As late as the mid-1970s, for example, data entry was still very expensive, and thus anything that would help reduce the number of times someone had to keypunch an address file, for example, into machine-readable form the better. At that time, keypunching and verification cost about 50 cents per thousand characters. For example, a 2 million character file (not really that large) cost $1,000 just to type into the system. Then there was the cost of inaccurate information stemming as much from faulty keystroking as from maintenance of multiple copies which data center managers recognized was expensive but could not always quantify in dollars. The concern for errors was high, however. One survey of programs written in COBOL* showed that about 40 percent of the Procedure Division's lines of code were there solely to check for errors.

The other economic issue involved the cost of storage. Tape and disk storage space cost well over a hundred times more per character in the late 1950s than in the early 1980s. Moreover, much less storage was available in earlier decades. For example, consider two of International Business Machines Corporation's (IBM's) disk storage products. In 1956, IBM announced RAMAC* which could store 5 million characters. The company's 3340 disk drives of the mid-1970s could store 70 million characters. The real surge in online storage capacities on disk did not come until the 1970s. The IBM 2311 and 2314—the standard disk drives of the S/360* era of the 1960s—had twenty-five times the capacity of RAMAC, or only about 125 million characters. That figure more than quadrupled by the early 1970s, and this exponential growth in capacity has continued to the present. In addition to volume, the early devices were vastly slower in physically reading and writing data to disk than subsequent technologies. Each new product introduction typically sped up the process by 2.5 times. Such growth in speed took place around 1956–1957, again in the mid-1960s, and yet another time in 1970–1971. Since then, other product introductions have continued to speed up access times. The point, however, was that devices were slow in the 1950s and early 1960s and tolerable in the mid-1960s. Not until the second half of the 1960s did technology and costs begin to make it possible to expand the amount of machine-readable storage that a data center could support realistically. When that happened, the need for DBMS surged.

One of the earliest efforts to create tools to manage data involved separating files from programs so that they could be shared. This was done as much for convenience as to reduce the expenses of file maintenance. The phrase used to describe this effort in the 1960s was "physical data independence." Data were always said to be physically independent of a particular program if they could be gotten to through various access methods and be structured in numerous ways. By 1970 "physical data independence" existed "if the program or *ad hoc* requests (were) relatively independent of the storage or access methods" (according to an IBM manual). Logical data independence (how software looked at data and not how data was stored physically in a computer's disk or tape)

Table 15
Database Managers: RPG Family, 1956–1965, Selected Years

Developer	Software	Computer	Year
GE	MARK I	IBM 702	1956
GE	MARK II	IBM 702	1957
SHARE	9 PAC	IBM 709	1959
SHARE	SURGE	IBM 704	1959
IBM	RPG	IBM 1401	1961
AIS	GIRLS	IBM 7090	1962
IBM	RPG	IBM S/360	1965
IBM	RPG II	IBM S/3, S/370	1970s

SOURCE: J. P. Fry and E. H. Sibley, "Evolution of Data-Base Management Systems," *Computing Surveys* 8, no. 1 (March 1976): 7–42.

meant "the ability to make logical change(s) to the data base without significantly affecting the programs which" accessed them.

The history of DBMS began in the 1950s when files of fixed-length fields were set up, drawing on the limits of the amount of data that could be punched out on an eighty-column computer card. These pieces of information on cards were called "flat files." "Records" represented collections of data (such as addresses), whereas "fields" were portions of records (such as street numbers, towns, or zip code portions). In the early days files were kept in sequential order and accessed one after the other, typically from cards or tape. By 1965 they were also on disk and could be gotten to directly, skipping over records that were not wanted. Where to go to get such files was accomplished by systems of indexes, a major portion of database management systems.

The first software tools involved data definition languages which enabled a programmer to describe databases that could be used across multiple applications. One of the earliest was developed for the SAGE* project of the 1950s called COMPOOL. Written at the Massachusetts Institute of Technology (MIT), it managed files for hundreds of programs. Many of the techniques developed for this package later found their way into languages such as JOVIAL.* By the mid–1950s languages contained facilities for managing files. Some of the more obvious included IBM's Commercial Translator* and Honeywell's† FACT. GECOM, developed by General Electric,† also fit into this category of programs. In the cases of GECOM and Commercial Translator, a programmer could define intrarecord structures. FACT went further by adding the ability to have hierarchical structures where a program sought a piece of information by going first to the head of a cluster of data and then working down through what looked like an organization chart of a company to the piece of information desired. One common family included RPG (see Table 15).

With the creation of COBOL* at the start of the 1960s, a more formal approach and the best of the past were merged together, creating the Data Division within

Table 16
Database Managers: MITRE/Auerbach Family, 1962–1970, Selected Years

Developer	Software	Computer	Year
MITRE	ADAM	IBM 7030	1962
MITRE	COLINGO	IBM 1401	1962
MITRE	C-10	IBM 1401	1965
SDC	LUCID	AN/FSQ32	1967
Auerbach	DM-1	U1218, H6000	1969
Western Electric	SC-1	IBM S/360	1970

SOURCE: J. P. Fry and E. H. Sibley, "Evolution of Data-Base Management Systems," *Computing Surveys* 8, no. 1 (March 1976): 7–42.

the language. For the first time and in dramatic form, data were separated from the actual procedures that would be employed on information within a language.

The impetus for many developmental activities during the 1950s, as with many other data processing technologies, came from requirements articulated by the U.S. military. The U.S. government's civilian agencies, representing some of the most important supporters of data processing, also encouraged work on DBMS. Along with the creation of new ways to define data in COBOL, other milestones were reached in the early years. For example, in 1961 a system called BASE-BALL came out that made it possible to access data through the use of English-like commands. Other similar developments included B-Tree and QUERY. From Radio Corporation of America (RCA) came Retrieval Command-Oriented Language in 1962 which had five commands to handle database management. By specifying a data descriptor with a particular query, a programmer could link together a program and relevant data.

An equally interesting development, sponsored by the U.S. Army during the 1950s, was called ACSI-MATIC. The system took advantage of disk-oriented systems and allowed one to make inferences. For example, if you are Jim's son, and Dora your sister, then Dora is Jim's daughter. The software used dynamic storage allocation and took requests in batch mode. This ambitious software was never completely implemented owing to the unreliability of disk storage devices of the late 1950s. The U.S. Air Force, also a strong supporter of software, had its own project called Experimental Transport Facility (ETF). This project was developed for the Air Force by the MITRE Corporation which later also created the Advanced Data Management System (ADAM) (See Table 16). ADAM was used on a STRETCH* computer (IBM 7030) in 1962 and was one of the most important of the early database packages.

Two other software tools in the early history of database include COLINGO (contemporaneous with ADAM) and LUCID. COLINGO was a tape-oriented database system coded in a manner similar to how COBOL managed logical data structures. It ran on an IBM 1401* computer, one of the most widely used systems of the early 1960s. C-10, a follow-up to COLINGO, operated on an

Table 17
Database Managers: MARK IV Family, 1962–1968, Selected Years

Developer	Software	Computer	Year
AIS	GIRLS	IBM 7090	1962
INFORMATICS	MARK I	IBM 1401/60	1962
INFORMATICS	MARK II	IBM 1401/60	1964
INFORMATICS	MARK III	IBM 1401/60	1966
Scientific Data Systems	MANAGE	XDS 940	1967
INFORMATICS	MARK IV	IBM S/360	1968
Application Software Inc.	ASI-ST	IBM S/360	1968

SOURCE: J. P. Fry and E. H. Sibley, "Evolution of Data-Base Management Systems," *Computer Surveys* 8, no. 1 (March 1976): 7–42.

IBM 1410 and incorporated many of the functions evident in ADAM. LUCID, created by the System Development Corporation, confirmed ADAM as a very influential, early piece of software. One other system, DATA MANAGER-1 (DM-1), also grew out of ADAM-like programs.

Beginning in 1964—the year IBM announced the System/360 and a collection of new hardware and software—and running to about the end of 1968, families of programming systems emerged with more sophisticated database management procedures. During these years three different lines of development were visible. Each identified many of the functions that became common in database systems of the 1970s.

The first of these groups involved the Mark IV* family. In 1962 the Generalized Information Retrieval and Listing System (GIRLS) appeared for use on an IBM 7090, while at the same time Mark I came out, soon to be followed by the Mark II (1964), Mark III (1966), and Mark IV (1968) (Table 17). The Mark IV ran on the widely used S/360, whereas the earlier versions ran on various pre-S/360 computers from IBM. The Mark series made possible multiple-segment types within hierarchical structures in almost unlimited quantities. These involved structuring data so that they were read on a "top down, left-right" sequence.

A second line of development involved the IDS family. Integrated Data Storage (IDS) represented General Electric's database management contributions of the 1960s and 1970s. The pieces of software included IDS/GECOM (1964), IDS/COBOL (1966), a data BASIC (1970), and IDS/II (1975) created at Honeywell† (Table 18). All of these systems took advantage of random-access hardware and high-level programming languages,* contributing new data-handling verbs that interfaced between information and high-level languages. They offered storage and program-level descriptors that were separate from each other (as required, for example, by the architecture of COBOL). Other features

Table 18
Database Managers: IDS Family, 1964–1975, Selected Years

Developer	Software	Computer	Year
GE	IDS/GECOM	GE 400	1964
GE	IDS/COBOL	H6000	1966
General Motors	APL	IBM S/360	1966
GE	dataBASIC	GE CPUs	1970
CODASYL	DDLC/DBLTG		1973
ANSI	COBOL 74		1974
Honeywell	IDS/II	H6000	1975
Honeywell	MDQS	H6000	1975

SOURCE: J. P. Fry and E. H. Sibley, "Evolution of Data-Base Management Systems," *Computing Surveys* 8, no. 1 (March 1976): 7–42.

included implicit insertion and deletion of groups, retrieval or modification of primary and secondary keys, paging of data, recovery and restart functions on an incremental basis, and shared access to databases. Not all of these functions were available in the earlier releases; rather, they were added during the 1960s and 1970s.

Similar tools appeared. From General Motors Research in 1966 came Associative PL/I (APL), which was used for managing data files in computer-aided design applications (today called CAD). In this same vein was dataBASIC at General Electric (1970) which made it possible for a nonprogrammer to reach homogeneous files in a time-sharing system using BASIC* as the programming language. By the early 1970s one final offspring of IDS-like systems came from Honeywell called the Management Data Query System (MDQS), consisting of query and report specification capabilities which in turn used sequential, index sequential, and more traditional IDS files.

A third trend, or family of systems, appeared during the mid- to late 1960s and was called the formatted file systems or GIS family (Table 19). Most of these systems were created to satisfy the specific requirements of various U.S. government agencies, particularly those concerned with command and control functions and intelligence gathering. Formatted files were first developed in the 1950s and are still being developed today. The earliest of these was Information Retrieval (IR) which ran on an IBM 704* for the first time in 1958. Tape Update for Formatted Files and Tape Updater and Generator appeared soon after and operated on the same kind of computer. The U.S. Air Force created its own called the SAC/AIDS Formatted File System which came into existence in 1961 for use on the 438L. For the first time a file format table appeared in a DBMS. This collection of software was one of the earliest "self-describing" database systems. Each piece of information had a data definer attached to it, which meant that this software could get every individual piece of information to it directly.

The U.S. Navy, on the other hand, used a multilevel hierarchical-structured

Table 19
Database Managers: Formatted File, GIS Family, 1958–1973, Selected Years

Developer	Software	Computer	Year
DTMB[a]	IRS	IBM 704	1958
DTMB	TUFF/TUG[b]	IBM 704/709	1959
SAC	FFS[c]	IBM 7090	1961
U.S. Navy	IPS	CDC 1604	1961
U.S. Navy	IPS	IBM 7090	1961
U.S. Navy	IPS	AN/UYK-1	1961
IDHS	FFS	IBM 1401	1963
FICEUR	FFS	IBM 1410	1963
DCA/NMCSSC	FFS	IBM 1410	1965
DIA-IDHS	FFS		1965
IBM	GIS	IBM S/360	1969
SAC/FICEUR	NIPS	IBM/360	1969
DIA	CDMS	IBM S/360	1969
IDHS	FSS	IBM S/360	1970
DIA	MIDMS	IBM S/360	1970
DIA	MIDMS	H6000	1973

[a]DTMB = the David Taylor Model Basin.
[b]TUFF and TUG = Tape Update for Formatted Files and Tape Updater and Generator, respectively.
[c]FFS = Formatted File System.

SOURCE: J. P. Fry and E. H. Sibley, "Evolution of Data-Base Management Systems," *Computing Surveys* 8, no. 1 (March 1976): 7–42.

database approach to go after sequentially stored files. Called Information Processing System (IPS), it first appeared in 1963 and ran on a CDC 1604 and later on an IBM 709/90 and the AN/FYK-1. Simultaneously, the Navy developed yet another variant. The Naval Fleet Intelligence Center (FICEUR) in Europe completed initial work on a system in 1963 which ran on an IBM 1410. SAC used a similar system, first on an IBM 1401 at its Pacific Air Force Headquarters, and later on an IBM S/360. FICEUR and SAC decided to merge their packages in 1965, creating the NMCS Information Processing System (NIPS). This software had logical file maintenance and an enhanced query language, and could be used in online systems. In 1968 it was moved from an IBM 1410 to an S/360 and was renamed NIPS-360.

Earlier it was noted that American intelligence organizations also had a need for more sophisticated computer-based information retrieval systems. Although little is known about their work during the 1960s, one project that has been publicly documented is the Intelligence Data-Handling Formatted Files System. It was characterized by an efficient processor that could handle relatively large files. This was first used by the Defense Intelligence Agency (DIA) in 1968 which called the software the COBOL Data Management System (CDMS) and renamed it the Machine Independent Data Management System (MIDMS) in

1970. It first ran on an IBM S/360 and by the end of 1973 on a Honeywell H6000 computer as well.

The years 1964–1968 marked a period of major advances in the development and use of higher level programming languages and complex operating systems rich with functions running on ever more powerful, yet less expensive, computer systems. Database management improvements also occurred. By the late 1960s many of the lines of research evident throughout the 1970s had been established, while the demand and applications for such software were clearly there. This was most precisely demonstrated by the fact that by around 1968 the overwhelming developments in database management systems had shifted from inhouse projects (primarily within military communities and their related consulting firms) to vendors who sold operating systems and compilers for high-level languages, and now saw the need to create commercially available DBMS packages. That trend has continued to the present day.

Three important lines of evolution now emerged: CODASYL, IMS, and Inverted File families. The first grew out of work done with IDS systems and also relied on APL-based DBMS. The CODASYL Programming Language Committee (Conference on Data Systems Languages, organized in 1958 with members from vendors and users of data processing systems to standardize the use of languages) decided in the early 1960s to enhance COBOL's data-handling features. It wanted to incorporate database methods into COBOL. This is an important episode in the history of software because the entire data processing industry in the United States had rallied collectively around COBOL as a universal language for commercial applications by the time IBM started shipping its S/360s in 1965. By the end of the decade, COBOL was the most widely used language for commercial applications. In the early 1960s many believed that incorporating database management capabilities in COBOL was vital. The CODASYL operated on the problem under various names: first as the List Processing Task Force and then as the Data Base Task Group (better known as DBTG). The change in names strongly suggested a shift in focus on the issue. The group did not issue its first recommendations until 1969 when it called for semantics and appropriate syntax for a data description language (soon referred to as DDL). The group envisioned a system with a network-structured database and statements that enhanced COBOL itself. It recommended that DDL be universal—that is, that it should function with all languages, not just COBOL. In a second set of recommendations, published in 1971, the group called for the division of all data descriptions into two parts. They recommended a Schema DDL to be written to define entire databases and then for a second, or Sub-schema facility, to define specific views of the entire database. All of these functions were to be compatible with the major programming languages of the day. That same year additional task forces were established to pursue both lines of development. These groups met during the 1970s and made specific recommendations on a continuing basis, many of which were incorporated into various database packages.

These suggestions involved the development of software offered for sale by

Table 20

Database Managers: CODASYL Family, 1964–1976, Selected Years

Developer	Software	Computer	Year
GE	IDS	GE 400	1964
General Motors	APL	IBM S/360	1966
CODASYL	(List Processing Task Group convenes)		1968
CODASYL	DBTG Specifications		1969
B. F. Goodrich & Cullinane	IDMS	IBM S/360	1970
Xerox	DMS	SIGMA 5, 7, 9	1970
CODASYL	DBTG Specifications		1971
Univac	DMS 1100	Univac 1100	1971
CODASYL	DDLC Specifications		1973
Xerox	EDMS	SIGMA 5, 7, 9	1973
Philips	PHOLAS	P 1000	1973
Honeywell	IDS-II	H6000	1975
CODASYL	DBLTG Specifications		1975
B.F. Goodrich & Cullinane	IDMS-II	PDP 11/45	1976
CODASYL	COBOL Specifications		1976

SOURCE: J. P. Fry and E. H. Sibley, "Evolution of Data-Base Management Systems," *Computing Surveys* 8, no. 1 (March 1976): 7–42.

Xerox,† Univac, Philips, and others. Perhaps the most widely used of all was the package developed by B. F. Goodrich and Cullinane and marketed by Cullinane beginning in 1970 as IDMS-I and, in 1976, IDMS-II—a rival to IBM's database management products of the period. Some of the key developments of that era included the UNIVAC DMS 1100 which came out in 1970 for use on the 1108 computer and Digital Equipment Corporation (DEC)'s† DBMS-10 which ran on its PDP-10 computer. IDMS first ran on the S/360 and later on the PDP–11/45. Other variants included Control Data Corporation's (CDC's)† Query/Update and Xerox's EDMS. For others see Table 20. Philips offered PHOLAS for sale, and Honeywell issued a new release of IDS called IDS/II in 1973.

Developments at IBM were ultimately to prove more important for the data processing industry because so many copies of that company's database management programs were used throughout the world. IBM embraced database as a strategic direction by the early 1970s and used its powerful marketing organization to sell the industry as a whole on the value of database systems. IBM's enormous success almost singlehandedly insured that by the early 1980s most major corporations and intermediate-sized companies and agencies would accept database as the primary way to handle data. Although many licenses for its software were sold to its customers, the installation of database software came slower than the acceptance of the wisdom of their use. Thus, by the early 1980s

Table 21
Database Managers: IMS Family, 1965–1979, Selected Years

Developer	Software	Computer	Year
North American Aviation Space Division	GUAM	IBM 7010	1965
Rockwell Int. & IBM	RATS	IBM 7010	1965
Rockwell Int. & IBM	DL/I	IBM S/360	1966
Rockwell Int.	ICS	IBM S/360	1966
IBM	IMS-1	IBM S/360	1969
IBM	IMS-2	IBM S/360	1969
IBM	IMS/VS	IBM S/370	1969
IBM	DL/I	IBM 4300	1979

SOURCE: J. P. Fry and E. H. Sibley, "Evolution of Data-Base Management Systems," *Computing Surveys* 8, no. 1 (March 1976): 7–42.

many companies were still using non-database management techniques for controlling machine-readable files. Nonetheless, IBM's endorsement of database had the same effect on the industry as a whole as did its blessing of other technologies: acceptance of database grew sharply as the 1970s progressed.

The single most important project within the company involved the creation of the IMS family of database management packages. Although it first appeared in 1969 as one of the first such systems that made widely available a hierarchical database package, work on earlier variants had been ongoing since the early 1960s. Originally motivated by efforts for the Apollo project to land a man on the moon, IMS could be traced to the aerospace industry. The Space Division of North American Aviation (later Rockwell International) created a package in 1965 called the Generalized Update Access Method (GUAM) which ran on an IBM 7010. This particular package was the predecessor to IBM's Data Language/One (DL/1)—the centerpiece of IBM's database offerings of the 1970s and early 1980s. DL/1 first ran on an S/360 in 1966 as a result of work done at Rockwell International.

One other line of IMS came from the use of the Engineering Document Information Collection Task (EDICT)—an application program—and the Logistics Inventory Management System (LIMS), both of which were supported by Remote-Access Terminal System (RATS). RATS grew out of IBM's and Rockwell's efforts in 1964–1965 and initially ran on the IBM 7010 (1965). In 1966 the two companies joined forces with Caterpillar Tractor Corporation to produce a sophisticated database system known as the Information Management System, or simply IMS (Table 21). Their objective was to create a package for use in the Apollo program. The three then separated, with Rockwell developing a new version of the code called the Information Control System/Data Language/I

(ICS/DL/I) which first ran in 1968. IBM continued with its own work, leading to further enhancements of its own software.

Because DL/I and IMS (variants of each other, with DL/I for use in small and large S/360-370 systems and IMS for use with CICS in large systems only) were so widely used by 1980 that their functions should be understood even if only superficially. IMS used a hierarchically structured concept for databases. A record logically had a single type of "root segment." The root segment (equivalent analogously to the highest block on a company's organization chart such as for the president) could have "child segments" (e.g., like a block to indicate a vice-president). These children in turn could also have as many child segments as needed, up to fifteen "segment types," with a maximum of 255 segment types within a database's "record type." All records had one root segment, and no child segment could exist without a "parent segment." IMS had various methods by which information could be organized physically on disk. These organizations were clustered into groups which used specific access methods to reach them. One of the four main methods employed, called the Hierarchic Indexed Sequential Access Method (HISAM), was the sequential access method used by IMS to get to a segment of a record. Hierarchic Indexed Direct Access Method (HIDAM) was the most widely used method for getting to data directly. It made available "pointers" that allowed a user to get to representations of hierarchical structures, or clusters of data, quickly, employing a sophisticated indexing methodology. IMS was a convenient tool for managing large and numerous databases in the 1970s and 1980s.

In its early years, DL/I was a data descriptor that could provide an organized, hierarchically structured database, rather than the very comprehensive system it became by the mid-1970s. At first, it also provided the necessary software to allow a program written in COBOL to get to data in this package, confirming the industry's desire to have that function. By 1968 this same task could be done with an online version, and by then, it too supported PL/I. The following year, the first release of IMS to run on an S/360 appeared. By the end of that year, a second became available called IMS-2. When IBM announced virtual storage in 1969, it produced yet a third version which it called IMS/VS and ran on the S/370 family of computers. That version became the basis for all releases and versions that appeared in the 1970s and 1980s.

The other major trend in database management packages involved the Inverted File Family (Table 22). As its name implies, data were organized so that a file's sequence was reversed. This became a method for organizing an indexing system using a keyword identifier for a record. Database packages using this approach first appeared in the 1960s. LUCID (1967) was frequently thought to be the source of this kind of database management technique because the same organization that created it—System Development Corporation—produced Time-Shared Data Management System (TDMS) in 1969. Work on TDMS was supported by the Advanced Research Projects Agency (ARPA) in order to have a database system that could operate in a time-sharing environment running on an

Table 22
Database Managers: Inverted File Family, 1967–1972, Selected Years

Developer	Software	Computer	Year
SDC	LUCID	AN/FSQ32	1967
SDC	TDMS	IBM S/360	1969
University of Texas	RFMS	CDC 6000	1969
MRI	S2000	CDC 6000	1970
CDC	MARS VI	CDC 6000	1970
MRI	S2000 (Level 1)	Univac 1100	1971
MRI	S2000 (Level 2)	IBM S/360	1972

SOURCE: J. P. Fry and E. H. Sibley, "Evolution of Data-Base Management Systems," *Computing Surveys* 8, no. 1 (March 1976): 7–42.

IBM S/360. TDMS was the first software to combine the use of an inverted file and a hierarchical data model in an interactive computing environment. That same year scientists at the University of Texas completed work on a package (under development since 1966) called the Remote File Management System (RFMS). It ran on a CDC 6000. RFMS was basically an inverted file database management system, although its specifications and design varied from TDMS. RFMS continued to evolve. Some of the people who had worked on it developed the System 2000 (for the MRI Systems Corporation) which appeared in 1970. Other releases continued the evolution to allow implementation on the Univac 1100 (1971) and on the IBM S/360 (1972). Thus, by the early 1970s, RFMS had a variety of host languages and self-contained instructions. However, use of these types of software was limited in comparison to IMS, for example, in the 1970s and 1980s.

Yet other developments took place in the general field of database management systems during the 1960s and 1970s. With the introduction of disk drives (DASD) in the 1960s, it became possible to develop all of the systems that directly accessed information conveniently and cost-effectively. As the cost of disk drives dropped, encouraging the growth in the amount of machine-readable data, the requirement for better data management systems rose.

One of the earliest and most obvious sources of demand for database systems outside of the military came from manufacturing. Bill-of-materials (BOM) applications were programs that kept track of the parts needed in order to manufacture products. BOM was an obvious application for database systems because it required that many pieces of data (on parts) be associated with each other (parts necessary to make up a product or subassembly and possible substitutions). In addition, a part might be a component of other parts or subassemblies. Perhaps the most important early work supporting this application was done at IBM in the 1960s. By 1965 the company had a package called Bill-Of-Materials Processor (BOMP) and, by the end of the decade, issued another version which replaced the first. This version, Data Base Organization and Maintenance Processor (DBOMP), was widely used in the early 1970s by manufacturing companies. Each release of the software had master and chain files. This approach also appeared in TOTAL, another widely used database management package of the 1970s. The Integrated Data Store (IDS) software developed at General Electric was also based on a network structure with records and chains of records. Its record chain was similar to a BOMP chain. IDS was also the basis of early work done by the CODASYL task force in the late 1960s and early 1970s.

TOTAL proved to be the most important of these database systems because of its wide acceptance, particularly within the United States. TOTAL, a direct access database package when introduced in 1969, developed into a system using chain-like pointers and a Schema-Sub-schema processor.

Another important package was Data Manager–1 (DM–1) which emerged from work done for the U.S. Army's ACSI-MATIC project and ADAM. It had service routines to move and store data which allowed for ad hoc queries or use with application programs. It initially ran on an Air Force U1218 computer, and on a Honeywell H6000. Western Electric's version, Control-1, ran on the S/360 in 1968. The other major computer vendor of the mid-1960s, Burroughs,† introduced the Data Management System II which ran on its B6700/B7700 computer in 1974. It worked with COBOL and relied on a process called set-theoretic terms for defining its data language. It was that company's primary database offering of the 1970s.

Although the history of database management systems is complicated and some of its outlines have been omitted here (e.g., the development of data dictionaries), by 1975 issues had emerged that reflected previous patterns and work for the future. The earliest systems were designed to help the relatively inexperienced nonprogrammer. These packages evolved into host-language systems by the late 1960s. During the 1970s developers created database packages that serviced the needs of professional programmers writing complex applications and ad hoc users who knew little about programming. Query packages of the 1980s appeared with embedded database management facilities. Research on artificial intelligence* in the 1960s and 1970s created pressure for even more sophisticated methods of managing data. The need for relational databases had also become evident by 1980.

Taking advantage of the fact that people could communicate within a "system" geographically scattered, using telecommunications, by 1980 users were clamoring for database tools that were not just resident in one computer. They wanted access to various databases located in multiple systems geographically dispersed. By the early 1980s, there was a demand for relational databases, and, by 1981 some vendors claimed that their products were relational. Much remained to be done, however, to make that statement a reality. One major breakthrough in the early 1980s that came close to being a relational database was IBM's SQL (Table 23).

Another trend of the past twenty years has involved making it possible for

Table 23
Early Relational Database-Styled Software, 1970s

Developer	Software
MIT Project Mac	MacAIMS
General Motors	RDMS
IBM	IS/I
University of California Berkeley	INGRES
University of Toronto	ZETA
IBM	System R
IBM	QBE
Relational Software, Inc.	ORACLE
IBM	SQL/DS[a]

[a]By the mid-1980s SQL was most closely a relational database manager. For a similar list, from which this table was drawn, see W. C. McGee, "Data Base Technology," *IBM Journal of Research and Development* 25, no. 5 (September 1981): 510. Copyright 1981 by International Business Machines Corporation; reprinted with permission.

programs written in widely used languages to access data housed in a database, thereby reducing the programmer's need to be concerned with the administration of information. As new languages appeared, database managers were better able to allow users to employ those idioms with existing files.

One final trend of this period involves the idea of data (information) as a corporate asset. The amount of machine-readable information kept growing during the 1960s and 1970s, as can be measured by monitoring the sale of disk drives. Statistics on their sale suggest the extent of the issue. The number of hard disk devices shipped in the United States in 1979 was 156 million. By 1984 this figure had climbed to over a million and was projected to rise to over 6 million by the end of 1989. This projection was made by INFOCORP, a publisher of a data processing industry newsletter. In terms of the amount of data stored (measured in gigaytes [Gb]), these statistics suggest that in 1979 organizations in the United States acquired an additional 18,887 Gb of online storage capability and that by the end of 1984 annual shipments had exceeded 56,400 Gb. One forecast has estimated sales of 306,700 Gb in 1989. These statistics reveal that both the quantity of information and the amount of money invested in such data were rising.

Many managers believe that the additional information, as used in applications relying on data processing technology, has made their organizations more effective and competitive. Employees are increasingly becoming "knowledge workers" in a service-oriented economy. Information is increasingly being viewed as an asset to be managed much as any other asset: money, people, buildings, or inventories. In addition, decisions are widely made on the basis of computer-based information. Thus, the need has arisen for tools to effectively manage the use, control, security, quality, and integrity of machine-readable data, and the acceptance of database management has become critical to any organization's success. By 1980 database management software had become as important as high-level programming languages and query packages, or operating systems.

For further information, see: J. P. Fry et al., "Data Management Systems Survey," *MITRE Corporation Report, MTP 329* (Santa Clara, Calif.: MITRE Corporation, January 1969) and J. P. Fry with Edgar H. Sibley, "Evolution of Data-Base Management Systems," *Computing Surveys* 8, no. 1 (March 1976): 7–42; James Martin, *Computer Data-Base Organization* (Englewood Cliffs, N.J.: Prentice-Hall, 1975); W. C. McGee, "Data Base Technology," *IBM Journal of Research and Development* 25, no. 5 (September 1981): 505–519; Saul Rosen, "Programming Systems and Languages—A Historical Survey," *Proceedings of the Spring Jt. Computer Conference, 1964* (Montvale, N.J.: AFIPS Press, 1964), 25: 1–25; Jean E. Sammet, *Programming Languages: History and Fundamentals* (Englewood Cliffs, N.J.: Prentice-Hall, 1969).

DEUCE. This British computer, first delivered in 1955, was an outgrowth of an earlier device called the Pilot ACE.* The DEUCE was constructed by the English Electric Company as one of its first electronic digital computers.* One of the first such machines was installed at the National Physical Laboratory (NPL), home of much of Britain's early research and development in the field of digital computers.

The English Electric Company had been a leading manufacturer of electronic equipment before expressing interest in computers. This new concern grew out of work it had conducted jointly with the NPL after World War II, as well as separate projects conducted by scientists at that laboratory. The company's chief executive officer, Sir George Nelson (and later Lord Nelson), served on NPL's Executive Council and thus was knowledgeable about its work. In January 1949 he formed a team that would work with NPL to produce the Pilot ACE, completed in 1952. At that time the company's Nelson Research Laboratories was more involved with computers. The company, first motivated to build a computer for its own use, soon realized the commercial possibilities of such a device. Hence, it built the DEUCE, finishing the first one in 1955.

The DEUCE depended on mercury delay lines along with a magnetic drum for memory.* It used punched cards for input while the processor operated at the speed of 1 megacycle, which was fast for its day. Although it proved very similar to the Pilot ACE, their differences were more important. For one thing, engineers, attempting to improve the ability to do maintenance on the equipment, built component subsystems. The machine had nearly twice the number of valves as the NPL device, and in its final configuration, it had 1,450 such devices. The initial models cost £50,000. The DEUCE 2 became available in late 1955, (working better than the first), and the DEUCE 2A appeared in 1957. These two models had additional input/output equipment. Between 1955 and 1964 the English Electric Company sold thirty-one copies of the DEUCE 1 and 2. Its success with DEUCE encouraged it to produce other computers such as the

KDN2 (1962), KDF6 and 7, and KDF10 (1961), which was then redone as the KDF8. The DEUCE series used a 32-bit word length. The early devices could store 8,192 words, which gave them one of the largest memory systems of all British computers of the late 1950s.

For further information, see: Simon Lavington, *Early British Computers* (Bedford, Mass.: Digital Press, 1980).

DIALOG. This programming language was developed in the mid-1960s to do numerical scientific problems at the Illinois Institute of Technology. In the late 1960s it ran on a UNIVAC* 1105 time-sharing monitor, although another version operated in batch mode on an IBM 7094. DIALOG was an online experimental language that could best be characterized as algebraic in expression. Unlike other languages of the time, however, it used a stylus and one push button to handle input. All programs and their results were displayed on a screen. Paper tape* served as the input medium, and results could be printed.

By using a stylus, a user would point to the characters desired which were displayed on a terminal's screen. DIALOG would interpret each character presented one at a time. The number of characters was typical of many programming languages*—twenty-six upper-case letters, all ten digits, and another twenty-four special characters (e.g., $+$, $-$, $/$, $=$, V, (), ?, etc.).

For further information, see: S. H. Cameron et al., "DIALOG: A Conversational Programming System with a Graphical Orientation," *Communications, ACM* 10, no. 6 (June 1967): 349–357.

DIFFERENCE ENGINE. This was the first computer-like device built by Charles Babbage.** Its partial construction during the 1820s represented a major technological leap forward over previous calculators and toward the development of concepts common in the design of today's computers. Its inventor, who later designed an analytical engine,* constructed a portion of the difference engine under a grant from the British government, making it one of the first scientific projects of technological importance subsidized by a nineteenth-century European government. Also as a consequence of its partial construction, Babbage made important contributions to machining processes. Ultimately, the effort advanced the scientific community's understanding of the potential of calculating devices.

Babbage's machine was able to add, computing polynomials up to six decimal places. Specifically, his machine could calculate more accurately than the existing navigational tables. The approach involved calculating differences in intervals, hence the name difference engine. Babbage's machine was made up of linked adding devices that automatically created successive algebraic values using the finite differences method. After several years of design and partial construction, by the end of 1822 he had built a portion of the machine to handle two orders of differences. This machine relied on work done by earlier inventors, for it consisted of a series wheels on rods serving as axes.

Babbage never completed the construction of this machine, although he continued refining its design throughout the 1820s. Through his experience with the difference engine, he sought to improve on the design in a more comprehensive fashion. Thus, in the late 1820s and for the rest of his life he turned his attention to the design of the analytical engine. The second device proved to be a far more sophisticated and complex series of machines capable of continuous mathematical processes complete with systems control and programmable-like qualities. He dropped extensive work on the difference engine in the early 1830s partly because tooling methods had progressed to the point where it was almost cheaper and easier to start from scratch rather than patch an older design. Many companies in the computer business would learn this lesson in the 1950s and 1960s.

The problem of leapfrogging technologies has therefore not been limited to the data processing industry of the twentieth century. Indeed, at the design level, Babbage recognized that the difference engine had only one function and so did not satisfy his perceived need for a machine that could do far more. It was not a device to automate mathematical processes in general—the fundamental requirement behind the research on the analytical engine. Yet the difference engine was important nonetheless, for Babbage learned a great deal from this early work, as would be reflected in his more important work on the analytical engine.

In his lifetime Babbage received considerable attention and publicity. He was socially well connected and had friendships with the leading scientists of his day. While building the difference engine for the British government, he received extensive press coverage on his progress. The work propelled him into the first ranks of contemporary British scientific leaders. His work encouraged others, particularly in France, to work on calculators and the general application of technology in industry. Construction of the difference engine also represented one of the first instances in the 1800s when the British government contributed to a highly technical research and development project. This device improved the accuracy of navigational tables, which reduced the high rate of maritime accidents caused by faulty data. This was an important advance, for at that time England had the largest maritime and naval fleets in the world. When no significant progress was made toward its completion, however, the government lost interest in the engine. Without funds, let alone desire, Babbage never completed the project. Today, portions of the machine can be found in British museums.

For further information, see: B. H. Babbage, *Babbage's Calculating Machine; or Difference Engine* (London: South Kensington Museum, 1872); L. H. D. Buxton, "Charles Babbage and His Difference Engine," *Transactions of the Newcomen Society* 14 (1934): 43–65; B. Collier, "The Little Engines That Could've; The Calculating Machines of Charles Babbage" (Ph.D. diss., Harvard University, 1971); D. Halacy, *Charles Babbage, Father of the Computer* (New York: Crowell-Collier, 1970); A. Hyman, *Charles*

Babbage: Pioneer of the Computer (Princeton, N.J.: Princeton University Press, 1982); M. Moseley, *Irascible Genius: A Life of Charles Babbage, Inventor* (London: Hutchinson, 1964).

DIFFERENTIAL ANALYZERS. This class of analog computers* was constructed primarily during the 1930s. The most famous were designed and built by Professor Vannevar Bush** at the Massachusetts Institute of Technology (MIT). Although these machines did not ultimately represent the main line of development within the world of computers (digital computers* later did), they nonetheless expanded scientific knowledge of the physics of computing and the principles associated with the relationship of problem-solving and mechanical constructs. Some engineers first learned about the principles of electromechanical computing through early work with differential analyzers. They next went on to design digital computers in the 1940s and 1950s.

The fundamental problem addressed by differential analyzers involved calculating the area beneath a specific curve. This integral of the given function, expressed as $\int f(x)\,dx$, becomes difficult to solve if the function, $f(x)$, is unknown in analytic form or, if it is known, to quote Michael R. Williams, it "is not well behaved." During the nineteenth century various mechanical means for achieving an answer were devised, the most famous of which was by James Thompson, brother of Lord Kelvin, who created a machine composed of a rotating disk, sphere, and cylinder that could do elementary integration. By about 1910 another class of devices began appearing called planimeters which, in effect, graphed out an answer. The device built by Bush was the most successful.

Bush wound string around a spinning shaft which was pulled lightly at one end causing it to tighten. The shaft then took hold of the string, pulling the other end with greater force. If one took the control end of the string and connected it to the output cylinder of an integrator, the other end could be employed to drive various gears in a series, which in turn transmitted results to other mechanical computational components. The device would graph on paper results that noted how much a cylinder had rotated, thus representing the value of an integral. That machine was called a differential analyzer. Between the late 1920s and throughout the 1930s Bush built three different machines. The first was mechanical, and the last included some electronics to power motors. One of the most widely known of his machines was called the Rockefeller Differential Analyzer #2 (RDA2). During World War II it calculated ballistic firing tables.

Bush's machines were made up of integrators, gears used for constant multiplication, and other gears for differentiating (performing addition and subtraction). These components were lashed together to solve a specific problem by the adjustment of gears on rotating shafts. Many of these gears were several feet long, able to pass along analog values (through the rotation of shafts) from one part of the device to another.

Five copies of Bush's machines were constructed for use at other universities. Both Manchester University in Great Britain and the Moore School of Electrical

Engineering† at the University of Pennsylvania had what was soon known as a Bush Analyzer; both institutions locales for the development of important digital computers in the 1940s. By the end of that decade, these mechanical differential analyzers had given way to electronic versions. Bush's first device found its way to Wayne University (later renamed Wayne State University) in 1954 and, in the 1960s, to a scrap metal dealer.

In the late 1930s differential analyzers had proven that mechanical devices could be employed to solve complex mathematical problems and, through calculation of ballistic tables, do practical work. Such machines were used effectively during World War II. That experience probably gave scientists, particularly those working within the U.S. government, considerable faith in the value of computers, leading them to support other projects such as the development of digital computers. The most famous of the early ones was the ENIAC,* which was developed at the Moore School. It became the first electronic digital computer ever built. The early analog devices represented an important step toward the modern computer.

For further information, see: Vannevar Bush, "The Differential Analyzer," *Journal of the Franklin Institute* 212, no. 4 (1931): 447–488, *Pieces of the Action* (New York: William Morrow and Co., 1971), and with S. H. Caldwell, "A New Type of Differential Analyzer," *Journal of the Franklin Institute* 240, no. 4 (1945): 255–326; Michael R. Williams, *A History of Computing Technology* (Englewood Cliffs, N.J.: Prentice-Hall, 1985).

DISK DRIVES, IBM. Primary leadership in the development of disk storage (also known as direct access storage) equipment has rested with International Business Machines Corporation (IBM) from the earliest years to the present. IBM's introduction of storage products led the data processing community over the past thirty years, while the number of patents issued to the company's engineers defined the nature of this technology. Disk drives constituted IBM's most important technological contribution to the industry in the 1950s and 1970s. During the 1960s its most impressive contribution was the family of computers known as the S/360,* which included an impressive series of disk storage products that built on the company's earlier successes in this field.

Between 1953 and 1962, few such devices were available, and, during the transition years between 1963 and 1966, many new innovations made it possible for IBM's customers to begin using disk drives in quantity. But the disk drive's great era came between 1967 and the mid-1980s. The cost of storage in these years, for example, dropped twentyfold, whereas the amount of storage on a typical system increased forty times over previous quantities. Put another way, after 1973 disk capacities grew a thousand times greater than those for computer memory,* and yet memories also experienced profound growth.

Disk technologies had as great an influence on data processing as computers. The ability to store information in machine-readable form on disk drives made it possible for users to write applications that required the direct accession of data. That allowed computer applications to move in the direction of real-time, interactive systems as we know them today. Thus, jobs that began as batch operations in the 1950s, relying on tape storage technologies and their sequential accessing of information, could now depend on equipment that permitted fast selection of information directly. In combination with terminals, users could query systems directly and almost simultaneously. Without disk technology, the growth in dependence on data processing witnessed in the late 1960s and throughout the 1970s would not have occurred.

In the first era of disk drives, from 1953 to 1963, experimentation with the auxiliary storage medium of a magnetic tape ultimately led to the development of disk drives. During this period storage was dominated by magnetic tape or punched cards. Batch processing became faster owing to improved computers and the introduction of usable operating systems.* Main memories on computers were dominated by magnetic drums whose capacity grew throughout the period. But it was not until 1957, when IBM began shipping the first movable head disk drive, that direct access to data became feasible. Improvements in operating systems allowed the overlapping of processing and input/output functions concurrently along with multiprogramming, the use of data channels, and the implementation of control units to manage growing amounts of auxiliary storage equipment. Those circumstances made the introduction of disk drives ripe.

The concept of a disk drive has remained essentially unchanged for the past thirty years. At the risk of oversimplification, it consisted of magnetically sensitive surfaces (platters much like phonograph records) in a stack housed in cases. Read/write arms, much like those on a record player, and usually one for each platter, would go out over the surface to a sector, drop down, and either record (write) or pick up (read) specific pieces of information. Software* would keep track of where the data were (each piece of information had an address) and maintain an index. That household function resided either in an operating system or a related software subsystem, or in microcode usually in a disk drive control unit. In practice, the sharing of responsibilities resided in both computers and control units. The disk drives were of two types: fixed-head equipment in which the platters were permanently encased in the disk drive; and later, equipment with removable disk packs that could be carried around and looked like high-tech hat boxes. By 1980 problems with reliability made the portable packs unpopular when compared to fixed devices. Furthermore, portability became available on another magnetic medium in the 1980s, diskettes (such as those used on personal computers), which looked like 45 rpm records.

Beginning in the late 1940s, various research centers, primarily at American universities, started experimentation with disk drives (storage also being called DASD). IBM became the first company to offer commercially available disk drives when, in 1957, it began shipping the IBM 350 and the IBM 650 RAMAC* system. It was the first movable-head disk drive, and it recorded data on moving at a density of 100 bpi (bits-per-inch). The disks rotated at about 1,200 rpm and were about 2 feet in diameter. One characteristic of the disk over the years would

Table 24
Select Technical Features of Early IBM Disk Drives, 1957–1963, Selected Years

Device	350	1405	1301	1311
Year first shipped	1957	1961	1962	1963
Feature				
Recording density (bpi)	100	220	520	1,025
Disk diameter (in.)	24	24	24	14
RPM	1,200	1,200	1,800	1,500
Fixed/removable	Fixed	Fixed	Fixed	Removable
Actuators/spindles	3	3	2	1
Avg. seek time (ms)	600	600	165	150
Data rate (Kbytes/s)	8.8	17.5	68	69
Cost per Mbyte/month ($)	100+	100+	75+	100+

be its continual shrinkage of size, but along the way disks would acquire nicknames. The 350 looked like a juke box, whereas the drives of the late 1960s were likened to pizza ovens.

The 350's disks rotated at about 1,200 rpm. The average seek time was 600 milliseconds (ms) per search, or 200 times the speed of contemporary tape. There were 100 tracks on each which could house 500 characters, thereby giving each disk the capacity to store 50,000 characters. The entire device could store 5 million characters. As Table 24 illustrates, improvements continued, and in 1962 IBM shipped the 1301 disk drive which could store 56 million characters in a device that housed 50 disks, each 24 inches in diameter. This additional capacity was made possible by increasing recording density which reached 520 bpi. Seek time dropped to 165 ms. By then formal standards for formatting data in fixed-length records insured order and improved the ability of the disk drive to move read/write arms to correct positions.

Problems with reliability led the company's engineers to continue further development. Disk management systems were enhancements of those used with tape systems and thus were inefficient in the new world of disks. During the 1960s such problems and those associated specifically with the hardware were solved or minimized to the point where IBM's customers felt confident about storing increasing amounts of data on disk. The breakdown of machines had been solved through engineering changes in existing drives or in subsequent products. Significant amounts of read/write errors persisted until the end of the 1960s. Finally, the cost of direct access storage remained too high when compared to tape, despite the fact that costs had dropped by a factor of two during this early period. During the early 1960s that expense declined so greatly that ultimately a piece of information on a disk pack cost almost the same as on tape. With that development, disk systems and online applications were born and exploded with growth.

IBM introduced its first removable disk pack in 1963, which remained nearly twenty times as expensive as reels of tape throughout the decade. The IBM S/360* had more advanced input/output management software and function, which made the effective use of DASD greater than ever before. Thus, tape began its role as a secondary form of storage, displaced as a primary form by DASD. Tape became the medium of choice for archival storage. In this period the IBM 1311 became a widely used disk drive, encouraging the company to spend money and resources on the development of additional disk products. Between 1963 and 1966 IBM changed the density at which data were recorded on disk from 1025 bpi to 2200 bpi. Costs dropped, while seek times went from an average of 150 ms to 75 ms. Rotational periods went from 57 ms to 25 ms. Data rates also improved (69 to 312 Kbps).

An important development took place in this period in the area of microprogrammed intelligent control units. This class of equipment made the management of storage devices easier while simplifying the attachment of disk drives to a computer system. Disk drives were attached to control units, which in turn were hooked to a data channel leading directly into the computer. A maximum number of disk drives were attachable to a control unit, constituting a "string" of disk. The control unit's capability to manage disk determined the size of a string. In time, two strings could be attached to a control unit, and later (by the 1970s) multiple control units could "hang off" a channel. The important control unit of the period was the IBM 2841 which appeared along with the S/360. That device opened a new era in the management of online data because it now became possible to manage many of the functions necessary to gather data into the computer or to remainder it back to disk storage.

One of IBM's most successful disk drives was the 2314 File Facility of 1966. This equipment, the first widely used disk drive, did more to initiate the use of online applications than any other piece of peripheral equipment. Its initial disk pack stored up to 25.87 megabytes [Mb] of data and used up to nine disk drives and the 2841 control unit. The disk drives could house up to 233 Mb of online data. New access methods (managed by software) appeared which remained in use for over a decade. Sequential Access Method (SAM) and Basic Direct Access Method (BDAM) dictated how information would be managed and by the end of the 1960s had been joined by the Indexed Sequential Access Method (ISAM), which was in use throughout the 1970s and into the 1980s, and the less used Basic Partitioned Access Method (BPAM). Along with other innovations, the system could also manage the allocation of storage space while tracking more efficiently the addresses of all data.

The years subsequent to 1967 were equally fruitful for disk drives (Table 25). Between 1967 and 1980, for example, the cost per Mb of disk storage dropped by a factor in excess of twenty, making online storage the dominant mode in the world of data processing. Vendors, led by IBM, introduced products that did not use removable disk packs made possible by better reliability and greater capacity. The 1970s began with the introduction of the IBM 3330 disk drive as

Table 25
Select Technical Features of IBM Disk Drives, 1966–1976, Selected Years

Device	2314	3330	3340	3350
Year first shipped	1966	1971	1973	1976
Feature				
Recording density (bpi)	2,200	4,040	5,636	6,425
Disk diameter (in.)	14	14	14	14
RPM	2,400	3,600	2,964	3,600
Fixed/removable	Removable------------a			Fixed
Actuators/spindles	1	1	1	1
Avg. seek time (ms)	60	30	25	25
Data rate (Kbytes/s)	312	806	885	1,198
Cost per Mbyte/month ($)	30+	5-10	9	5

[a]Fixed disk pack version called the 3344 was available.

part of the new family of S/370* computers. The 3330 was the industry's dominant device in the early to mid-1970s. The company also brought out the IBM 2305 as a fixed-head disk file to replace drum storage on computers which had been popular in the late 1950s and throughout the majority of the 1960s. By the early 1970s virtual memory could be implemented on 3330s.

The 3330 continued the growth in capacity already evident, going from 29 to 100 Mb, while densities doubled again to 4,040 bpi. Seek times dropped from 75 ms to 30 ms, while the rotational period went from 25 to 16.7 ms. The cost of storage dropped per byte by a factor of three and by another factor of two with the introduction of the 3330-11 in 1974. That device had dual density, which meant one could record data on both sides of a platter. Up to that time, all recording had been on only one side. Additional enhancements in microprogramming along with more reliable hardware made the 3330 the workhorse of the industry in the mid-1970s, encouraging "look-alike" products from other companies, such as from Memorex.†

The next major evolution was the arrival of the Winchester technology of the IBM 3340. Besides physically smaller devices with greater capacities, the Winchester allowed IBM to make better read/write heads that flew above the disks at a height of only 20 microinches. One could now start or stop rotation while the head was in contact with the platter, a time saver and a convenience. Such innovations improved speeds and allowed disk technology to keep up with the greater speeds and capacities of new computers. The 3340 was introduced in 1973 and remained in IBM's product line throughout the 1970s along with a variety of 3330s and 3350s. The Winchester represented a great improvement in reliability over older technologies. For one thing, regularly scheduled

Table 26
Select Technical Features of Recent IBM Disk Drives, 1979 and 1981

Device	3310	3370	3380
Year first shipped	1979	1979	1981

Feature			
Recording density (bpi)	8,530	12,134	15,200
Disk diameter (in.)	8.3	14	14
RPM	3,125	2,964	3,620
Fixed/removable	Fixed	Fixed	Fixed
Actuators/spindles	1	2	2
Avg. seek time (ms)	27	20	16
Data rate (Kbytes/s)	1,031	1,859	3,000
Cost per Mbyte/month ($)	1	1	0.90

maintenance on disk drives was no longer necessary with the Winchester. The data module (disk pack) could store 35 Mb of data, and up to four could be used on a disk drive with minimal read/write errors. The 3350 had nonremovable disk storage (first shipped in 1976) and also relied on Winchester technology. It was a logical growth path for users of 3330s, while the 3344 did the same for 3340s. The cost of storing data on a 3350 over the 3330-11 dropped by a factor of nearly two. That innovation also meant that the cost of storing data had declined by a factor of 70 over the previous two decades.

The IBM 3370, announced in 1979, with its more compact film head technology, and the small 3310 (also announced in 1979) broadened the product line for intermediate and very small systems (Table 26). By then demand for online storage across the industry was growing at a compound rate of nearly 45 percent annually, a rate that continued into the 1980s. In 1980 IBM introduced the IBM 3380. Storage grew to 625 Mb per actuator (read/write arm and the associated surface area it managed) and to 1,250 Mb per spindle. Dramatic improvements were made in reliability while costs per byte continued to decline. The use of an LSI microprocessor to manage these disk drives, embedded in the

Table 27
Disk Storage Capacities on Select IBM Products, 1957–1981, Selected Years

Device	Millions of Bytes	Year First Shipped
350	04.40	1957
2314	25.87	1966
3330	100.00	1971
3350	317.50	1976
3380	1,260.00	1981

Table 28
Cost Performance: Space Per $1 Rental of Select IBM Disk Drives, 1957–1981,
Selected Years

350	6.8 thousand bytes	1957
2314	38.2 thousand bytes	1966
3330	245.6 thousand bytes	1971
3350	470.0 thousand bytes	1976
3380	1.19 million bytes	1981

3880 control unit, significantly expanded the manageability of data. It controlled fixed-block and count-key-data (CKD) formats, did maintenance functions, and interfaced with channels more efficiently than had earlier control units. The device was so reliable and so inexpensive when compared to its predecessors that it became the industry standard almost within a year of introduction. Competitors had difficulty developing their own versions of the 3380, giving IBM nearly a five-year lead over all other vendors. Thus, IBM had much the same advantages it had when it introduced RAMAC and then the 3330.

Table 27 summarizes the disk's continuing growth trend over the past thirty years. Table 28 suggests the relative cost of storage, indicating the incentive users had to expand their use of disk storage as long as the reliability of hardware proved acceptable. Innovations in products emerged from technological developments. However, the preponderance of evidence that came out of the lawsuit proceedings between IBM and the U.S. government during the 1970s indicates that competitive pressures from other vendors, particularly by 1970, had provided IBM with the greatest incentives to bring out new products.

For further information, see: Charles J. Bashe et al., *IBM's Early Computers* (Cambridge, Mass.: MIT Press, 1986); Franklin M. Fisher et al., *IBM and the U.S. Data Processing Industry: An Economic History* (New York: Praeger Publishers, 1983); J. M. Harket et al., "A Quarter Century of Disk File Innovation," *IBM Journal of Research and Development* 25, no. 5 (September 1981): 677–689; Emerson W. Pugh, *Memories That Shaped an Industry: Decisions Leading to IBM System/360* (Cambridge, Mass.: MIT Press, 1984); L. D. Stevens, "The Evolution of Magnetic Storage," *IBM Journal of Research and Development* 25, no. 5 (September 1981): 663–675.

E

EDSAC. Also known as the Cambridge EDSAC, the Electronic Delay Storage Automatic Calculator was the product of work at Cambridge University done by Maurice V. Wilkes.** He built one of the earliest British computers to employ the design principles developed at the Moore School of Electrical Engineering† at the University of Pennsylvania, home of the ENIAC.* This British stored-program computer was one of the first such machines constructed in Europe in the late 1940s.

Wilkes attended a class at the University of Pennsylvania during the summer of 1946 where he was exposed to the work of J. Presper Eckert** and John W. Mauchly** on the ENIAC and the EDVAC.* Upon his return to England, he assembled a team to build a machine. Wilkes intended his device to be a binary stored-program computer with programming capability. He also wanted to use it in a variety of scientific applications common to any university. The team decided that it was more important to construct a functioning machine than to build a better designed one than the Americans had at the University of Pennsylvania. Their main concern was to quickly build a computer that could be used by researchers at the university.

Wilkes elected to use mercury delay lines for memory.* Each of the mercury-filled components (often called tubes) was about 5 feet long and could house 576 binary digits. Memory on the computer was made up of thirty-two delay lines. Others were added to serve as registers within the processing unit itself (that is, in that portion of the computer that conducted calculations). The net result was that EDSAC had the capability of storing 512 words, each made up of 36 bits. Wilkes employed electromechanical paper tape* readers for input and a teleprinter for output. Some 3,000 thermionic valves were also installed, giving this computer its very large, room-size look. The designers chose to make it a relatively slow operating machine in order to reduce the sophistication and complexity of design. Thus, the pulse repetition rate was some 500 kc/second

Table 29
Technical Features of EDSAC (Cambridge)

Word length (bits)	36
Instruction length	18
Instruction format	1-address
Instruction set	18 operations
Store size, fast	512 words
Store type, fast	delay
Store size, backing	none
Average add time	1.4 ms
Average multiply time	5.4 ms
Input medium	5 track paper tape reader
Output medium	teleprinter
Digit period	2 microseconds
Main valve type	EF 54
Approximate number of valves	3,000
Approximate number of GE diodes	none

SOURCE: Reprinted with permission from *Early British Computers*, Simon Lavington, First Edition. Copyright © Digital Press/Digital Equipment Corporation (Bedford, MA), 1980.

instead of the 1 Mc/second common to other devices of the late 1940s. Addition, for instance, could be done in 1.4 milliseconds, which was one-third the speed of the Pilot ACE* and even slower than the BINAC* or EDVAC.* Wilkes described the machine as "a serial binary computer with an ultrasonic memory." The technical features of the EDSAC are listed in Table 29.

Each instruction cost a half word, while each instruction set was a single address. This machine was capable of seventeen instructions. It could add, subtract, and multiply but not divide; all data had to be numerical and there were no alpha characters. Paper tape eventually provided both input and output. This device would operate as follows: a group of routine commands (called "initial orders" at the time) were loaded into memory and were then executed. Wilkes was especially pleased that his machine had the capability of relying on initial orders to determine the manner in which all instructions would be punched out to paper tape, a convenience over earlier devices where every step required human intervention.

EDSAC ran its first program on May 6, 1949, and was in routine operation by early 1950. It was finally powered off in July 1958. During the period it operated, students at Cambridge used the computer. It was frequently easier to work with EDSAC than with other contemporary devices because its programming was by way of symbolic assembly language. Hence, a user could create instructions using alphabetic characters which were then punched on tape and read into the computer. The machine would then take these orders and convert them into machine-level instructions which were binary and which could be understood by the computer itself. Wilkes and his team also developed a good collection of subroutines on paper tape. This meant that widely used programming instructions did not have to be written each time a student or professor needed them; they would simply use a set already programmed—a standard feature in modern languages.

EDSAC was the product of a clever scientist with a practical bent of mind. Wilkes, who still lectured on computer science as late as the mid-1980s, never lost sight of the need to make computers easier to use. His machine encouraged a generation of British scientists and engineers to use and build computers. In the late 1940s and in the 1950s, EDSAC became the reason for having a number of national computer conferences at Cambridge University. One of the byproducts of these efforts was Wilkes's early efforts (in 1951) to encourage computer designers to incorporate microprogramming into the design of their systems. In the early 1950s microprocessing was, to use Wilkes's own definition used in 1980 (since he is the originator of the term), a specific way "to provide a systematic approach and an orderly approach to designing the control section of any computing system." To Wilkes the concept of computer control was the interpretation followed by the execution of instructions to the machine. Although the definition and scope of microprogramming changed over the next three decades, all systems today have embedded either in operating systems or in the hardware standard sets of instructions that are often used in order to manage transactions through the computer.

By the time EDSAC II was in use in 1957, Wilkes had made another contribution in the form of training a generation of computer scientists while encouraging the J. Lyons and Co., Ltd., to become interested in building and selling computers in Great Britain. This company went on to develop LEO for its own use in 1951; this early computer was used in England for commercial applications. By the end of 1953 this version of EDSAC was performing well. The encouragement thus born of success led to the establishment of LEO Computers,* Ltd., on November 4, 1954, which marketed computers in Great Britain during the 1950s.

For further information, see: Simon Lavington, *Early British Computers* (Bedford, Mass.: Digital Press, 1980); Maurice V. Wilkes: *Automatic Digital Computers* (New York: John Wiley and Sons, 1956); "Early Programming Developments in Cambridge," in N. Metropolis et al., eds., *A History of Computing in the Twentieth Century* (New York: Academic Press, 1980): 497–501; "The EDSAC, An Electronic Calculating Machine," *Journal of Scientific Instruments* 26 (December 1949): 385–391; and *Memoirs of a Computer Pioneer* (Cambridge, Mass.: MIT Press, 1985).

EDVAC. EDVAC was the follow-on computer to the ENIAC* and became the first electronic digital computer* which applied the concept of the stored program to its design. That concept suggested that programs run by a computer should be stored within it along with the data to be acted on. This approach represented a major step forward in the development of the modern computer. The EDVAC, the design for which began in 1944, was completed in 1951 as a sequel to ENIAC. It became one of the most important of the early computers and represented another chapter in the lives of J. Presper Eckert** and John Mauchly,** two giants in the early history of computing. Both worked on this device while at the Moore School of Electrical Engineering† in the mid-1940s. A third giant in the early history of the modern computer, John von Neumann,** also made significant contributions to its design, particularly concerning the stored-program concept.

ENIAC was less than an ideal machine because of its lack of internal program control; it was the one function most needed to make it a more productive tool. That requirement became one of the central issues in the development of the EDVAC. Other concerns were the ENIAC's limited storage and the lack of reliability inherent in the use of thousands of vacuum tubes. In August 1944 Herman H. Goldstine** recommended that the U.S. Army contract with the Moore School for the construction of a new machine with the stored-program capability. It was to be called the Electronic Discrete Variable Automatic Computer—EDVAC. In October 1944 a contract was signed by the Army Ordnance Department authorizing the expenditure of $105,600 for research and development leading to the improvement of ENIAC's technology. Meanwhile, work continued on the ENIAC along with initial design of the EDVAC. The principal early designers were Goldstine and von Neumann, and, later, Arthur W. Burks.** They concentrated on the logic and arithmetic issues concerned with the new design.

Von Neumann, already a distinguished mathematician with considerable influence with the government, supported the work of the Moore School and was particularly in favor of the stored-program approach. He made a number of recommendations concerning the design of the EDVAC over a series of visits to the University of Pennsylvania. Because of his influence, the U.S. Army was willing to sign a contract with the Moore School for the construction of EDVAC. Finally, his influence forced the staff at the Moore School to implement more formal project management procedures than might otherwise have been the case, thereby increasing the possibility of the new project being completed. But his greatest contribution was in helping to define the concept of the stored program and making it a reality in computer design, a central feature of all computers made from the late 1940s to the present. Indeed, computers today are sometimes still called von Neumann computers because of his work in describing the stored-program concept for the EDVAC project and how it might be applied to computer design.

During 1944–1945 the concept developed at the Moore School and thus was not solely the product of von Neumann's genius. In June 1945 he prepared one of the most important papers in the history of data processing, entitled "First Draft of a Report on the EDVAC," a widely read document by scientists interested in the construction of computers in the late 1940s. It summarized the thinking of the staff at the Moore School regarding the design characteristics of a stored-program machine, formalizing and putting into writing what became the blueprint for a new generation of computers. His contributions lay in formalizing the collective thoughts of the staff while adding new suggestions born of his own skills as a logistician and mathematician. Years later, during Eckert and Mauchly's patent fights with others, the paper played a central role in identifying the true originator of the concept.

The issue was a simple one. Because the paper carried von Neumann's name on the title page, the implication was that he had developed the concept. And he did little to state otherwise. Others argued, however, that the concept represented the collective research of those at the Moore School. The argument was not settled for over a quarter of a century. The fact that this paper was widely read damaged Eckert and Mauchly's claim that their patents were valid since the court argued that on some points von Neumann's paper had made many of the key design features of computers of the late 1940s common knowledge and in the public domain. The final consensus of historians is that the stored-program idea emerged from the collective efforts of researchers at the Moore School, including von Neumann. The controversy over the authorship of design features erupted early, however, when Eckert and Mauchly became embroiled with the University of Pennsylvania over who had the right to apply for patent rights on the ENIAC—they or the university. The arguments created tensions at the Moore School just at the time when work on the EDVAC was beginning and clouded patent issues concerning the new machine. After the quarrel began, the University of Pennsylvania required that all members of the Moore School sign over to the university any rights for patents that might emerge as a result of their research. As a consequence, Eckert and Mauchly left the Moore School and started their own firm. Other engineers also left, thereby terminating an important era of computer development at the University of Pennsylvania.

During the early stages of the controversy, Eckert and Mauchly nonetheless worked on the EDVAC, employing many of the concepts described in von Neumann's unpublished paper. They worked on the machine until they left the University of Pennsylvania on March 31, 1946. Burks went to the Institute for Advanced Study with von Neumann, while two other key members of the project, Robert Shaw and C. Bradford Sheppard, joined Eckert and Mauchly.

The EDVAC that was finally built stored programs the same way it stored data. The machine had only 4,000 tubes instead of the nearly 18,000 in the ENIAC, and it used 10,000 crystal diodes (Table 30). EDVAC's design relied on a 1,024-word recirculating memory-delay line memory which made memory* much larger than ENIAC's, thereby permitting larger projects to be run through the computer. Words were 44 bits long, and EDVAC used four-address type instructions. It also used four 1-bit addresses. Fixed and floating-point operations were possible within its arithmetic unit. Both cards and punched tape provided input and output. All of these features, when coupled to the actual experiences gained with components in building the ENIAC, made possible a superior design of EDVAC. It was delivered to the Ballistic Research Laboratories at Aberdeen, Maryland, in 1949, but it was not fully operational until 1951. Delays were caused by normal startup problems as well as by the large turnover in staff at

Table 30
Technical Features of EDVAC

MANUFACTURER: Moore School of Electrical Engineering, University of Pennsylvania.

OPERATING AGENCY: U.S. Army Ordnance Corps Ballistic Research Lab, APG.

GENERAL SYSTEM: Application solutions of ballistic equations, bombing and firing tables, fire control, data reductions, related scientific problems.

Timing   Synchronous
Operation   Sequential

A general-purpose computer that may be used for solving many varieties of mathematical problems.

NUMERICAL SYSTEM: Internal number system: binary. Binary digits per word: 44. Binary digits per instruction: 4 bits/command, 10 bits each address. Instruction per word: 1. Total number of instructions decoded: 16. Total number of instructions used: 12. Arithmetic system: fixed-point. Instruction type: four-address code. Number range: $-(1-2^{-43}) \leq x \leq (1-2^{-43})$.

ARITHMETIC UNIT: Add time (including storage access): 864 microsecond (minimum 192, maximum 1,536). Multiply time (including storage access): 2,880 microsecond (minimum 2,208 maximum 3,600). Construction: vacuum tubes and diode gates. Number of rapid access word registers: 4. Basic pulse repetition rate: 1.0 megacycle/second. Arithmetic mode: serial.

STORAGE:

Media	Words	Microsec. Access
Mercury Acoustic		
Delay Line	1,024	48-384
Magnetic Drum	4,608	17,000

Includes relay hunting and closure. The information transfer to and from the drum is at one megacycle per second. The block length is optional from 1 to 384 words per transfer instructions.

INPUT:

Media	Speed
Photoelectric Tape Reader	942 sexadecimal characters/second
	78 words/second
Card Reader (IBM)	15/rows/second

OUTPUT:

Media	Speed
Paper Tape perf.	6 sexadecimal characters/second
	30 words/minute
Teletypewriter	6 sexadecimal characters/second
	30 words/minute
Card Punch (IBM)	100 cards/minute
	800 words/minute

Table 30 *(cont.)*

NUMBER OF CIRCUIT ELEMENTS:

Tubes:	3,563
Tube types:	19
Crystal diodes:	8,000
Magnetic elements:	1,325 (relays, coils, and trans.)
Capacitors:	5,500 approximately
Resistors:	12,000 approximately
Neons:	320 approximately

CHECKING FEATURES: Fixed Comparison—Two arithmetic units perform computation simultaneously. Discrepancies halt machine. Paper tape reader error detection.

PHYSICAL FACTORS: Power consumption: computer 50 kW. Space occupied: computer: 490 sq. ft. Total weight: computer: 17,300 lbs. Power consumption: air conditioning: 25 kW. Space occupied: air conditioning: 6 sq ft. Total weight: air conditioning: 4,345 lbs. Capacity: air conditioning: 20 tons.

MANUFACTURING RECORD: Number produced: 1.

COST: Rental rates for additional equipment: IBM card reader $82.50 per month; IBM card punch $77.00 per month. Approximate cost of basic system: $467,000.

PERSONNEL REQUIREMENTS: Daily operation: three 8-hour shifts, number of technicians: 8, 7 days/week. No engineers are assigned to operation of the computer, but are used for design and development of improvements for the computer. The technicians consult with engineers when a total breakdown occurs.

RELIABILITY AND OPERATING EXPERIENCE: Average error-free running period: 8 hours. Operation ratio: 0.79. Good time: 130.5 hours (Figures for 1955). Attempted to run: 166 hours/week. Number of different kinds of plug-in units: 3. Number of separate cabinets (excluding power and air conditioning): 12. Operating ratio figures for 1954: Operating ratio: 0.79. Good time: 129 hours. Attempted to run: 163 hours/week.

ADDITIONAL FEATURES AND REMARKS: Oscilloscope and neon indicator for viewing contents of any storage location at any time. Exceeded capacity option: halt, ignore, transfer control, or go to selected location. Unused instruction (command) halt. Storage or previously executed instruction and which storage location it came from, for viewing during code checking. Storage of current instruction and storage location from which it originated. Address halt when prescribed address appears in anh of four addresses of instruction to be executed by cpmputer. Tape reader error detection.

SOURCE: Specifications are from Martin Weik, BRL Report No. 971. Reprinted with permission from *From Eniac to Univac*, Nancy Stern, First Edition. Copyright © Digital Press/ Digital Equipment Corporation (Bedford, MA), 1981.

the Moore School, primarily in 1946, which meant a whole new staff had to be trained.

During 1952 the EDVAC was in normal productive use. By the mid-1950s it ran an average of eight hours error-free at a time and, 78 percent of the time run times were meeting design standards. In short, there were about 130 hours of good run time per week. It was finally shut down in December 1962 and dismantled, with components shipped to various museums.

EDVAC proved that the stored-program principle could be made part of a computer. Other computers were designed in the late 1940s, and in the years that followed the concept was utilized and improved. It was one of the first applications of the principle. Its successful design meant that the range of applications a computer could undertake would vastly exceed that of the ENIAC or any other earlier device. Moreover, the EDVAC allowed the engineering staff to improve on the physical design of the ENIAC, augmenting its reliability by using fewer vacuum tubes and installing better performing components. The EDVAC project provides a classic example of modern technology evolving to higher levels of efficiency and reliability. It was a direct descendent of the UNIVAC,* the first widely used, commercially available stored-program computer of the 1950s. The UNIVAC was designed largely by Eckert and Mauchly.

For further information, see: H. D. Huskey, "EDVAC," in A. Ralston and C. L. Meek, eds., *Encyclopedia of Computer Science* (New York: Petrocelli/Charter, 1976): 534–535; B. Randell, "Digital Computers: Origins," in ibid.: 486–490; Nancy Stern, *From ENIAC to UNIVAC: An Appraisal of the Eckert-Mauchly Computers* (Bedford, Mass.: Digital Press, 1981); W. W. Stifler, Jr., ed., *High Speed Computing Devices* (New York: McGraw-Hill, 1950).

ENIAC. ENIAC, the first fully operational large-scale electronic digital computer,* was built at the Moore School of Electrical Engineering† at the University of Pennsylvania between 1943 and 1946. It virtually launched the modern era of the computer, even though other machines were under construction at the time. Its builders, John Presper Eckert, Jr.,** and John William Mauchly,** learned enough from this project to go on to build the EDVAC,* BINAC,* EDSAC,* and finally the UNIVAC,* the first computer to be widely accepted for commercial use. The ENIAC, therefore, was the direct precursor of the many computers available during the 1950s. It proved that computational devices could work at electronic speeds which, for the mid-1940s, meant it could calculate some 500 times faster than available electromechanical machines. People could now compute vast quantities of data faster than ever before and use technology to expand the capabilities of the human mind. Some argue that the ENIAC was the single most important computer ever built; others rank it at least with the IBM System 360.*

The immediate reason for constructing the ENIAC was World War II. The U.S. military establishment, especially the Army, needed a device to calculate and print firing tables. That task required that vast amounts of arithmetic calculations be performed quickly and accurately. To create such a machine, a team was put together at the Moore School.

Firing tables and ballistics in general had long stimulated scientists to find mechanical and, later, electronic ways to calculate numbers. A firing table would instruct a gunner at what angle to set a cannon in order to deliver a shell of a particular size and weight at a target that was often farther away than he could see. All the tables listed firing angles and ranges in numerical form. More complete tables showed the velocity of a shell and the velocity of fall at impact, data that were needed when trying to determine the shell's capability of penetrating armored targets. Such calculations were done with differential equations, taking into account Sir Isaac Newton's law which states that all objects influenced by the earth's gravity will fall with a constant acceleration. In addition to these concerns, scientists had to consider air resistance in preparing a firing table. Differential equations were used to calculate the rate at which air resistance grew as the speed of the shell's trip increased.

Traditional methods of calculation involved using arithmetic to tabulate subtotals of the equations, storing that information, and then continuing other rounds of calculations until each element of the table was arrived at. All of these tasks were very time consuming and frequently inaccurate to generate by hand or even with the use of a desk calculator. Thus, many war-related projects involved computers for the creation of firing tables. Another project that addressed this need was Howard H. Aiken's** Mark I,* constructed by International Business Machines Corporation (IBM)† and Harvard University at the start of the war. Vannevar Bush's** differential analyzer* at the Massachusetts Institute of Technology (MIT) in the 1930s had the capability of calculating differential equations and was rebuilt in 1942 to prepare tables. The Moore School had a copy of Bush's computer for that purpose. The Moore School also employed women to hand calculate trajectories, each trajectory taking several hours to develop. Bush's machine could do it in twenty minutes, but even that was too slow.

These concerns motivated Mauchly to consider designing a machine that would do the work faster based on vacuum tube technology. Through the device he sought to replicate many of the numerical methods used in the manual calculation of trajectories rather than relying on analog approaches that were the bases of differential analyzers. His difference in approach to calculating finite numbers was one of the early origins of the digital computer* (a name that was not used at that time, however). The machine he built was called the Electronic Numerical Integrator and Computer, or simply ENIAC by Colonel Paul Gillon, one of the early proponents (1942) of more advanced technologies to help ballistics. He was then working at the Ballistic Research Laboratory (BRL).

Mauchly, Ekert, and John Grist Brainerd** drew up a proposal to construct such a machine in April 1943. Mauchly was a new arrival at the Moore School where he served as an instructor. His primary interest was in meteorology and he had seen the need for a device that would help make weather predictions. The scientific community already knew that certain calculations could be performed to increase the accuracy of weather prediction, but doing these by hand and rapidly was almost impossible. Before attempting the design of a machine, Mauchly decided to attend a class on electronics at the Moore School, and while there he joined the staff. The ENIAC was his original idea. The second man most closely associated with the construction of this machine was Eckert who had been at the Moore School longer and had worked with some existing computational devices but had not yet developed any serious research interests in them.

In late 1940 in Philadelphia, Mauchly met John V. Atanasoff,** a professor from Iowa State University who was also working on a computational project. This meeting later led to a controversy over who was the actual inventor of the first digital computer, a discussion that remains unresolved today. Atanasoff's machine was designed to solve simultaneous linear equations using electronic circuits. Rather than relays or analog methods, he used numerical processes, which made his device extremely advanced by the standards of his day. Mauchly visited Atanasoff at Iowa State University and saw the partially built machine. This visit has led many to believe that Mauchly got his ideas for the ENIAC from Atanasoff. If so, the patents emerging from the ENIAC would represent a usurption of rights belonging to Atanasoff. Atanasoff could then claim the title of father of the modern digital computer.

The issue was not settled legally until 1974 when a judgment was made against Sperry Univac Corporation.* In that patent case, Judge Earl Larson ruled that the patents which Eckert and Mauchly had taken out at the end of World War II on the ENIAC were invalid. Since they had transferred these patents to Univac, that company lost title to them. The judge further ruled that Atanasoff had invented the first digital computer.

Although Mauchly had been impressed with Atanasoff's work, much of the design of the ENIAC actually grew out of Mauchly's work with his colleagues at the Moore School. There were distinct differences between the two machines, perhaps the most obvious being that the ENIAC could be programmed. Thus, mathematical functions could be altered, that is, the sequence of steps taken to calculate a solution could be changed on the ENIAC. This was not the case with Atanasoff's machine. A second difference was that the ENIAC did not use binary systems, whereas Atanasoff's did. A third was that the ENIAC worked. Atanasoff's machine never went into full operation, although portions could be demonstrated. Both groups—at Iowa State University and at the Moore School— contributed to the development of digital computers. The ENIAC was more important, however, for it became the first well-known digital computer to be used in the United States.

The ENIAC was proposed on April 9, 1943, and soon after a contract was signed with the U.S. Army to construct it. The computer was completed by November 1945, and the first set of problems was run on it in December. In

November 1946 it was moved to the Aberdeen Proving Ground and was finally dismantled by early October 1955. It cost $500,000 to build. The ENIAC used vacuum tubes for computing and relied on electromechanical technology for peripheral equipment. It employed card readers and punches and telephone relay buffer storage. The ENIAC was a large machine, using 18,000 vacuum tubes and 1,500 relays. It consisted of forty panels, each of which was 9 feet x 2 feet x 1 foot and was set up in a U arrangement. Peripheral equipment could be deployed around the computer as needed. For a summary of the history and characteristics of the ENIAC, see Table 31.

The word length used on the computer was ten decimal digits, and it had a fixed decimal point. Multiplication was done by table lookup. Additions and subtractions were performed by using accumulators. Its memory* had the capability of storing 20 numbers in its accumulators and 100 numbers of read-only in memory. Punched cards stored additional data, and the machine was equipped with twenty ten-digit constant switches. The ENIAC was a fast machine by the standards of the early 1940s; for example, 5,000 additions or 357 multiplications could be done in one second. All programming was done by wiring plugboards and using switches. It could do conditional branching and handle subroutines. This machine could also be programmed to perform parallel sets of operations. Typically, programming the machine to do a problem could take one to two days. Input and output were handled by cards, and some output could go to printers.

From the beginning, ENIAC's inventors felt they could do better. Their machine was difficult to program, and its input/output procedures were slow. In addition, it had limited memory and capacity—problems that were fixed in the machines that followed—and it was overdependent on vacuum tubes which were very unreliable compared to electronic components. Tubes constantly failed. On balance, however, the machine proved more reliable than existing relay computers because of its speed. Thus, for instance, one hour's worth of work on the ENIAC was equivalent to fifteen days of calculations on the Bell Labs* Model V. The machine could run about twenty hours before vacuum tubes began failing in quantity.

The machine was used for various government projects, including the Manhattan Project and the U.S. Army's preparation of firing tables. More important, it taught a small community of computer scientists what the next set of computers needed to perform better. Future machines were built with decreasing numbers of vacuum tubes, memories grew in size, digital rather than analog designs became the norm, and electromechanical technologies were abandoned as too slow and cumbersome. Binary scales dominated after ENIAC and high-speed read/write memories were developed. The concept of a computer storing both its data and programs led to the modern computer of today. The ENIAC gave birth to the first generation of commercially available computers in the late 1940s.

Table 31
Technical Features of ENIAC

MANUFACTURER: Moore School of Electrical Engineering, University of Pennsylvania.

OPERATING AGENCY: U.S. Army Ordnance Corps Ballistic Research Lab, APG.

GENERAL SYSTEM: Application solutions of ballistic equations, fire control problems, data reduction, and related scientific problems.

Timing   Synchronous
Operation   Sequential

NUMERICAL SYSTEM: Internal decimal in basic computer, binary coded decimal in magnetic storage. Decimal digits per word: 10 plus sign. Decimal digits per instruction: 2. Instructions per word: 5 or 6. Total number of instructions decoded: 100. Total number of instructions used: 97. Arithmetic system: fixed-point. Instruction type: one-address code. Number range: $10^{-10}-1$ to $1-10^{-10}$.

ARITHMETIC UNIT: Add time (excluding storage access): 200 microsecond. Multiply time (excluding storage access): 2,800 microsecond. Divide time (excluding storage access): 24,000 microsecond. Construction: vacuum tubes. Number of rapid access word registers: 120. Basic pulse repetition rate: 60-125 kilocycles/second variable. Arithmetic mode: parallel. Information is transferred in parallel as a serial train of pulses.

STORAGE:	Media	Words	Microsec. Access
	Vacuum tubes	20	200
	Magnetic core	100	200
	Function table	304	Lines of 12 decimal digits + sign on each constant set switch
	Plugboard	96	Lines of 12 decimal digits + sign each (IBM)
	Relays	8	Buffer capable of storing contents of one card

INPUT:	Media	Speed
	IBM cards	125 cards/minute

Each card has eight ten-decimal-digit words plus signs.

OUTPUT:	Media	Speed
	IBM cards	100 cards/minute

200 milliseconds of card cycle are available for other computer operations. At 50 cards/minute rate, 800 milliseconds are available for computer operations per card cycle.

Table 31 *(cont.)*

NUMBER OF CIRCUIT ELEMENTS:
        Tubes: 17,468
        Tube types: 16
        Crystal diodes: 7,200
        Magnetic elements: 4,100

PHYSICAL FACTORS: Power consumption: computer: 174 kW. Space occupied: computer: 1,800 sq. ft. Air conditioning: forced outside air.

MANUFACTURING RECORD: Number produced: 1.

COST: Additional equipment: magnetic storage $29,706.50. Rental rates for additional equipment: IBM card reader $82.50 per month, IBM card punch $77.00 per month. Approximate cost of basic system: $750,000.

PERSONNEL REQUIREMENTS: Daily operations: three 8-hour shifts, number of technicians: 6, 7 days/week. A minimum requirement for operation and servicing on a 24-hour/day, 7 day/week basis. No engineers are assigned to operation of machine, but they are used for design, development of improvements, and consultants when total breakdowns occur.

RELIABILITY AND OPERATING EXPERIENCE: Data unit passed acceptance test, 1946. Average error-free running period: 5.6 hours. Operating ratio: 0.69. Good time: 113 hours. (Figures for 1955). Attempted to run 164 hours/week. Number of different plug-in units: 44. Number of separate cabinets (excluding power and air conditioning): 42. Operating ratio figures for 1954: Operating ratio of 0.70. Good time: 116 hours. Attempted to run: 166 hours/week.

ADDITIONAL FEATURES AND REMARKS: There are four modes of operation: continuous, pulse time, add time, or instruction time. A manual preset stop box is available. Count instructions and transfer instructions are used.

SOURCE: Specifications are from Martin Weik, BRL Report No. 971. Reprinted with permission from *From Eniac to Univac*, Nancy Stern, First Edition. Copyright © Digital Press/ Digital Equipment Corporation (Bedford, MA), 1981.

For further information, see: J. G. Brainerd, "Genesis of the ENIAC," *Technology and Culture* 17, no. 3 (July 1976): 482–488; A. W. Burks and A. R. Burks, "The ENIAC: First General-Purpose Electronic Computer," *Annals of the History of Computing* 3, no. 4 (October 1981): 310–399; Paul E. Ceruzzi, *Reckoners: The Prehistory of the Digital Computer, from Relays to the Stored Program Concept, 1935–1945* (Westport, Conn.: Greenwood Press, 1983); J. P.Eckert, Jr., "The ENIAC" in N. Metropolis et al., eds., *A History of Computing in the Twentieth Century* (New York: Academic Press, 1980): 525–539; W. D. Gardner, "Will the Inventor of the First Digital Computer Please Stand Up?," *Datamation* 20, no. 2 (February 1974): 84, 88–90; John W. Mauchly, "The ENIAC," in N. Metropolis et al., eds., *A History of Computing in the Twentieth Century* (New York: Academic Press, 1980): 541–550 and his "Mauchly: Unpublished Remarks," *Annals of the History of Computing* 4, no. 3 (July 1982): 245–256; Joel Shurkin, *Engines of the Mind: A History of the Computer* (New York: W. W. Norton and Co., 1984); Nancy Stern, *From ENIAC to UNIVAC: An Appraisal of the Eckert-Mauchly Computers* (Bedford, Mass.: Digital Press, 1981).

ENIGMA. This German machine was used to convert messages from the German language into coded form for the purpose of transmitting them by telephone or radio. It was used to preserve the secrecy of such messages for military and political purposes beginning in the 1920s. This cryptographic device played its most important role during World War II. ENIGMA and the work of cryptoanalysts proved critical to the outcome of the war, particularly in the European theater.

The technology inherent in this machine was created in the early 1920s, and late in the decade two versions of the device were available. The commercial version became available by mid-1927, whereas the military version had been used for several years but did not go into regular use until 1928. The ENIGMA, also seen in published accounts as "Enigma" or simply Enigma, was no larger than a typewriter and was housed in a wooden box. It employed all twenty-six letters of the alphabet on a keyboard similar to that of a typewriter. Twenty-six lights in the lid represented each letter. Three rotating wheels could take any of twenty-six different positions making twenty-six electrical contacts on either side of a flat wheel. Wires running through the wheels connected contacts on either side of each. Any electrical signal on one side of a wheel could contact any other on the opposite side. Users would press a key for each letter in the coded message. Each depressed key would cause a wheel to turn 1/26 around. The same key caused electricity to flow to a contact, through the wheel, to another point on the next wheel. Thus, electricity passed through the series of wheels in the machine or back, causing one of the bulbs to light. Obviously, the position or the wheels and the path of the electrical signal established which light came on. The letter symbolized by that light was the symbol as code for a letter whose key had been originally depressed. By typing a coded message, the correct German-language message was displayed on the lid in the form of lighted lamps. In order to work, however, both sender and receiver had to have all wheels set identically. A major part of the work done by the British involved understanding how to set their wheels, tracking German patterns.

The task for anyone using these devices to break German codes was not as easy as this description might imply. Because the German military had developed a plugboard to alter codes frequently, they had more than 100 billion possible settings. Furthermore, they changed the settings frequently to provide security for ciphers. Commercial users did not often change their settings, possibly sending changes to various branch offices by mail on a weekly or monthly basis. By World War II, German military settings changed at least once daily. Various schemes were worked out to cause changes in the cipher (settings of the wheels), even making them unique for simply one message.

The first consequence of the machine came on July 15, 1928, when the Germans began transmitting coded messages created by the machine over the radio. Polish radio operators informed their nation's Cipher Bureau, which then bought a commercial copy of the ENIGMA, added the plugboard to it which the German military had developed to provide various combinations of ciphers, and then went to work establishing methods for breaking German codes. By then Poland recognized that Germany was an enemy and, therefore, felt a growing pressure to gather intelligence. Failing to succeed with this immediate project, the Bureau dropped the effort. But with mounting political tensions in Eastern Europe, in 1932 the government asked a mathematician, Marian Rejewski, to renew the effort. Within a short period of time, and with the help of other mathematicians, Rejewski worked out the pattern by which the Germans set the wheels. Thus, the Poles could now decipher approximately 75 percent of all messages the Germans sent on a routine basis. Throughout the late 1930s they developed devices to help manage the decoding of German messages. When it became obvious that Germany would go to war in 1939, the Poles informed their allies about this project. On July 25 and 26, 1939, the Poles informed the British and the French of their ability to crack German codes. Both the British and French intelligence communities had failed to break German ciphers in the 1930s and were therefore very surprised and pleased to discover what the Poles were doing on a regular basis. To this day most diplomatic historians remain unaware of Poland's success and so, in their analysis of Anglo-French-Polish-German relations in the final months before the start of the war, have overlooked ENIGMA. Poland's gift to the British, for example, was complete. It gave Great Britain a German military ENIGMA along with descriptions of Polish devices (*bombas*—bombs) used to aid in cracking codes. With this equipment and descriptions of German operational modes, the British were able to regularly break German codes while sharing critical information with their allies.

Although the Germans did not think anyone had breached ENIGMA, after the war started, they began to use a more complex device called the *Geheimschreiber* (secret writer). This escalation in ciphering led the British decoders located at Bletchley Park to build what became known as the COLOSSUS* computer to decipher new messages. It appears that the British built advanced versions of the Polish *bombas* which they too called ENIGMAS, introducing some confusion in terms for later historians.

The German Navy developed a newer, more complex ENIGMA that had more wheels, and the British apparently responded with additional devices. Almost every major military message that the Germans transmitted during World War II was apparently decoded or at least subject to enemy exposure. This activity was so important to the British that they would sometimes opt not to take advantage of some of the information gained in order to preserve the anonymity of their decoding. For example, even when it became clear that Coventry would be bombed, Winston Churchill painfully allowed the bombing to occur in order to protect the system. On a more positive side, he also had access to all German military messages surrounding the Allied invasion of France on D-day. Major German offensives during the war and many political dealings were known to the Allies mostly because of Britain's decoding activities. The Germans never discovered that their ENIGMA had been breached. One of the most exciting and as yet incomplete stories about the war to be uncovered in recent years has been the story of British intelligence.

The story of ENIGMA is far from over. The British government has in effect a fifty-year ban on access to official records, and exceptions are made only on specific files. Thus, for example, the Public Records Office opened up diplomatic papers beginning in the early 1970s, and by 1980 new revisionist monographs appeared. That same government has also maintained a lock on decoding activities of World War II, despite the fact that it has also sponsored the publication of histories on related topics, following the example of other Allied governments in publishing accounts of the activities of various agencies during the war. Historians of data processing have had to content themselves with slips of information and memoirs. Much has been written on Poland's role, but that unfortunately involved only the 1930s. By the end of the first month of World War II Germany had conquered Poland. While the fiftieth anniversary of the end of the war will arrive in 1995, the British may well continue to hold back data inasmuch as the technique of decoding messages has not changed materially; only the technology used to do it has altered. To reveal how such work was conducted in the 1940s might threaten British policies and practices which are in place today and can be expected to remain in some form in the 1990s.

For further information, see: P. Beesly, *Very Special Intelligence: The Story of the Admiralty's Operational Intelligence Centre, 1939–1945* (London: Hamish Hamilton, 1977); Ralph Bennet, *Ultra in the West* (London: Hutchinson, 1979); C. A. Deavours and J. Reeds, "The Enigma, Part I: Historical Perspectives," *Cryptologia* (October 1977): 381–391; Jozef Garlinski, *The Enigma War* (New York: Charles Scribner's Sons, 1979); I. J. Good, "Early Work on Computers at Bletchley," *Annals of the History of Computing* 1, no. 1 (July 1979): 38–48; F. H. Hinsley et al., *British Intelligence in the Second World War*, 3 vols. (London: HMSO, 1979–1984); R. V. Jones, *The Wizard War* (New York:

Table 32
Technical Features of ERA 1101

Serial/parallel	Parallel
Word length (bits)	24
Instruction length	24
Instruction format	1-address
Main store size	16,384
Main store type	Drum
Backing store type	None
Average add time	96 microsec.[a]
Average multiply time	352 microsec.[a]
Basic clock frequency	400 KHz
Approximate number of valves	2,700
Approximate number of GE diodes	2,385

[a]Times were minimum because drum periods were 10 milliseconds, while maximum times could be very long in comparison.

SOURCE: Reprinted with permission from *Early British Computers*, Simon Lavington, First Edition. Copyright © Digital Press/Digital Equipment Corporation (Bedford, MA), 1980.

Coward, McCann and Geoghegen, 1978); D. Kahn, *The Code Breakers* (New York: Macmillan Co., 1967), "The Geheimschreiber," *Cryptologia* 3, no. 4 (October 1979): 210–214, "Why Germany Lost the Code War," ibid., 6, no. 1 (January 1982): 26–31; Wladyslaw Kozaczuk, *Enigma: How the German Cipher Was Broken and How It Was Read by the Allies in World War II* (Federick, Md.: University Publishers of America, 1984); Marian Rejewski, "An Application of the Theory of Permutations in Breaking the Enigma Cipher," *Applicationes Mathematique* 16, no. 4 (1980): 543–559; "How Polish Mathematicians Deciphered the Enigma," *Annals of the History of Computing* 3, no. 3 (July 1981): 213–234; Gordon Welchman, *The Hut Six Story: Breaking the Enigma Codes* (New York: McGraw-Hill, 1982); Richard Woytak, "A Conversation with Marian Rejewski," *Cryptologia* 6, no. 1 (January 1982): 50–60.

ERA 1101. This parallel processor of the early 1950s was originally built for the U.S. military community but later became available for commercial users. It was constructed by Engineering Research Associates,† which was made up of ex-naval officers who had worked on various computer-like projects during World War II, primarily in code-breaking. The company, also called ERA, was established with the encouragement of the U.S. Navy to continue the development of specialized computational devices. One of these was the ATLAS I,* an early stored-program computer. The company obtained permission to make a version of this computer to sell to commercial and public institutions. This machine became known as the ERA 1101 and was marketed between 1948 and 1958. The first copy of the system was delivered in December 1950 to the U.S. Bureau of the Census. This computer was shipped several months before the UNIVAC I,* perhaps one of the most widely publicized computers of the 1950s.

The ERA 1101 had a relatively conventional design. It used vacuum tubes and a rotating drum for main memory.* Table 32 documents its other technical specifications, suggesting that it was built in line with other available processors of the day. Yet ERA did relatively well with the machine, selling approximately three copies. The company also produced the 1102, three of which were also sold. The ERA 1102 was a general-purpose computer that could handle applications calling for instrumentation. It sold for approximately $575,000 in 1952 and, like the ERA 1101, used commercially available peripheral equipment: American Telephone and Telegraph (AT&T)† teletype punches, Ferranti's† paper tape* reader, multiple Frieden Flexowriters, and for the 1102, an FAI digital plotter. But in the early 1950s customers were already able to acquire less expensive machines than the 1101, for example, and thus it had to be withdrawn from marketing. A second derivative of the 1101 was the 1103, which was first shipped in 1953 and was less expensive. Twenty of these systems were eventually sold by Remington Rand Corporation† which had acquired ERA by that time. These systems were used for both commercial and scientific computing.

For further information, see: Franklin M. Fisher et al., *IBM and the U.S. Data Processing Industry: An Economic History* (New York: Praeger Publishers, 1983); E. Tomash and A. A. Cohen, "The Birth of an ERA: Engineering Research Associates, Inc.," *Annals of the History of Computing* 1, no. 2 (October 1979): 93–97.

F

FACT. Fully Automatic Compiling Technique, an early programming language, was designed for business data processing on computers made by the Minneapolis-Honeywell Regulator Company (later known as the Honeywell Corporation†). It was developed at the same time as COBOL,* the data processing industry's most widely used programming language for commercial applications in the late 1960s and throughout the 1970s. The language was available only on a limited number of computers and thus had little impact within the industry.

In the first quarter of 1959 the Datamatic Division of Honeywell signed a contract for the development of FACT with the Computer Sciences Corporation.† By that time large computer vendors had recognized the need for such a language. International Business Machines Corporation (IBM)† had the Commercial Translator,* representatives of the industry as a whole were working on COBOL, and Honeywell needed a language for its own computers. The tiny Computer Sciences Corporation (it had a staff of ten employees) began work on a compiler to operate on the Honeywell 800 computer—the only machine FACT ever ran on because it was dropped after that computer was no longer sold. Knowledge of FACT became public in the fall of 1959. The news surprised the committee developing COBOL since a representative from Honeywell was a member and had never mentioned it before. Honeywell went ahead with FACT and produced several releases, some of which had features developed within COBOL and IBM's Commercial Translator.

When finally completed, FACT became the largest and most complicated package developed as a compiler up to that time. It consisted of several hundred thousand instructions. Yet this massive piece of code died because users of compilers were adopting COBOL in response to government pressure to standardize on COBOL as well as support for COBOL from the industry as a whole. As a result, COBOL eliminated the need for FACT, and eventually

Honeywell had to shift its attention and support to the more widely accepted language. By 1970 only a handful of data centers were using FACT, all on old 800 computers.

As a language, FACT had some interesting features. For example, it handled input/output functions in a more flexible manner than did the early releases of COBOL. FACT assumed that data came into a computer via cards creating an input tape. It would perform considerable amounts of error checking during that process and had verbs similar to COBOL's. Some of its features would become common in other programming languages* during the 1960s: validity checks on arithmetic, synonyms for key verbs, and an automatic search for values of arguments.

For further information, see: R. F. Clippinger, "FACT—A Business Compiler: Description and Comparison with COBOL and Commercial Translator," in R. Goodman, ed., *Annual Review in Automatic Programming* (New York: Pergamon Press, 1961) 2: 231–292; *The Honeywell-800 Business Compiler: A Preliminary Description* (Newton Highlands, Mass.: Datamatic Division. Minneapolis-Honeywell Regulator Company, 1959); Jean E. Sammet, *Programming Languages: History and Fundamentals* (Englewood Cliffs, N.J.: Prentice-Hall, 1969).

FLAP. FLAP was one of over a dozen programming languages* created in the United States to handle algebraic manipulations during the 1960s. Written in LISP 1.5,* it was created at the U.S. Naval Weapons Laboratory at Dahlgren, Virginia, and initially ran on an IBM 7090. The intent of the language was to allow a user to work with various types of symbolic mathematical data, including differential equations, vectors, and matrices. Its nomenclature, but not its notation, came from LISP. According to one specialist on early programming languages, FLAP offered a number of mathematical systems which were modes of arithmetic.

FLAP was used at the weapons laboratory to find ways to manipulate families of matrices and partial differential equations. It was also employed to analyze algebraic structures of indefinite integrals and their related operators. These included both Laplace and Fourier transforms. There is no evidence that this particular language was used anywhere else other than at the facility at Dahlgren during the 1960s and early 1970s.

For further information, see: Jean E. Sammet, *Programming Languages: History and Fundamentals* (Englewood Cliffs, N.J.: Prentice-Hall, 1969).

FLOW-MATIC. One of the earliest programming languages* to appear for the purpose of solving problems in business was FLOW-MATIC. It was developed at Remington Rand† in the mid-1950s, and, unlike most programming languages of the time which catered to scientific and numerical problems, it was designed to handle business applications. It was an important language because it ran on computers sold by one of the major vendors of the mid-1950s and led the way in defining the characteristics of business-oriented languages, most notably COBOL.*

In January 1955 Grace B. M. Hopper,** then a programming language manager at Remington Rand, had in hand a set of specifications for a computing language to perform business problems using her company's computers. Initially, the designers attempted to introduce standard abbreviations for business terms (e.g., GROSS PAY and COMPUTE) but were not very successful. Yet, they were able to employ the concept of using nouns as descriptors for specific data. Prior to that, the common practice had been to use symbols which, of course, either had to be learned or looked up to be understood. The initial manual for this new language was available by July 1957 and was offered for customer use in early 1958. Hopper devised one of the earliest computer languages to use English-like notation, a feature that would characterize many languages by the 1970s. Hopper and her company next played an important role in persuading users to apply such a language to business problems.

This language, also known within Remington Rand as B-Ø, formally introduced English words to describe data and commands. The second important notion that they worked on was the idea that data could be designed and created separately from that portion of a program that executed instructions. Hence, a programmer could write a lengthy and even sophisticated description of some file (e.g., a customer address file) apart from any program that used such information.

Third, FLOW-MATIC applied the concept of code generators in a compiler for a business language. The idea was to use a compiler to translate instructions into machine-readable code in which the size of a word or the point of positioning a decimal period would fluctuate and yet be correctly interpreted by the software.* Hopper and her team therefore developed a compiler that could take business application programs and create object codes for computers. Furthermore, this language had a compiler that would translate code into object form just before execution rather than completely force the translation of an entire program prior to its being run by the computer. This feature allowed large programs, for example, to operate on small computers with limited amounts of memory.* This was an important feature since all computers in the 1950s had very small memories, often smaller than would be found today in sophisticated hand-held calculators.

FLOW-MATIC contributed to the development of languages in other ways. For example, together with a sequel called AIMACO, it provided much food for thought to members of the short-range committee designing COBOL. By the late 1960s COBOL had become the industry's most widely used programming language for business applications. In the late 1950s FLOW-MATIC was the only business-oriented programming language of any consequence available. It proved successful with Remington Rand's business customers who initially had little desire to use the more widely available symbolic languages designed for scientific numerical problem-solving. Some later did, employing FORTRAN* until COBOL became an industry standard in the mid-1960s. In the 1950s the writers of business applications were less interested in stating formulas or employing algebraic notation. Rather, they wanted to use languages that had statements in English. Instead of coding $A + B = C$, they wanted to code Cost + Profit = Price. FLOW-MATIC dramatically contributed to this requirement.

For further information, see: Jean E. Sammet, *Programming Languages: History and Fundamentals* (Englewood Cliffs, N.J.: Prentice-Hall, 1969); A. Taylor, "The FLOW-MATIC and MATH-MATIC Automatic Programming Systems," in R. Goodman, ed. *Annual Review in Automatic Programming* (New York: Pergamon Press, 1960) 1: 196–206.

FORMAC. Formula Manipulation Compiler was one of the early formal algebraic manipulation languages to appear and represented International Business Machines Corporation's (IBM's)† first major work in this area. It was developed under the guidance of Jean E. Sammet,** a programming manager and the author of an important history of programming languages.* FORMAC was the first such language widely used and had the richest set of functions available in the early 1960s.

Together with Robert G. Tobey, Sammet worked on the language at IBM's Advanced Programming Department in Boston, beginning work in July 1962. By the end of that year they had specifications for a system that would provide formal algebraic features to an already existing programming language designed to do scientific computing, the widely used FORTRAN.* The original version of FORMAC ran on an IBM 7090/94 computer under IBSYS/IBJOB control software.* It was initially developed for experimentation, and the first version became operational in April 1964. Then IBM released the software for limited sales in November 1964. In the late 1960s a different release permitted the software to operate on the System/360.* The pre-System/360 version relied on FORTRAN, and the second on PL/I,* called PL/I-FORMAC, and was made available in November 1967.

The FORTRAN-based earliest releases borrowed heavily from FORTRAN IV. The developers intended FORMAC to deal strictly with a specific type of mathematical problem that needed algebraic manipulations. They designed it to handle those problems that were frequently solved before by tedious, manual algebraic calculations. The language satisfied this need and proved successful. One of the most interesting features of FORMAC was the variable that allowed FORTRAN to have formal algebraic data variables. Doing algebra using a computer became easier. Other commands were added to handle various algebraic functions while offering a wide variety of arithmetic functions (mixed mode, floating point, rational arithmetic, etc.). The version designed to run with PL/I was similar in function, although changes were made to conform to the different language. This version was also used widely during the era of the S/360 (1964–1970) and later the S/370* (1970s).

FORMAC was probably the first programming tool to show how to add another language, or collection of commands, to an existing language designed for numerical scientific computing. That is, FORMAC illustrated how to enhance an existing language for scientific computing with features to handle formal algebraic manipulation. It also proved that it was possible to do formal algebraic manipulation using computers; prior to FORMAC that was only a theoretical possibility, and earlier experiments with such languages as ALGY* were of limited success. Furthermore, it proved to be a language that was relatively simple to learn and use. As a consequence, both mathematicians and engineers employed FORMAC to solve analytical problems. The language contributed to a series of algorithms for performing automatic simplifications. It encouraged users to do more analytic problem-solving than the more popular approach of the 1950s of relying on numerical analysis.

FORMAC also contributed to a better understanding of computing power. When using FORMAC it became very evident that the greatest limiting factor in solving problems from the perspective of computing power was the limitation of memory* that was characteristic of technologies of the 1960s, not slow run times. Until the early 1960s the issue of whether memory or cycles was a more sensitive issue in determining throughput remained unanswered. FORMAC's experience suggested that computers would need more memory in order to solve problems effectively and that to solve them faster would require memory, not necessarily more cycles. In other words, problems could be solved faster by adding memory rather than simply increasing a computer's horsepower. That lesson was confirmed by the experiences of many computer manufacturers in the 1970s when memories increased sharply in size as the cost of chips* declined.

For further information, see: J. E. Sammet, *Programming Languages: History and Fundamentals* (Englewood Cliffs, N.J.: Prentice-Hall, 1969) and her study with E. Bond, "Introduction to FORMAC," *IEEE Transactions, Electrical Computing*, EC-13, no. 4 (August 1964): 386–394; R. G. Tobey, "Eliminating Monotonous Mathematics with FORMAC," *Communications, ACM* 9, no. 10 (October 1966): 742–751.

FORMULA ALGOL. This multipurpose programming language, a variation of the better known ALGOL,* was developed at the Carnegie Institute of Technology in 1963 to run on a CDC G-20 computer. No evidence exists that it was ever used elsewhere. The objective of this language was to add the ability to manipulate formulas using ALGOL, the language many scientists within the data processing community thought might become a universal tool in the early 1960s. Because Formula ALGOL was strictly an experimental language, it underwent numerous changes. It also ran on a computer that by the mid-1960s was rapidly becoming outdated. Yet while used at Carnegie, the language did become more broad-based in function, even acquiring features to do both string and list processing.

The language had some unique characteristics. For example, it used a *compiler-compiler* set of techniques. Furthermore, the language had a broader collection of character sets than ALGOL and executed Boolean expressions. Despite these positive enhancements, the impact of this language was minor largely because it was used at only one place. Yet is was unique for its time in that it had algebraic formula manipulation, could do list processing or manipulation of strings, and conduct matching of patterns. Jean E. Sammet, a student of old programming languages,* characterizes its use as "awkward," but, on the other hand, Formula ALGOL had powerful functions for an early multipurpose language.

For further information, see: A. J. Perlis and R. Iturriaga, "An Extension to ALGOL for Manipulating Formulae," *Communications, ACM* 7, no. 2 (February 1964): 127–130; Jean E. Sammet, *Programming Languages: History and Fundamentals* (Englewood Cliffs, N.J.: Prentice-Hall, 1969).

FORTRAN. FORTRAN was the first major high-level programming language* created. Also known as the IBM Mathematical Formula Translating System, it became one of the most widely used languages in the late 1950s and early 1960s and continues to be used extensively today primarily by scientists and engineers. This language was of crucial importance for the expansion of computer use and the development of a generation of programmers. No other piece of software* had as great an impact in the early years of digital computers.* The language proved that efficient object code could be created through automatic programming by using a compiler, thereby increasing the productivity of the too few programmers then available. Within five years of its introduction, the FORTRAN language had become an industry standard, used in almost every large data center in the United States and increasingly in Europe. It proved that high-level programming languages were efficient, better than what had existed before, and thus encouraged the development and use of others such as BASIC,* COBOL* and PL/I.*

In 1954 International Business Machines Corporation (IBM)† decided to develop what eventually became known as FORTRAN. Prior to this language, programming on such machines as the IBM 604* or the IBM CPC* had been done by arranging wires on plugboards. By the early 1950s programming consisted of writing code at the machine level or in assembly language. That meant generating lines of code in terms that were directly understandable by computers. The process was very complicated, and few understood how to do it well. Within IBM alone, for example, there may have been as many as 150 such programmers/engineers in the early 1950s. Because of the enormous cost of equipment, efficiency in coding was critical; that is, a premium was placed on writing commands which a computer understood in the shortest terms possible, using the least amount of computer resources, especially memory.* The high cost of memory was an issue, but even more so were the small memories on those early computers which forced programs to be tight and limited.

The programming community had widespread doubts about whether or not automation would create more efficient programming than what programmers could accomplish. No language would be acceptable if it created inefficiencies greater than what the data processing world already had. In 1954 a data center's programming costs were typically as much as the expense of the computer. Between 25 percent and 50 percent of a computer's time was spent in debugging (finding errors in) programs. As much as 75 percent of the cost of operating a computer was therefore devoted to programming activities, not in running programs. Yet, all the benefits of computing came from using programs, not writing or debugging them. As the cost of hardware declined, the problem of rising programming costs as an increasing percentage of the total costs of data processing obviously became more severe. This circumstance of changing economics provided the business case for the development of a high-level language. IBM's software community intended to improve the productivity of programmers while reducing the cost of developing completed, usable programs.

IBM was interested in developing programming tools as a means of encouraging the use of its 704* computer. This device had floating-point arithmetic and index registers, features attractive to those who wanted to write more complex programs for scientific and engineering problems. The majority of the applications being written in the early 1950s were scientific, often solving problems stated in mathematical terms. Various computing specialists were developing languages, translators, subroutines, and other aids to computing that would later become standard features of all programming languages. But no major, widely accepted language existed.

In late 1953 John Backus** at IBM proposed that his company develop a high-level language that would allow programmers to use familiar algebraic notations that a compiler (translator) could then convert into terms understood by a computer for execution. His proposal was twofold: first, to develop an easy-to-learn and easy-to-use language and, second, to provide a compiler to turn programs written in this language into machine-readable form accurately and efficiently. He urged that a simple language be developed but with a compiler that was very efficient in generating object code (machine-readable) instructions.

Backus's recommendations were accepted, and on November 10, 1954, the preliminary design for FORTRAN was released in a report. In October 1956 his group published a user's manual and within six months a primer. IBM made the language available in early 1957, initially for use on the 704 computer. At first, potential users expressed real concern as to whether the object code generated by FORTRAN's compiler would be as efficient (meaning as tightly written) as what their own programmers could produce. That concern dissipated with the early use of FORTRAN. In June 1958 a new version called FORTRAN II appeared for the 704 with additional enhancements. This version gave programmers the ability to have subroutines (collections of frequently used portions of a program) and COMMON statements (which allowed communications between subroutines). Until FORTRAN II appeared, subroutines had never been part of a high-level language. This version of the language also allowed programs to communicate with assembly language programs for the first time. That was particularly useful, for it suggested the possibility of linkages between sets of programs without having to rewrite assembly code which, as a lower level language, was harder to work with than FORTRAN.

In the late 1950s variations of FORTRAN began to appear, as did a movement to reduce the number of variations dependent on specific types of computers. The idea developed that FORTRAN had to evolve to the point where it did not matter which computer it ran on and that commands would be the same from one vendor's version of FORTRAN to another. A major step in this direction was taken when IBM introduced FORTRAN IV. This version significantly reduced the number of machine-dependent features of earlier releases while improving the language and its compiler.

Along with the development of the language, work on the compiler in various releases took place throughout the late 1950s. Backus and others had worked on the compiler as early as 1955, and its first release had become available at the same time as the language in early 1957. As an early compiler, it had to establish the standard for efficient object code, an objective it met when its operation proved more effective than what programmers could generate on their own. The ability of this compiler to translate a programmer's statement into machine-readable code, and then the results of the computer's calculations into terms a programmer could understand, became the benchmark for future compilers. The impact of this efficient compiler was quickly felt. By the end of the decade, programming was no longer limited to programmers. Scientists, engineers, and mathematicians learned FORTRAN, which they found so similar to their own understanding of algebra. Computers were no longer the exclusive preserve of programmers. For a sample FORTRAN program, see Table 33.

Despite their initial reservations, managers within the data processing community soon saw the benefits of FORTRAN and a movement began to standardize all versions of the language. SHARE,* a data processing user group, for example, argued in favor of standardizing all releases and features of FORTRAN. In 1962 an industry study was begun on that issue. The move to standardize FORTRAN began almost immediately after the language first appeared and came to a head in 1962, paralleling similar movements that had always existed to codify the grammar of such human languages as English and French. The movement was a broad-based one. Already by November 1958 one survey of twenty-six data centers using 704 computers indicated that more than half used FORTRAN for over 50 percent of their programming and that some relied on the language for 80 percent of all their work. Put another way, of the sixty-six organizations using 704s in the United States at that time, over half had their total collection of programs written in FORTRAN. Usage of the language continued to rise into the early 1960s, making concern about the future evolution of the language a critical issue for the industry.

In May 1962 the American Standards Association, speaking for data processing users, established a committee to define standards for FORTRAN, and eventually it developed two standards of the language. The first, simply called FORTRAN,

Table 33
Sample Program: FORTRAN

PROBLEM:[a] Construct a subroutine with parameters A and B
such that A and B are integers and $2 \leq A \leq B$. For every odd
integer K with $A \leq K \leq B$, compute $f(K) = (3K + \sin(K))^{1/2}$ if K
is a prime, and $f(K) = (4K + \cos(K))^{1/2}$ if K is not a prime.
For each K, print K, the value of $f(K)$, and the word PRIME
or NONPRIME as the case may be.

Assume there exists a subroutine or function PRIME (K)
which determines whether or not K is a prime, and assume that
library routines for square root, sine, and cosine are available.

PROGRAM:

```
      SUBROUTINE PROBLEM (A,B)
      INTEGER A, B
      J = 2*(A/2) + 1
      DO 10 K = J, B, 2
      T = K
      IF (PRIME(K) .EQ. 1) GO TO 2
      E = SQRT (4.*T + COS(T))
      WRITE (1, 5) K, E
      GO TO 10
    2 E = SQRT (3.*T + SIN(T))
      WRITE (1, 6) K, E
   10 CONTINUE
    5 FORMAT (16, F8.2, 4X, 8H NONPRIME)
    6 FORMAT (16, F8.2, 4X, 5H PRIME)
      RETURN
      END
```

[a]This particular problem was used by a noted authority on the history of programming to illustrate
the kind of numerical calculations well suited to FORTRAN.

SOURCE: Jean E. Sammet, *Programming Languages: History & Fundamentals*, © 1969, p. 151.
Reprinted by permission of Prentice-Hall, Englewood Cliffs, New Jersey.

reflected the characteristics of IBM's FORTRAN IV. The second, named Basic
FORTRAN, mirrored the features of IBM's improved version of FORTRAN II.
However, FORTRAN II (Basic FORTRAN) was considered a subset of the first
standard. With the completion of this committee's work, the data processing
industry now had its first high-level language with standards to which computer
manufacturers and users could conform. From then on, standardizing high-level
languages became routine within the industry. After standardization, the use of
FORTRAN[a] increased even faster than it had in the late 1950s since users now
expected their investments in programs to be protected from future developments
in software and hardware that might otherwise make older programs obsolete.
(For sample FORTRAN statements, see Table 34.)

With the establishment of standards in the early 1960s, the great period in the
evolution of FORTRAN came to an end. There would be many new releases of

Table 34
Sample FORTRAN Statements Circa 1961 Operating on IBM Computers

Statement	650	650 FORTRANSIT	1620	705	Basic 7070/7074	704
ACCEPT n, list			X			X
ACCEPT TAPE n, list			X	X	X	X
BACKSPACE i TO n						X
CALL NAME $(a_1, a_2, \ldots, a_n)$						X
COMMON $(a_1, a_2, \ldots, a_n)$						X
CONTINUE	X	X	X	X	X	X
DIMENSION $(v_1, v_2, \ldots, v_n)$	X	X	X	X	X	X
DO n $i = m_1, m_2, m_3$	1	X	X	X	2	2
END $(I_1, I_2, I_3, I_4, I_5)$	1	1	2		2	2
END FILE i				X	X	X
EQUIVALENCE $(a, b, c, \ldots)$, $(d, e, f, \ldots)$,		X				X
FORMAT $(s_1, s_2, \ldots, s_n)$			X	X	X	X
FREQUENCY $(n(i, j, lll), m\,(k, l, \ldots), \ldots)$					X	X
FUNCTION name $(a_1, a_2, \ldots, a_n)$						X
GO TO n	X	X	X	X	X	X
GO TO n, $(n_1, n_2, \ldots, n_m)$				X		X
GO TO $(n_1, n_2, \ldots, n_m)$	X	X	X	X	X	X
IF ACCUMULATOR OVERFLOW n_1, n_2				X	X	X
IF DIVIDE CHECK n_1, n_2				X	X	X
IF QUOTIENT OVERFLOW n_1, n_2				X	X	X
IF (a) n_1, n_2, n_3	X		X	X		X
IF (SENSE LIGHT i) n_1, n_2				X		X
IF (SENSE WHICH i) n_1, n_2				X		X
PAUSE n	5	5	4	5	5	5

Table 34
Sample FORTRAN Statements Circa 1961 Operating on IBM Computers

Statement				
PRINT n, list			X	X
PUNCH n, list	6	6	X	X
PUNCH TAPE n, list	6	X		X
READ, n, list	6	X	X	X
READ DRUM, i, j, list			3	X
READ INPUT TAPE i, n, list		X	X	X
READ TAPE i, list		X	X	X
RETURN				X
REWIND i		X	X	X
SENSE LIGHT i		X	X	X
STOP n	4	4	X	X
SUBROUTINE name $(a_1, \ldots, a_n)$	5		X	X
TYPE n, list		X		X
WRITE DRUM i, j, list		3	3	
WRITE OUTPUT TAPE i, n, list		X	X	X
WRITE TAPE i, list		X	X	X

Notes: (1) l, not permitted; (2) l, optional and may be ignored; (3) may be included but will be ignored; (4) the n is not permitted; (5) the n is optional and may be ignored; (6) the n is optional and is ignored. The IBM 709/7090 had all the above commands except END command was governed by note (2).

SOURCE: This list of FORTRAN statements was published in a different format in International Business Machines Corporation, *FORTRAN: General Information Manual* (1961). It was also reprinted in Jean E. Sammet, *Programming Languages: History and Fundamentals* (Englewood Cliffs, N.J.: Prentice-Hall, 1969): 149. Courtesy of International Business Machines Corporation.

the language and its compilers in subsequent decades, with more functions and greater ability to take advantage of new hardware, but these were always modifications of an existing productivity tool. As late as the mid-1980s companies like IBM were introducing new releases of the language because it had so many customers with such large libraries of FORTRAN programs who had to be supported. Even the latest operating systems*, such as IBM's MVS/XA, had to support FORTRAN if the company's large processors were to be sold to engineers and scientists.

The use of the language and what it taught the industry proved far more significant than its technical details. FORTRAN became the preferred language for scientific and engineering users. Throughout the 1960s, 1970s, and 1980s the language improved while a whole generation of programmers and members of the scientific community grew up with it. It was the first language that could be used on a variety of computers. In 1985 one could run FORTRAN programmers on large IBM 308X processors or on some microcomputers. Almost every computer vendor supported FORTRAN. It was easier to learn than any previous language and some that appeared afterwards. The language optimized object code and therefore used hardware more efficiently than earlier and even some later languages. Its efficiency was critical between the late 1950s and the early 1970s when the cost of hardware, as a percentage of the total cost of computing, remained very high. In the late 1950s the computer often accounted for 50 percent or more of a data center's budget, while in the early 1970s, it took up a third of all operating expenses.

With this language, users no longer had to remain with a particular generation of computers; they could now move on to new technologies, confident that their FORTRAN programs would operate on different machines. For computer vendors newer machines could be introduced without posing a severe conversion problem for potential customers, as long as the new devices supported FORTRAN. Ultimately, therefore, FORTRAN encouraged the use of data processing, particularly by people outside of the data processing industry. The greatest number of users were typically *not* members of a data center. This stood in sharp contrast to COBOL which became the most widely used language by programmers within the data processing industry. The specific percentage of all programs in each decade written in FORTRAN is still unknown.

For further information, see: John W. Backus, "The History of FORTRAN I, II, and III," *Annals of the History of Computing* 1, no. 1 (July 1979): 21–37; Robert W. Bemer, "Computing Prior to FORTRAN," ibid. 6, no. 1 (January 1984): 16–18; Herbert S. Bright, "FORTRAN Comes to Westinghouse-Bettis, 1957," ibid. 1, no. 1 (July 1979): 72–74; J.A.N. Lee, "An Annotated Bibliography of FORTRAN," ibid. 6, no. 1 (January 1984): 49–58; Jean E. Sammet, *Programming Languages: History and Fundamentals* (Englewood Cliffs, N.J.: Prentice-Hall, 1969); Richard L. Wexelblat, ed., *History of Programming Languages* (New York: Academic Press, 1981). The entire issue of the *Annals of the History of Computing* 6, no. 1 (January 1984) is devoted to FORTRAN on the occasion of its twenty-fifth anniversary as a programming language.

FORTRANSIT. This programming system was developed in the late 1950s to take advantage of functions available in both of the International Business Machines Corporation's (IBM's)† scientific languages: FORTRAN* and IT.* In essence it was a subset of IBM's very popular FORTRAN. A user would write programs in FORTRAN, and the code would convert it to IT, which in turn put programs into the SOAP assembler used on an IBM 650* computer. The computer accepted the code when SOAP translated it into machine code. Someone wanting to use FORTRAN on a 650 would therefore have used FORTRANSIT. As best can be determined, it was used by only a few people and was one of the minor programming languages* of the late 1950s and of almost no significance.

For further information, see: Jean E. Sammet, *Programming Languages: History and Fundamentals* (Englewood Cliffs, N.J.: Prentice-Hall, 1969).

G

GECOM. GECOM, one of the first programming languages* for business data processing, was developed by General Electric (GE)† for use on its GE-225 computer. Like other languages designed to provide a compiler for use in writing commercial applications in the early 1960s, it could run only on specific computers and suffered from the fact that COBOL* had come out as a universal language (i.e., it could run on many types of computers) and was supported by a wide spectrum of the data processing industry.

GE's original intent was to produce a compiler similar to COBOL-61 for its computers. The final product, however, had some important differences from COBOL. It borrowed some syntax and functions from ALGOL*—another universal language primarily supported in Europe at the time. The General Compiler (GECOM) was designed less as a source language than as a method for compiling since source code could be in ALGOL, COBOL, FRING, or TABSOL. The result was a language whose syntax did not look much like that of any other. With the demise of the GE-225 in the mid-1960s, GECOM's usefulness diminished. The GE-225, a small computer introduced in 1961 as a scientific processor, became more popular for commercial applications in part because of GECOM's attractiveness. By 1963, however, GE was introducing newer, larger computers while customers were beginning to take COBOL more seriously. Those two circumstances relegated GECOM to history.

For further information, see: C. Katz, "GECOM: The General Compiler," in *Symbolic Languages in Data Processing* (New York: Gordon and Breach, 1962): 495–500; Jean E. Sammet, "A Method of Combining ALGOL and COBOL," *Proceedings, WJCC* 19 (1961): 379–387.

GPSS. The General Purpose Simulation System (GPSS) was one of the first simulation programming tools developed for use with computers. It was introduced in the early 1960s for the purpose of simulating discrete systems, which could be described as a sequence of changes taking place instantaneously over a predetermined period of time. By the early 1970s the programming system had become a popular tool for simulating telecommunication networks, airline reservation applications, and their costs, among other applications. Its popularity encouraged the creation of other programming packages to do simulations by the 1970s. The idea of a computer being asked to model circumstances—a widely employed use of computers today—grew out of such early successes with simulations as that provided by GPSS. Although it was not the first such tool, it became the most important one available during the 1960s and early 1970s.

Its primary developer was a systems engineer at International Business Machines Corporation (IBM),† Geoffrey Gordon. At the time GPSS was developed, he was manager of Simulation Development within IBM's Advanced Development Division. During the 1950s he had worked with both analog* and digital computers* and had gained considerable experience in developing simulators. During the late 1950s he was employed by Bell Laboratories† where he learned how to simulate electronic devices and telecommunication networks. In 1959, now at IBM, he worked on the design of the Sequence Diagram Simulator, a tool for examining advanced switching system designs. The following year he worked on what became known as the Gordon Simulator at IBM's research facility at White Plains, New York.

Gordon's work at White Plains led him and his management to conclude that they needed a better, easy-to-use tool. His solution would become known as GPSS. Gordon's design was a simple one. A user of GPSS would draw a block diagram to describe that which was to be simulated. The key data elements that represented pieces of traffic (called transaction equipment) were acted on by other variables (such as facilities, storages, or logic switches), and an item he called blocks described the logic of a system. Table 35 illustrates the types of blocks that could be used, and Table 36 is the example Gordon used to illustrate how GPSS functioned in simplistic form. Other components in his package did statistical measurements, simple arithmetic, and chaining of information. Each block also had associated subroutines. The system as a whole was thus a collection of blocks while his programs created transactions that were moved into blocks. The output would be the answers to "what if" questions concerning the impact of variables. For example, if one increased the speed with which information could run down a telecommunications line, how much more data could be moved and at what cost could be compared, for example, against having more telecommunication lines handling the same load by using slower speeds.

Early in 1961 Gordon began to revise earlier simulators used at IBM, writing the new one in assembly language and employing new algorithms. The project was made public that September. His program ran on IBM computers—704,* 709,* and 7090—and the company wanted to sell the software* as a product. Originally called the General Purpose Simulator (GPS), its name was soon changed to General Purpose Systems Simulator, and finally, with release III, to its current name, the General Purpose Simulation System. The first version could handle only twenty-five blocks, GPSS II—released in 1963—expanded it to

Table 35
Some of the GPSS Block Types and Corresponding Operations

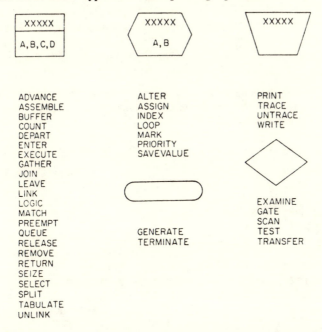

ADVANCE	ALTER	PRINT
ASSEMBLE	ASSIGN	TRACE
BUFFER	INDEX	UNTRACE
COUNT	LOOP	WRITE
DEPART	MARK	
ENTER	PRIORITY	
EXECUTE	SAVEVALUE	
GATHER		
JOIN		
LEAVE		
LINK		
LOGIC		EXAMINE
MATCH		GATE
PREEMPT		SCAN
QUEUE	GENERATE	TEST
RELEASE	TERMINATE	TRANSFER
REMOVE		
RETURN		
SEIZE		
SELECT		
SPLIT		
TABULATE		
UNLINK		

SOURCE: IBM, *General Purpose Simulation System S/360 Introductory User's Manual* (White Plains, N.Y.: International Business Machines Corporation, 1967): 73.

thirty-three. GPSS relied less on tables than on list processing to be more efficient than earlier versions.

Other enhancements were made to this batch package with GPSS III (1965). It also continued to be a card-oriented system, but rather than have just fixed form input statements, it was changed to a free format. After the announcement of the IBM System 360,* GPSS/360 appeared in 1967. It could run under either of the two major operating systems of this family of computers: OS/360 or DOS/360. In 1970 GPSS appeared which allowed part of the simulation model to be stored on disk rather than in the main memory* of a computer. It permitted the size of words to be full, half, or byte size, and was capable of taking advantage of floating-point features. A user could also allocate storage with this program. By the late 1970s an online version of the package had been released.

GPSS had begun as a simulator of telecommunication lines, telling users what the traffic on specific lines would be, given certain volumes and line speeds. This allowed engineers to optimize costs versus line loads through modeling while forecasting such variables and their behavior as line speeds on response times and suggesting optimal costs. The package was also used for other applications such as simulating a stock exchange system, traffic control in a city, the operations of a steel mill, and even the organization of a data center. Airline

Table 36
Sample Program: GPSS

PROBLEM:
Ships arrive at harbor in a specified arrival pattern.

If pier is free, dock ship and go to next block. If busy, join waiting line, if any.

Unload cargo at a specified rate. Unloading time is a function of amount of cargo. When finished, continue.

Record total time ship spent in harbor.

Ship leaves harbor.

PROGRAM:

```
LOCATION OPERATION
 * SIMPLE HARBOR SYSTEM
 * BLOCK DEFINITION CARDS
           GENERATE   32,5        ONE SHIP EVERY 32+5 HOURS
           QUEUE      1           JOIN QUEUE, WAIT FOR PIER
           SEIZE      1           OBTAIN PIER WHEN FREE
           DEPART     1           LEAVE QUEUE (NO LONGER WAITING)
           ADVANCE    25,20       HOLD PIER 25+20 HOURS
           RELEASE    1           FREE PIER FOR NEXT SHIP
           TABULATE   10          ENTER TRANSIT TIME IN TABLE 10
           TERMINATE  1           REMOVE SHIP FROM SYSTEM
 *
 * TABLE DEFINITION CARD
  10      TABLE      M1,10,5,20   DEFINE TRANSIT TIME TABLE
 *
 * CONTROL CARD
           START      100         RUN FOR 100 TERMINATIONS
```

This illustrates how to apply GPSS to a harbor problem, suggesting what will happen when ships arrive and unload.

SOURCE: IBM, *General Purpose Simulation System S/360 Introductory User's Manual* (White Plains, N.Y.: International Business Machines Corporation, 1967): 1, 6.

reservation systems and telephone networks were common subjects for this package. Although other software tools were available to do this kind of work by the early 1980s, GPSS continued to enjoy popularity for very complex models.

Besides IBM's version of GPSS, other companies brought out similar products because the use of computers for simulations was an excellent way to sell more equipment. Not surprisingly, computer manufacturing companies led the way with the introduction of their own versions of the software: Univac, Control Data Corporation (CDC),† Honeywell,† and Radio Corporation of America (RCA)† in 1968 and Digital Equipment Corporation (DEC)† later. The earliest known user of GPSS outside of IBM was the Norden Division of United Technologies Corporation which simulated massive military defense systems in 1962. The group using this package developed an interactive version called GPSS/360-Norden (1968). IBM's interactive GPSS appeared in 1976 and was programmed in APL.*

Like many software tools, GPSS's early versions were coded by a small group in short periods of time. The first release was written by one man (Gordon). By 1976, however, a group of programmers worked on the package, enhancing it continuously. By then, GPSS could use up to forty-nine blocks—almost twice the number as the original version—while the number of commands also increased. Modifications and enhancements made the breadth and capabilities of GPSS more useful for ever more complex models.

For further information, see: R. Efron and G. Gordon, "A General Purpose Digital Simulator and Examples of Its Application: Part 1—Description of the Simulator," *IBM Systems Journal* 3, no. 1 (1964): 21–34 and a sequel to this article by C. R. Velasco, "A General Purpose Digital Simulator and Examples of Its Application: Part II—Simulation of a Telephone Intercept System," ibid.: 35–40; G. Gordon, *The Application of GPSS V to Discrete System Simulation* (Englewood Cliffs, N.J.: Prentice-Hall, 1975), "The Development of the General Purpose Simulation System (GPSS)," in Richard L. Wexelblat, ed., *History of Programming Languages* (New York: Academic Press, 1981): 403–434, and his *System Simulation*, 2d ed. (Englewood Cliffs, N.J.: Prentice-Hall, 1978); E. C. Smith, Jr., "Simulation in Systems Engineering," *IBM Systems Journal* 1, no. 1 (1962): 33–50.

GRAY CODE DISK. This device was used by the U.S. Department of State and the military community on the eve of World War II in cryptography. It looked like a metal disk with code bar in the middle called a "cyclic permutation code." The Gray Code was an arrangement of data in the form of ones and zeros—the notation which in several years would be the standard used for electronic digital computers.* The combination of information was such that each group varied from the one next to it by no more than one digit. The purpose of such a device in coding messages was to reduce the opportunity for errors while increasing the speed of transmission. The Gray Code Disk reduced errors in sensing messages at the receiving end over manual systems. Its most dramatic use came on the eve of Japan's attack on the U.S. naval base at Pearl Harbor on December 7, 1941. During that critical period it was employed to transmit some of the results of intelligence gathered on Japanese movements.

For further information, see: David Kahn, *The Codebreakers* (London: Weidenfeld and Nicholson, 1968).

GREAT BRASS BRAIN. This tide predictor* was built by E. G. Fischer and R. A. Harris for the U.S. Coast and Geodetic Survey on the eve of World War I. It took fifteen years to build and when completed could compute using thirty-seven different variables. Results were displayed on dials. The machine added waves together to make a tide prediction. It remained in use for more than a generation.

For further information, see: C. H. Claudy, "A Great Brass Brain," *Scientific American* 110 (March 7, 1914): 197–198; R. A. Harris, "The Coast and Geodetic Survey Tide Predicting Machine," ibid., 110 (June 13, 1914): 485.

H

HARVARD MARK I, II, III, IV. These large calculating machines were built at Harvard University during the 1940s by Professor Howard H. Aiken.** He was the director of the Computational Laboratory at Harvard where, with the assistance of engineering support from International Business Machines Corporation (IBM),† he built the Mark I and, with the financial backing of the U.S. Navy and Air Force, the other three systems. He built electromechanical devices that presaged the arrival of the electronic digital computer.* His four machines ranged from the completely mechanical to magnetic ferrite-core memories. Intermediate machines also used electromagnetic relays as well and therefore reflected the evolution of such hardware in the 1940s.

During the 1930s Aiken realized the need for a powerful calculator that could help remove some of the tedium associated with calculating the results of nonlinear differential equations. He also recognized that existing punched card technology, even if rigged in a more sophisticated manner, could not handle the work. After canvassing several firms for support in designing and funding his ideas for an advanced calculator, in 1937 he contacted IBM. Along with support from the U.S. Navy, he was able to obtain the help he needed to build a calculator. IBM made engineers and parts available through its manufacturing facility at Endicott, New York. Beginning in 1939, Aiken began spending his summers there working on what would eventually be known as the Harvard Mark I. Benjamin M. Durfree, Francis E. Hamilton, and Clair D. Lake,** all employees of IBM, contributed to the design and construction of his equipment.

The first machine in the Mark series became operational in January 1943 at IBM's plant. The device was broken down and reassembled and made operational in May 1944 at Harvard. Known officially as the Harvard-IBM Automatic Sequence Controlled Calculator (ASCC), it soon acquired the nickname Mark I. When unveiled publicly in 1944, Thomas J. Watson, Sr.,** chief executive officer of IBM, concluded that Aiken had not given his company sufficient credit for work done on the machine and consequently refused to lend any further support to his work. That dispute probably cost Aiken the long-term opportunity to do more than he actually did in the development of the modern computer. Had he been able to maintain his connection with IBM, other projects might have come along for him when IBM decided to construct computers and sell them commercially. He also lost the opportunity to share ideas with the engineers at Endicott whom he had worked with that might have influenced future devices at Harvard. After the break with IBM, Aiken gained financial support from the U.S. military community. IBM's engineers ultimately built highly successful calculators and then computers, taking full advantage of their experiences with Aiken.

The Mark I was a very long machine that weighed five tons. Each of its registers was an adding machine constructed out of existing parts from IBM. They functioned individually as accumulator and arithmetic unit except that numbers (data) could be moved from one to another for addition. Punched paper tape* fed the machine the sequence of steps it was to take in performing a job. The physical transfer of numbers from one register to another was performed through electrical signals. The seventy-two mechanical registers could individually store up to twenty-three decimal digits. An additional digit per register carried the number's sign (plus or minus symbolized by 0 or 9, respectively). The registers themselves were actual IBM accounting machine registers driven by a 5 hp motor.

The mechanical accumulators were also enhanced with sixty constant registers. These other registers had switches turned on a panel as the means of loading numbers into them. Three paper tape readers were employed to read long "loops" of tabular data. A twenty-four-channel paper tape reader served as the control tape. The configuration included two card readers, one card punch, and two electric typewriters to print output; all of these were standard IBM products of the period.

To feed instructions into Mark I, commands were punched onto wide strips of tape (made out of uncut accounting machine card stock). Holes sensed by the card readers transmitted signals to the computer-like device. The control tape reader was always connected to the drive shaft which powered accumulators being used. Reading an instruction off the tape took approximately 0.3 seconds. Aiken and IBM's engineers structured arithmetic units to do normal functions (division and multiplication, for example) and more complex operations such as calculating $\sin (x)$, 10^x, and $\log (x)$. The machine could even be used to calculate trigonometric results. Yet it was a stiff device for programmers. They could hardly modify the sequence of instructions fed to it based on the answers received during the process of calculating the results to a problem. This problem was partially resolved later in the decade by adding the Subsidiary Sequence Mechanism, which was three additional control tape readers that made it possible to increase an operator's control over the machine's operations.

In its initial form in 1943/1944, the Mark I was 51 feet long and 8 feet high and had two additional panels each 6 feet long going out perpendicularly from the back of the machine. IBM's engineers had encased the device in attractive glass fronts and metal cabinets. It was a huge machine, one of the largest calculators ever made. Michael R. Williams calls it a "mechanical monster." The complicated-looking machine was made up of over 75,000 different parts and included such items as relays, binary and ten-position switches, rotating wheels, and cams. Users had access to 1,400 ten-position rotary switches, the majority of which were located on the panel housing the sixty constant registers. The Mark I's builders also laid some 500 miles of wiring in it to handle the flow of electricity.

Its initial work involved a project for the U.S. Navy's Bureau of Ships; other tasks were also military in nature for the duration of World War II. One of the early users of the machine was Lieutenant Grace B. M. Hopper,** who could rightfully be dubbed the first woman programmer. She went on to have a career in the U.S. Navy that extended into the second half of the 1980s, making her the oldest naval officer on active duty. She also played an instrumental role in the development of COBOL,* one of the most important programming languages* ever devised. The Mark I gave many young engineers and mathematicians their first experience with computing on a grand scale. After the war, the device was used primarily for mathematical purposes, such as calculating tables of Bessel functions. Although superseded by more sophisticated devices, it continued to function until 1959. Components of the Mark I then were shared among Harvard, IBM, and the Smithsonian Institution.

The press in the 1940s called the Mark I a mechanical brain, and indeed it was quite sophisticated for 1944, gaining almost instant recognition as an important achievement. Almost everyone in the United States who was working on computers in the 1940s visited the machine to see how it had been constructed. It gave a generation of engineers at IBM considerable experience with computers, endowing them with the background and determination it would take to build their own devices in the late 1940s. In turn, it gave IBM the technical expertise it needed to enter the world of computers. Finally, the Mark I supported Aiken's effort in training students in computing. Eventually, sixteen of them earned Ph.D.'s while working in his laboratory; most opted for careers in the nascent data processing industry or academia where they conducted related research. These students probably represented Aiken's most important contribution to the history of computing and data processing.

While students were being trained, Aiken did not stop with the Mark I. Casting about for a strategy which he could incorporate into an enhanced version of the machine, he settled on using existing technology (electrically managed mechanical components) which could be funded from the outside and had a proven record. Support came in 1945 from the U.S. Navy which needed a machine for the Naval Proving Ground in Dahlgren, Virginia. Aiken built it out of electromagnetic relays, and by July 1947 it could perform limited functions. He used relays instead of mechanical registers because they would make the

Mark II much faster. His use of a six-pole, double-throw contact, high-speed relay, made available by the U.S. Navy when it was difficult to acquire otherwise, could make and break contacts within 0.01 seconds. Thirteen thousand of these $15 units signaled a major advance in his design. In addition, the machine could operate almost as two independent devices. In the 1980s such a machine would have been called a diatic processor.

The Mark II had 100 registers dedicated to storing numbers; these were shared evenly by both parts of the processor. Each half of the machine had two multiplication units, another for addition, two paper tape* readers (to handle instruction tapes), and four tape readers for data handling. A ten-digit floating-point number could be housed in a numerical register. Stored digits employed a binary-code decimal (BCD). A dedicated relay carried the number's "plus" or "minus" sign. This machine had "hardwired" instructions (today called subroutines) which performed specific calculations. In theory the Mark II could perform various calculations concurrently, but in practice, that was difficult to program with tapes. For example, an answer might be generated at the wrong time for another part of the machine to use. (It lacked interlocks as we know them today.) Its control system, however, did provide support for this kind of activity. The machine as a whole was more efficient than the Mark I. It could perform an addition in 125 milliseconds, a multiplication in 750 milliseconds, and a hardwired transaction in 5 to 12 seconds.

Aiken had focused considerable attention on the speed of his first two machines but now saw that he needed to improve the accuracy of its work if he was to enjoy the real benefits of speed. Therefore, in the case of both the Mark III and Mark IV, he paid more attention to the use and quality of the machine. Aiken completed the construction of the initial configuration of the Mark III in September 1949. It used eight magnetic drums for main storage, each of which was 8 inches in diameter and 40 inches long, and rotated at 7,000 rpm. Instructions were recorded on the surface of another drum twice the size in diameter, 30 inches long, but it rotated at a speed of only 1,730 rpm. The drum for instructions housed nearly 4,000 commands for a program, thereby making the Mark III Aiken's first stored-program computer.

Aiken was fundamentally opposed to the concept of designing digital electronic computers based on the stored-program concept. He rejected the idea of storing within a computer the programs it would need to interact with data to execute an application (job). As a result, while the rest of the data processing industry moved into digital computers that used stored programs, his work remained outside that mainstream. For that reason, his work as an inventor of computers became irrelevant in the 1950s.

This is not to say that the Mark III did not function as promised; it was a clever device. He rigged it with a "mathematical button board" that allowed a user to issue commands to the computer by the mere pressing of a button. Each button had a mathematical notation on it. When pressed, a button would activate specific commands within the Mark III, causing it to execute a function. An even greater change in the Mark II occurred when Aiken abandoned paper tape as the way of operating the control system which drove the machinery. In Mark III, commands resided in the machine on drums, not outside on tape. An operator could thus rely on his panel of buttons to a greater extent than users of earlier machines. The result was a more flexible computer.

Although the Mark III housed up to 4,350 16-bit numbers, processing was relatively slow. The machine was partly electronic and electromechanical. It contained over 5,000 vacuum tubes and some 2,000 relays. It also experienced considerable problems with reliability after it left Harvard in March 1950 at its new home at Dahlgren, Virginia. Most of its problems were traced to the constant heating and cooling of parts that took place when the machine was turned off over weekends. Most computers of the 1950s solved this problem largely through better components as well as through the simple expedient of leaving them on all the time. Even today almost every data center in the world leaves computers running seven days a week.

The Mark IV was Aiken's last machine and the final one in the series built at Harvard. The primary physical difference between it and the Mark III was its memory* which this time was made out of magnetic cores with 200 registers. That change alone increased the speed of the machine which was used by the U.S. Air Force. The Mark IV became operational in 1952. It remained at Harvard where, with its molded metal casings, dials, and switches, it looked more like a computer from the 1950s. The Mark IV was also smaller than its predecessors and thus did not crowd the room.

Of all of Aiken's machines, only the Mark I was a state-of-the-art device for its time. Other machines were under construction with better components elsewhere, or they were faster and more efficient. That was certainly the case by the time Mark IV was constructed. Nonetheless, they represented a class of machines interim between analog differential analyzers* and the true stored-program computers of the 1950s. The first generation of computer scientists learned many valuable lessons from the Mark series. Aiken hosted numerous conferences on computing at Harvard where engineers from universities and laboratories shared information about their work while visiting and analyzing the Marks. That pattern lasted from the mid-1940s through the 1950s. Aiken built no more large computers in the 1950s; instead he trained computer engineers. He left Harvard in 1961 to form his own firm in Florida, Aiken Industries, and did some teaching.

For further information, see: Howard H. Aiken, "Proposed Automatic Calculating Machine," *IEEE Spectrum* (August 1964): 62–69 (reprints his original proposal of 1937); Paul E. Ceruzzi, *Reckoners: The Prehistory of the Digital Computer, from Relays to the Stored Program Concept, 1935–1945* (Westport, Conn.: Greenwood Press, 1983); Michael R. Williams, *A History of Computing Technology* (Englewood Cliffs, N.J.: Prentice-Hall, 1985).

HEATH ROBINSON. See ROBINSON MACHINES

IAS COMPUTER. This computer was built by John von Neumann** and his staff of engineers at the Institute for Advanced Study (IAS) in Princeton, New Jersey, in the late 1940s. The IAS machine provided him with the opportunity to implement his concept of the stored-program computer. The result was the construction of a new class of machines, five of which were built by other locations based on his design. The IAS computer was one of the more important of the early computer projects in the United States.

Von Neumann funded his projects through the same military organizations that had supported the construction of the ENIAC* and the EDVAC* during World War II. He obtained additional support from the U.S. Atomic Energy Commission and the Radio Corporation of America (RCA).† In the spring of 1946, when he began to form the organization to build his digital computer,* von Neumann sought J. Presper Eckert,** who had been so successful with the ENIAC at the Moore School of Electrical Engineering,† but was unable to persuade him to join the staff. Yet by June von Neumann had five engineers working for him. The plan called for design specifications to be drawn up at the Institute and then copied at the Los Alamos Laboratory, University of Illinois, Oak Ridge National Laboratory, Argonne National Laboratory, and the RAND Corporation. Each would then construct its own IAS computer. In reality each location varied the machines slightly based on modifications made by local engineers. But they were basically similar.

The design strategy called for using off-the-shelf parts to speed up construction. That worked except for the computer's memory* which was originally slated to be RCA's selectron. That technology did not work out, however, and after attempting to develop their own, the team discovered the British Williams tube memory which proved satisfactory and so was adopted. These engineers also decided to install forty parallel data paths rather than simply build a serial machine. The addition of so many paths required careful design to insure that all components would work in harmony and that electrical impulses and the flow of data would be synchronized correctly. This was not an easy feat in the early days of computing. The solution to the problem was to use "flip-flops," which are more formally known as trigger circuits. These circuits prevented data from going from one point to another until needed and therefore made it possible for data to flow in a parallel fashion throughout the system—hence the name parallel processor to describe this kind of machine. By the summer of 1947 each part of the processor itself had been designed and built. When they were all lashed together, however, they did not always work well. This forced the team to formulate policies regarding circuit designing that were needed for data processing. These policies included testing procedures for components and establishing levels of tolerance in deviations for parts from performance specifications.

The next major concern was to enhance the machine's storage capability. RCA's capacity was limited to 4,000 words, which the group thought too little. Next they examined the possibility of using magnetic wire memory (also known as the "bicycle wheel" endless loop wire memory), but that was not reliable enough. Even by the spring of 1948, RCA's selectron was not progressing technically, and so it was ruled out. At that point, the engineers discovered work being done at the University of Manchester in Great Britain by Frederick C. Williams** and decided to pursue his ideas further. The team elected to put together forty cathode ray terminals (CRTs) to form a parallel memory, with each CRT capable of housing 1 bit of a computerized word. After experimenting with this approach the memory worked, even though it offered only 1,024 words of additional memory to the system. But at last the end could be seen, and in January 1951 von Neumann anticipated that soon the IAS computer would receive its first program.

The first real test of the machine's processing came in midyear with calculations related to the construction of the hydrogen bomb. The machine had to run sixty days, around the clock, in order for the program to be executed. The programs were designed at Los Alamos, and two IAS computers were used to insure that the answers received were correct. Finally, on June 10, 1952, the IAS computer was dedicated. Features of the machine as it looked then are described in Table 37.

The computer continued to be modified. High-speed card input/output equipment was added, along with faster tape drives. A magnetic drum to handle more memory was also added. Meanwhile, other copies of the computer were being constructed. RAND's machine was called the JOHNNIAC* after von Neumann; MANIAC* was built at Los Alamos; Argonne's was called AVIDAC; and the other two built there were named ORACLE and GEORGE. The University of Illinois constructed ORDVAC* for the Aberdeen Proving Ground and the ILLIAC I* for use on campus. All were in use by the end of 1954, with most of them available as early as the end of 1952. One of the more famous

Table 37
Technical Features of IAS Computer

Serial/parallel	Parallel
Word length (bits)	40
Instruction length	20
Instruction format	1-address
Main store size	4,096
Main store type	CRT
Backing store type	Drum
Average add time	62 microsec.
Average multiply time	713 microsec.
Basic clock frequency	Asynchronous
Approximate number of valves	2,300
Approximate number of GE diodes	None

SOURCE: Reprinted with permission from *Early British Computers*, Simon Lavington, First Edition. Copyright © Digital Press/Digital Equipment Corporation (Bedford, MA), 1980.

copies built outside of the United States was the SILLIAC made in Sidney, Australia. These machines were used for a variety of projects during the 1950s. The JOHNNIAC, for example, did processing for many important military projects and provided the computing necessary for the creation of early high-level programming languages,* database management systems,* and research in artificial intelligence.*

These machines were impressive. First (in a departure from earlier devices), they used fewer vacuum tubes. In its case only 2,300 were employed as opposed to some 18,000 on the ENIAC and 3,500 on EDVAC. The machine was no longer a giant but a mere 6 feet long, 8 feet high, and 2 feet wide. The use of tubes that were two-thirds smaller than those in the ENIAC or EDVAC helped to keep the size small. Furthermore, because instructions were only 20 bits long, two could go into every memory word, again reducing size and effort from the past. It also caused the machine to run somewhat faster than earlier devices. IAS used single-address instruction codes and relied heavily on random-access memory and therefore did not need to optimize coding schemes as much as the American EDVAC or the British Pilot ACE.* An addition took some 60 microseconds to do, while multiplication required about 300 microseconds. Put into perspective, the IAS computer was one of the fastest, most powerful computers of the early 1950s. All parallel computers built in the next fifteen years relied heavily on the technologies and lessons learned from the IAS.

For further information, see: J. Biglow, "Computer Development at IAS Princeton," in N. Metropolis et al., eds., *A History of Computing in the Twentieth Century* (New York: Academic Press, 1980): 291–310; Michael R. Williams, *A History of Computing Technology* (Englewood Cliffs, N.J.: Prentice-Hall, 1985).

IBM 305 RAMAC. See RAMAC

IBM 405. The official name of this machine was the IBM Type 405 Alphabetical Accounting Machine. It was the most successful card punch product that the International Business Machines Corporation (IBM)† probably ever had. Introduced in 1934, it quickly became the single most important item in the company's sales manual until after World War II. In addition to helping fuel the company's growth and success, the IBM 405 served to train electrical engineers at IBM while causing other products to be introduced.

Although the machine underwent various modifications during its lifetime, its essential characteristics did not change. When fully configured, it could support sixteen accumulators, which could be divided evenly by lengths of two, four, six, and eight columns. Two or more accumulators could be joined to form a single unit, thereby allowing a user to allocate the use of accumulators better than with older or competitive machines. The device grew out of the Type 285 tabulator; hence, the IBM 405 frequently was also called a tabulator. Because it was manufactured by the Electrical Accounting Machine (EAM) Division, some called it an EAM (pronounced *ee ay em*) machine.

The IBM 405 was a small black hulk with a removable control panel on the left. The panel housed over 1,600 functional hubs, all laid out in a rectangular grid. The device could tabulate at a rated speed of 150 cards per minute and print at 80 cards per minute. The printing portion of the machine had eighty-eight type bars, forty-three of which were used for alphanumeric characters and the other forty-five for digits. The machine was extremely reliable and fast and, therefore, popular with IBM's customers. For example, at the start of 1944, 10,000 IBM tabulators were out on rental (they were not offered for purchase); nearly 6,400 were Type 405s and another 3,000 were Type 285s.

The IBM 405 played an early role in the evolution of IBM's computer products when, in 1947, Northrop Aviation† linked a 405 to an IBM Type 601 electric multiplier together with cables to run continuous jobs from one machine through the other with little human intervention. Northrop was very happy with the results, nicknamed their combination "Betsy," and often referred to it as a "poor man's ENIAC"* at industry meetings. That success led other customers to ask for the "combination." In 1949 IBM responded with the IBM Card-Programmed Electronic Calculator (CPC),* which represented a major step forward for the firm into the world of electronic computing, verging on computers themselves.

Earlier, in July 1948, IBM had already announced a successor to the 405 called the Type 402, an improved and more modern-looking accounting machine. Also available was a variety of the 402 without alphabetic printing called the Type 417. The 402 read cards at 150 per minute and printed at a rated speed of 100 lines per minute. Although the 402 was announced late in the history of card punch equipment, by being a part of CPC's configuration it enjoyed a good life: nearly 700 CPCs were built and shipped during the 1950s.

IBM manufactured other 400 series machines, and at least one other important device fit into the product line. The IBM Type 407 was an accounting machine introduced in 1949 and had the dubious distinction of being the last important electromechanical accounting machine ever introduced by IBM. It boasted a wheel printer under development since the late 1930s which turned out to be so outstanding that it became a standard to which all printers made by IBM throughout the 1950s were compared. It also had a good production speed, printing at 150 lines per minute. Before its arrival, no IBM device printed alphanumeric data at a rate faster than 100 lines per minute. Therefore, its introduction was greeted as important news within the office machine market.

For further information, see: Charles J. Bashe et al., *IBM's Early Computers* (Cambridge, Mass.: MIT Press, 1986).

IBM 603. The IBM 603 was the first electronic calculator introduced by the International Business Machines Corporation (IBM)† in 1946. It derived the product of two six-decimal numbers which it punched on a card in a cross-footing keypunch and at the rate of 100 cards per minute. After 100 copies were built, it was replaced with a more advanced model called the IBM 604.* Its official title was the IBM 603 Electronic Multiplier. From then on all IBM products would contain electronics; the day of the mechanical card punch device had ended. The 603 reinforced the determination of Thomas Watson, Jr.,** then a rising young executive within his father's company, to push for more electronically based products.

The 603 was one of a series of machines designated as a 600 device. The IBM 602, announced at approximately the same time as the 603, was also a calculator, although it was electromechanical. It could perform division and multiplication and, like the 603, was frequently used by scientists and engineers. The company marketed both machines primarily to commercial customers. The first of the 600 series was the IBM Type 600, introduced in 1931, as a multiplying punch. It read two factors on cards, multiplied them, and punched out the results on a card. In 1933 a more advanced model appeared called the IBM Type 601. It could do automatic cross-footing multiplication, thus summing a table of numbers by column and row. As of the end of 1943, there were some 2,000 601s on rent.

The entire series of machines represented a class of multiplication devices available from IBM between the early 1930s and the mid-1950s which essentially remained unchanged over time. Electronics was added, some devices sped up, and flexibility in function enhanced, but they did what they had always been designed to do: process stacks of cards through simple arithmetic into other stacks of output cards. Yet they provided much of IBM's revenues for nearly two decades before engineers moved fully into the world of electronics and later computers.

For further information, see: Charles J. Bashe et al., *IBM's Early Computers* (Cambridge, Mass.: MIT Press, 1986).

IBM 604. Also known as the Electronic Calculating Punch or the IBM Type 604 (all International Business Machines Corporation [IBM]† products of the 1930s and 1940s were called Type, followed by a number), this was one of IBM's first electronic calculating machines to depart from more traditional punched card technology. It was a step toward the world of electronics and, subsequently, computers. Experience with the 604 and related products enabled IBM's engineers to build the IBM CPC* and follow-on computational devices.

The 604 developed from an earlier device, the IBM 603, which was IBM's first electronic calculator. It could generate the product of two six-decimal numbers and punch the results out on cards at the rate of 100 cards per minute, the standard speed of all such devices in the 1930s and early 1940s. The first machine was demonstrated in 1946, and initial orders for it amounted to 150. After about 100 were made, IBM brought out the more powerful 604. The new device used a plugboard to control its functions. The 604 had 1,400 vacuum-tube circuits, qualifying it as an early first-generation computational device. The tubes made it a faster machine than the 603. In effect, it had a memory* made up of fifty decimal digits, hence twice the capacity of the 603. Each of the four basic arithmetic operations—addition, division, multiplication, and subtraction—could be performed at the rate of 100 cards per minute. The 604 could transfer programs from one machine to another, and repetitious steps could be programmed. Because circuits for the first time were mounted on a removable board, of which there were many in the machine, maintenance improved dramatically. All the repairman needed to do was swap a "bad" board for a new one and the machine would once again function. This feature subsequently became standard on all competitive and IBM equipment. The concept of swapping parts remains a design strategy of all computer-related equipment to this day. In the 1950s boards with transistors could be swapped while, by the end of the 1960s, subassemblies were made up of chips.* That whole approach could be traced directly back to the 604.

The machine became available in the third quarter of 1948, and, before it was retired as a product in the late 1950s, approximately 5,000 copies had been sold, making it one of IBM's most successful offerings of the 1940s and 1950s. It also became a critical piece of CPC's configuration. It is difficult today to appreciate the technical enhancements that made this product a success. At the time of its announcement, no competitive machine had as much function or speed for the cost. It initially could do between twenty and forty plugboard-controlled program steps, a number that soon rose to sixty. Yet it should not be characterized as a computer because programs in the machine were linked to its reader-punch which in turn functioned as Charles J. Bashe noted, "within the time provided by that unit rather than directing input-output along with internal operations." In other words, it took advantage of electronics to improve the speed and capacity of otherwise familiar punched card calculators.

For the first time, some of the problems that all computer manufacturers would face existed as the 604 went through development and manufacture. Lessons learned with the 604 were applied successfully with future products at IBM. For example, questions had to be answered concerning how to manufacture thousands of units in a cost-effective manner. Policies and practices were drawn up relating to maintenance (hence, for example, the plugboard approach). Such concerns also influenced the decision to minimize the number of circuits necessary. Plugboards made it possible for computers to be packaged in three dimensions rather than in two, increasing the utilization of space in electronic equipment. Such devices took up less space than would otherwise have been the case. The 604 used smaller vacuum tubes which also saved space. Through the 604 engineers learned how to standardize designs, encouraging them to take that approach again with subsequent machines. The machine introduced IBM's sales force to the benefits of electronics, thereby making them more receptive to future electronically based products in the 1950s.

For further information, see: Charles J. Bashe et al., *IBM's Early Computers* (Cambridge, Mass.: MIT Press, 1986); René Moreau, *The Computer Comes of Age: The People, the Hardware, and the Software* (Cambridge, Mass.: MIT Press, 1984).

IBM 650. The IBM 650 was the single most widely purchased first-generation computer in the United States and represented one of International Business Machines Corporation's (IBM's)† earliest entries into the data processing industry. Also known as the 650 RAMAC*, 470 were installed by the end of 1956, and in the following year the figure had grown to 803, far exceeding the number of installations by any other vendor. This medium-priced computer, which rented for between $3,000 and $4,000 per month, was intended to expand IBM's market for computerized products. The 650 contributed successfully to the expansion in the number of customers using computers.

Considerable debate had been generated within IBM during the Korean War regarding the potential size of the market for computers in general, as well as the amount a user would pay for such equipment. Finally, in the spring of 1953 IBM decided to build and announce what would become the 650, forecasting that it might sell over 200 of them. Known as a drum calculator because of the technology making up its memory, IBM unveiled the product line in 1953 with first shipments in 1954. Two models were based on rotating magnetic drum technology for main memory.* The total size of memory in real storage ranged from 10,000 to 20,000 characters. Over 1,800 of these machines were ultimately sold, making it an outstanding commercial success in an industry where the sale of less than 100 machines was considered good. Even the then mighty UNIVAC* never exceeded several hundred installations.

The 650, although a small device, was nonetheless important because it was a general-purpose commercial system and was mass produced. Both circumstances were firsts for IBM and taught the company valuable lessons about market demand and manufacturing techniques that would contribute to the firm's successes in the decade to follow. The system's wide popularity could be traced

Table 38
Technical Features of IBM 650

Word length	
Alphabetic	None
Numerical	10 decimals
Number of instructions	89
Memory capacity	60 words
Memory access time	
(microseconds)	96
Drum	1,000–2,000
Add time (fixed point)	0.63
Magnetic tape: transfer rate	
(characters/second)	15K
Card reader (cards/minute)	150–200
Card punch (cards/minute)	100
Printer (lines/minute)	IBM Tabulator

Data as of 1958.

SOURCE: R. Moreau, *The Computer Comes of Age: The People, the Hardware, and the Software*
(Cambridge, Mass.: MIT Press, 1984): 70, Table 2.4. Copyright 1984 by MIT Press,
all rights reserved; reprinted with permission.

in part to its cost/performance relative to existing options but also to its ease of use. It was a system based on card input/output and came at a time when most accounting departments still relied on card I/O. Its peripherals were very reliable by the standards of the day, and the system was most suitable for bread-and-butter commercial (mainly accounting) applications. It was also popular with noncommercial users. Chrysler's two 650s for research and development allowed engineers to design gas turbine engines while modeling other automative activities. Yet a third machine supported the company's more traditional accounting functions. Caterpillar Tractor did inventory control on theirs, Stanford University performed statistical analysis, while both Purdue University and the Massachusetts Institute of Technology (MIT) conducted academic work on 650s.

The machine was considered relatively easy to use and quick to install. Furthermore, it took up less room than other vendors' machines. A typical 650 system consisted of three devices or more, well packaged into attractive "boxes" on rollers. The key to its configuration was the integration of punched card input and output into a single unit (Table 38). It was a binary decimal machine that used a two-address instruction set. It could multiply in 9 milliseconds and divide in 12 milliseconds. Storage access time was approximately 2.4 milliseconds. It could read cards at 200 per minute and punch them out at 100 per minute. RAMAK was the name given to its disk memory. A printer was later attachable to the system. IBM also introduced tape drives* for the 650. The system also had software* called SOAP, which was a set of Assembler programs designed to permit a user to employ a symbolic language rather than code machine instructions, thereby saving considerable time and effort in programming the machine. The 650 represented one of the first important steps toward moving programming up from the machine-level instruction sets of the 1940s and early 1950s to what would soon be known as "higher level" languages in which compilers translated instructions written in a language into terms that were understood by the hardware. That process of translation reduced the degree of complexity in programming while also shrinking the number of lines of code necessary to execute an instruction.

The 650 computer enjoyed a healthy life from 1954 through 1959, after which competitive pressures, the march of technology, and the advent of second-generation machines made it obsolete. In its day, it was the single most important computer product in IBM's inventory and the most widely used machine in the industry.

For further information, see: Charles J. Bashe et al., *IBM's Early Computers* (Cambridge, Mass.: MIT Press, 1986); Franklin M. Fisher et al., *IBM and the U.S. Data Processing Industry: An Economic History* (New York: Praeger Publishers, 1983); Cuthbert C. Hurd, "Computer Development at IBM" in N. Metropolis et al., eds., *A History of Computing in the Twentieth Century* (New York: Academic Press, 1980): 389–418; René Moreau, *The Computer Comes of Age: The People, the Hardware, and the Software* (Cambridge, Mass.: MIT Press, 1984).

IBM 701. Also known as the Defense Calculator, the IBM 701 was a first-generation computer from International Business Machines Corporation (IBM).† The company had already gained some experience with high-speed computational devices, such as with the SSEC* and the IBM 603* and, in 1948, began selling the IBM 604.* Engineers at IBM had watched John von Neumann's** construction of the IAS Computer* at the Institute for Advanced Studies (IAS) during the late 1940s with considerable interest. They learned enough to give them ideas that led to the IBM 701—the first computer built by IBM for commercial customers. Earlier machines either had been special-purpose devices, only quasicomputers, or had been built for the military. The 701 is generally viewed as the company's first commercial computer.

Thomas J. Watson, Jr.,** an early and enthusiastic supporter of the use of electronics and computers, wanted to develop a high-speed calculator for use by the U.S. military community during the Korean War. His father, Thomas J. Watson, Sr.,** then chairman of the board at IBM, strongly supported efforts to build a calculator for the military. The result was the 701 or, because of the war and its original purpose, the Defense Calculator. When the senior Watson expressed the desire to have the machine built by the spring of 1952, IBM engineers began exploring ideas on how to construct it simply and quickly. The parallel processor under development at IAS provided an excellent model and significantly influenced the Defense Calculator's design. It was to be a binary device that could perform 16,000 additions each second or 2,000 multiplications in the same amount of time. IBM's machine also used Williams tubes for memory because of their great reliability compared to such traditional tools as vacuum

Table 39
Technical Features of IBM 701

Add time	62 microseconds
Multiplication time	500 microseconds
Word length	36 bits (2 instructions per word)
Memory type	Williams tube
Memory size	2,048 (36 bits)
Secondary memory	Magnetic drum (8,192 words) Magnetic tape (IBM 727s)
Card reader	150 cards/minute
Printer	75 lines/minute, 120 characters/line

These features were typical of the machine constructed in 1953. The version that appeared in 1956 used ferrite-core memory (IBM 701M).

tubes or relays. A variety of IBM card punch equipment would serve as the machine's peripherals. (For a summary of the 701's features, see Table 39.)

The combination of existing technologies and products made it possible for IBM's engineers to demonstrate the machine for the first time on April 7, 1953. (Coincidentally, it would be on another April 7, in 1964, that the company would introduce a family of computers called the System 360* which would totally transform IBM into a computer company and would forever kill card punch equipment as the primary technology for processing in the industry.) The Defense Calculator set a pace that encouraged management, and in time nineteen of them were built.

A number of the Defense Calculator's technical distinctions represented departures from the IAS Computer. First, the IBM machine used a 36-bit word length rather than 40, which made it compatible with other developments underway in the company with a new tape drive. Second, each of its addresses referred to an 18-bit half-word. That feature simplified programming for the machine. Third, numbers were presented as absolute values, complete with separate sign indications that permitted simplified logic processes and hence lowered manufacturing and operating costs. Overall, users of the device found that these changes made it easier to operate than the IAS Computer.

The Defense Calculator's configuration by March 1952 included two tape units, a magnetic drum unit, Williams tube memory,* an L-shaped arithmetic and control unit, an operator's console (known as an operator's panel), a card reader, a printer, and a card punch. The entire configuration was driven by three power units.

Originally constructed for the military, IBM next decided to produce a commercial version which it eventually called 701. The incentive to produce a commercial model was powerful. First, customers wanted more computing capability and the Defense Calculator offered the promise of such a product. Second, because of its design, it took 75 percent less space than the old SSEC while doing work at a speed twenty-five times greater. In May 1952 IBM began notifying its sales branches that such a machine, called the IBM Electronic Data Processing Machine, was under development. In December 1952 the first commercial copy of the 701 left the plant in Poughkeepsie, New York, for IBM's headquarters at 590 Madison Ave in New York City. The commercial version had a variety of peripheral equipment that increased its overall complexity. These included an electronic analytical control unit (Type 701), two electrostatic storage units (Type 706), a Type 711 punched card reader, a Type 716 alphabetical printer, a Type 721 punched card recorder, two magnetic-tape and recorder units (Type 726), and a magnetic drum reader and recorder (Type 731). As early as June 1952, the computer as a whole was being called the 701, taking its name from the first unit in the configuration. Built with tubes and a memory of 2,048 words, IBM rented the system for $17,600 per month. In April 1953 it was announced to the public. By June 1954 eighteen machines had been built and shipped (the eighteenth went to Lockheed Aircraft). The majority of its users employed the machine for scientific applications.

The 701 was an important machine for IBM for reasons that have not always been obvious. First, it caused the company to establish formal procedures for announcing new products in the computer field while defining more precisely the relationship between IBM and customers, particularly concerning maintenance. The company was therefore forced to define how much of the development work had to completed before a machine could be announced to the public, thereby increasing the odds of it being shipped on time and as announced. Second, the 701 gave IBM's management the confidence to build other computers rather than simply to ask whether there was a market for such machines. After the 701 IBM knew that it could introduce other computational machines and, therefore, could ask when and with what configurations. By the time the nineteenth 701 rolled off the assembly line in March 1955, IBM had already brought other computers to the market. The 701 gracefully moved IBM a long way from its punched card perspective to an electronic computing mindset.

For further information, see: Charles J. Bashe et al., *IBM's Early Computers* (Cambridge, Mass.: MIT Press, 1986); Byron E. Phelps, ''Early Electronic Computer Developments at IBM,'' *Annals of the History of Computing* 2, no. 3 (July 1980): 253–267.

IBM 704, 705. The IBM 704 was the first widely sold core memory* computer from International Business Machines Corporation (IBM).† First delivered in 1955, it exemplified state-of-the-art technology, features of which continued to exist in computers until at least the mid-1960s. These computers were also some of the largest available for commercial users in the 1950s. More important for IBM, these machines allowed the company to take the initiative away from Sperry Rand† and other vendors in leading the data processing industry into the computer revolution.

IBM converted much of the technology developed for the SAGE* project into

Table 40
Technical Features of IBM 704

Technology	Valve
Word length (bits)	36
Store size, fast	Up to 32,768
Store type, fast	Magnetic core
Store size, backing	8,192 per drum
Store type, backing	Drum
FXPT add time[a]	24 microsec.
FLPT add time[b]	228 microsec.
FLPT multiplication time	192 microsec.
Date first delivered	1956 in U.K.
	1955 in U.S.

[a]Fixed-point arithmetic.
[b]Floating-point arithmetic.

SOURCE: Reprinted with permission from *Early British Computers*, Simon Lavington, First Edition. Copyright © Digital Press/Digital Equipment Corporation (Bedford, MA), 1980.

components of the 704 and 705 computers, replacing the smaller 701* and 702. The 704 was between two and three times the speed of the 701 and over twenty times that of the UNIVAC I.* The computer was also priced competitively. IBM announced it in May 1954 and delivered copies in 1955; the 705 was unveiled in October 1954, and initial shipments began in 1956.

The technology employed in these machines was significantly different from that of the past. First, IBM employed magnetic core memories in these machines. Their design was originally conceived by Gene M. Amdahl,** later considered one of the key architects of the IBM System 360* family of computers, and in the 1970s, founder and chief executive officer of the Amdahl Corporation,† one of the most important manufacturers of computers marketed against IBM's System 370.* With the 704 and 705, Amdahl wanted a large amount of memory* and thus designed the machines so that one could expand memory incrementally to 4,096 bytes, making these the largest memories available on commercial machines. Such quantities of storage in a computer meant that programs could either be larger than before or that more data could be processed. By 1957 up to 32,768 positions of memory were available on such computers (Table 40), providing ample proof of Amdahl's wisdom in designing expandable computers. Yet even in 1957, that amount of memory was considered massive. By 1970 such memory sizes were thought to be small. Second, the 704 and 705 operated faster than other available systems, which also encouraged large users to process more than before.

Other innovations included floating-point arithmetic—the first time such a feature was available on a commercial computer. At the risk of oversimplification, it may be stated that floating point made the calculation of large numbers possible, such as those involved in complex engineering and scientific applications. The 704 was the first IBM computer to have index registers, which meant that three registers could help track the progress of a calculation in which a sequence of steps was repeated many times. It counted the number of occasions when a series of steps was executed and stopped the process when the desired number of repetitions had been completed. These features were well received. Programmers wrote subroutines and kept them resident in computers for use many times without having to write the same portion of a program each time such a piece of code was needed in an application. Larger engineering and scientific applications were also developed. These capabilities represented important productivity boosts for programmers and encouraged the acquisition of more IBM systems.

These computers were the first from IBM with any software* that even approached the concept of an operating system.* Such software serves as a "traffic patrol," directing the flow of information and processes while coordinating the activities of the computer. All processors have such a function, usually called a systems control program, or simply an SCP. Prior to the 704, the computer had to be instructed on every step to take in handling its functions along with writing application software. The 704 had a crude operating system to the extent that it had some software to control the input of data, to monitor the progress of calculations, and to deliver output. Such functions have been expanded by all vendors in all systems to the present day. The experience of the 704 and 705 suggests that operators would be more productive with SCPs. The evolution to today's massive operating systems had begun.

These early machines also offered a programming language—FORTRAN*— to their users. This language, made available in 1957 on IBM's computers, was one of the most important programming languages* ever introduced in the history of data processing. It quickly became the language of choice for engineers and scientists until the late 1970s when others became more convenient and powerful to use. The introduction of FORTRAN alone would have insured the 704s and 705s a place in the history of computing without any consideration having to be made for their large memories. Yet core memory was easier to manufacture and maintain, which helped drive down the cost of computing.

The 704 and 705 made possible larger programs and more systems. New users of such technology included aircraft manufacturers, oil and chemical companies, universities, and government agencies such as the Social Security Administration. The 701 and 704 processors were generally sold to scientific users, whereas the 702 and 705 machines were intended primarily for scientific applications. Customers did not observe such distinctions, however. In fact, nearly half of 704 users also ran scientific and engineering work on them. This blurring of scientific and commercial applications on 704s and 705s eventually contributed to IBM's decision in 1962 to design the S/360 computers to handle both types of processing within one machine.

The 704 and 705 products came at an opportune time for IBM. The use of computers was already spreading, and the Remington UNIVAC systems threatened to prohibit IBM from playing a major role in the new technological revolution. Almost on the verge of dominating the industry with UNIVAC, Remington lost the lead to IBM in the mid-1950s largely because of the timely introduction of the 704s and 705s. The installed computers tell the story. In 1955 more UNIVACs were being used than 700 series machines. By the end of 1956 IBM had seventy-six installed as opposed to forty-six UNIVACs. On order were 193 704s and 705s to Remington Rand's sixty-five processors.

For further information, see: C. J. Bashe et al., "The Architecture of IBM's Early Computers," *IBM Journal of Research and Development* 25, no. 5 (September 1981): 363–375 and *IBM's Early Computers* (Cambridge, Mass: MIT Press, 1986); Franklin M. Fisher et al., *IBM and the U.S. Data Processing Industry: An Economic History* (New York: Praeger Publishers, 1983); René Moreau, *The Computer Comes of Age: The People, the Hardware, and the Software* (Cambridge, Mass.: MIT Press, 1984); Robert Sobel, *IBM: Colossus in Transition* (New York: Times Books, 1981). See also the entire issue of *Annals of the History of Computing* 5, no. 2 (April 1983) devoted to the IBM 701, with articles by many of the engineers and scientists who developed computers in the mid-1950s.

IBM 1401. This second-generation computer from International Business Machines Corporation (IBM)† became a series of machines and peripheral equipment introduced in the late 1950s and early 1960s, first at the small end of the computer market and later expanded to other peripherals and computers. IBM sold between 15,000 and 20,000 1401s, exceeding by many thousands the number of processors sold by a particular model type of all earlier systems. Through 1960 in the United States, only 6,000 general-purpose computers had been sold—the year the 1401 was shipped to customers. The 1401 was also initially a replacement for the 650* computer, which had been sold in greater quantities than any other first-generation system in the world.

The 1401 was first announced in October 1959. The first model immediately set significant price/performance standards over earlier technologies, while offering one of the least expensive entry-level computers available. Its subsequent success was due to its low cost, important increases in speed and reliability, the introduction of new peripheral equipment, ease of use, and reliance on card input/output procedures (in the initial models). It performed up to seven times the number of instructions as a 650. Eventually, it could be configured in nearly one dozen models with an equally varied set of peripherals. Other computers were later introduced which were compatible with the 1401, allowing a user to upgrade with less conversion effort than would have been the case in moving from a 650 to a 1401.

Its technological improvements quickly destroyed the need for first-generation vacuum tube systems. Even electromechanical card I/O-based systems were rapidly transformed over the next several years. The five models of the 1401 (401, 1440, 1460, 1410, 7010) were based on transistorized technology and magnetic core memory which together provided greater reliability and cost reductions over previous systems. It used 6-bit BCD code and parity checking bits. Instructions could be formatted in variable form ranging from a single character to eight, while data fields could also be variable.

A wide variety of peripherals was announced with the 1401 but none as important as the 1403 printer, a variety of which would remain as the standard printer on the S/360* (1403 Model N1 printer). This was the most popular printer ever introduced in the history of data processing; variations of the original 1403 were still in use in 1979 as the preferred printer by many on the small-to intermediate-size S/370s.* Known as a chain printer, it relied on a belt of characters that moved at high speeds, placing the character to be printed at exactly the correct position on the paper for impact printing. It could operate at a maximum speed of 1,000 lines per minute and produced straight lines of print years before that capability became the norm in the industry. Prior to the 1403, wavy lines were common. Initial models operated at only 600 lines per minute, but even that was attractive when compared to the standard 150 lines per minute for first-generation systems. Costs also dropped. At the end of the life of the 1401 computer IBM believed that, in addition to general price/performance improvements over older systems, the success of this new computer was largely due to the attractiveness of the 1403 printer. The 1403 appeared at a time when a great deal of output was increasingly shifting away from cards toward printed reports and at the same time that a number of new systems were being installed for such report-dependent applications as payroll, inventory control, and accounting.

Other major components of the 1401 family included the 1402 card reader/ punch which could read 800 cards per minute or punch at 250 per minute. The 729 magnetic tape drives later provided users with seven-track tape-handling, ranging in speed from 15,000 to 62,000 characters per second. Density of data on tape ranged from 200 characters per inch to 556 and eventually 800. The 7330 tape drive functioned at 7,200 characters per second and was very inexpensive. This system eventually also had disk storage in the form of the 1405 fixed-disk units; the units stored either 50,000 or 100,000 records, each with character lengths of 200. The 1311 disk storage appeared as a replacement for the 1405 with the removable disk pack which was so evident throughout the 1960s and part of the 1970s. A pack could house 2 million characters as 20,000 hundred-character records. Even MICR devices were also available in the 1401 configurations.

IBM sold these systems to operate as stand-alone computers or to work in tandem with its 700 or 7000 series of processors. In large data centers, the 1401 might be seen controlling peripheral equipment for much larger computers, such as the 7090. Some of IBM's competitors produced systems that were compatible with the successful 1401, such as the Honeywell† 200 system and the General Electric (GE)† 400 series.

The 1401 had several programming languages* as well. First, it had the standard ability to program in machine language. Second, an assembly language called Symbolic Programming System (SPS) provided a higher level tool. Third,

Table 41
Technical Features of IBM 1401

Type	Decimal
Word length	Variable
Number of words	1.4K–16K
Memory type	Ferrite
Access time (microseconds)	11.5
Times for operations (microsec.)	
Addition	230
Multiplication	2,100
Division	2,600
Number of instructions (in code)	43
Number of addresses per instruction	1–2
Number of index registers	3

Typical characteristics as of about 1960.

SOURCE: R. Moreau, *The Computer Comes of Age: The People, the Hardware, and the Software* (Cambridge, Mass.: MIT Press, 1984): 132–133. Copyright 1984 by MIT Press, all rights reserved; reprinted with permission.

a language that became known as Autocoder and that survived throughout the bulk of the 1960s came close to resembling Assembler, one of the more widely used languages of the S/360 days. Later FORTRAN* and COBOL* could be used on the 1401 computers. A Report Program Generator (RPG) eventually made itself into the configuration and may represent the first significant software* tool in the industry for producing reports. RPG remained in vogue for the next twenty years. Some operating systems* software emerged to do job control, but no real system control program was developed for the 1401. (See Table 41 for a list of 1401's characteristics.)

The configurations varied over the years as new models appeared. Thus, the initial availability of memory* ranged from 1,400 characters up to 4,000. Later models of larger processors supported up to 16,000 characters of storage. Circuitry improved from model to model. For example, the small 1410 and the larger 1460 replaced earlier versions of the 1401. The 1410 was larger, had more processing power, and worked faster. Its memory could be expanded to 80,000 characters. The largest of the 1401 product series was the 7010, which was totally compatible with the 1410; it could support 100,000 characters of memory.

The 1401 series was one of the last major sets of computer products to appear from IBM before the S/360. The 1401 was announced on October 5, 1959, the 1410 on September 12, 1960, the 1440 on October 11, 1962, and the 7010 on October 29, 1962. The last model, the 1460, was unveiled on May 16, 1963, hardly a year before the introduction of the S/360 on April 7, 1964. The S/360 was introduced partly to replace a product set (which included all of the 1401 computers whose useful life was projected only to 1965.

An average 1401 configuration would have rented for approximately $8,000 per month, and a larger system might have gone as high as $12,000. Entry-level systems rented for one-third that price. A 1410 system averaged about $11,000 per month and ran as high as $18,000. The System/360 models leased for similar amounts of dollars but always had much greater capacity and speed.

For further information, see: C. J. Bashe et al., "The Architecture of IBM's Early Computers," *IBM Journal of Research and Development* 25, no. 5 (September 1981): 363–375 and *IBM's Early Computers* (Cambridge, Mass.: MIT Press, 1986); Franklin M. Fisher et al., *IBM and the U.S. Data Processing Industry: An Economic History* (New York: Praeger Publishers, 1983); René Moreau, *The Computer Comes of Age: The People, the Hardware, and the Software* (Cambridge, Mass.: MIT Press, 1984); Robert Sobel, *IBM: Colossus in Transition* (New York: Times Books, 1981).

IBM 1620. This was one of International Business Machines Corporation's (IBM's)† earliest computers and was designed to be a small, low-cost, scientific processor. It was developed during the late 1950s in response to demands from the company's sales organization which was experiencing increased competitive pressures. Vendors were catering to the needs of scientists and engineers with processors that could not be matched with IBM's existing product line as of the late 1950s. The task of developing such a machine went to Wayne Winger, who began to design the computer in the spring of 1958.

Winger took advantage of many new technological developments while stripping the new machine down to the bare essentials in hopes of keeping the potential lease cost of the computer to between $1,600 and $2,000 per month. He as well as his managers believed that a machine half the size and cost of an IBM 650* could satisfy the specific need for a new scientific processor. Winger and his staff elected to use core memory* for the new machine (code-named CADET) which allowed him to take advantage of the speed and capacity of this kind of technology. When announced in October 1959, the IBM 1620 System was a serial, decimal processor that had a fixed twelve-digit instruction set. It used variable-length data words. Addition and multiplication were done through the use of stored tables; subtraction was done similarly to addition, although it relied on circuitry as well. Initially, there was no DIVIDE function; that would come later. To perform division, therefore, one either programmed the function or used multiplication reciprocals. The only input/output equipment was a paper tape* reader and punch. A typewriter became the data-entry component of the system.

The 1620 could store and retrieve up to 20,000 digits, all of it in core memory. The small device rented for $1,600 or could be purchased for $74,500. Users later recalled that it had a useful instruction set, and the memory was substantial for the period. It was announced with FORTRAN*, which had already become one of the most popular programming languages* for engineers and scientists. The machine was supported both by customers and IBM sufficiently to warrant enhancement with additional capacity and peripherals. Thus, memory grew to 60,000 characters while punched card I/O was added. Software was improved along with providing a DIVIDE instruction. By the end of 1963, it even had disk drives* while a faster model of the base machine appeared.

Some observers have noted that IBM was late in responding to the needs of the scientific community (a problem the company faced again in the 1970s). When it did introduce the 1620, however, it sold well. By the end of 1963 IBM had shipped over 1,000 systems. The reliability of the machine made it an ideal candidate for process-control by IBM. When equipped with analog-to-digital conversion capability, it became the IBM 1710 Control System, announced in March 1961.

The IBM 1620 exposed the company to the scientific user in a favorable light. The use of typical second- and third-generation technologies led to such great improvements in performance that the firm could use the computer for control systems. That a customer could employ a commercially available system to control the operations of a manufacturing or processing plant, with the attendant financial risks should the machine break down, was a real indicator of the 1620's reliability. Under the title 1710, hundreds of systems were sold in addition to the 1,000 1620s cited above. It was also one of a series of transistor computers which came out of IBM, suggestive of a new era in computing. Other solid-state machines of the late 1950s or early 1960s included the 7070, 7090, 1401,* 7080, STRETCH* (7030), 7074, and the 1410.

For further information, see: Charles J. Bashe et al., *IBM's Early Computers* (Cambridge, Mass.: MIT Press, 1986).

IBM AN/FSQ-7. See SAGE

IBM CPC. The IBM Card-Programmed Electronic Calculator, one of International Business Machines Corporation's (IBM's)† first computer-like devices, was introduced in 1949. It led the company to move from calculating equipment toward computers in general. As early as 1946 and continuously throughout 1947, Wallace J. Eckert,** at the Watson Scientific Computing Laboratory at Columbia University, had been experimenting with linking card punch equipment to relays in an attempt to take card input and process calculations electronically, much like a miniature sequence calculator driven by instructions originating on cards and involving multiple card punch equipment. Northrop Aviation† next linked an IBM 601 electric multiplier to an IBM 405* accounting machine with cables. This innovation allowed users to send jobs from one machine to another automatically, that is, electronically rather than by having an operator carry output cards from one machine manually to a second device to process next in a sequence of steps. The net result was faster computation. During the winter of 1947–1948, Northrop's data processing community began to present papers on the efficiencies of their modifications (called combination at the time) of IBM's machines, naming their new configuration "Betsy." They also tagged it a poor man's ENIAC.* By the end of 1948, nearly a dozen of IBM's customers had asked to obtain a "combination" such as Northrop used.

Table 42
Technical Features of IBM CPC

Type 604 calculator	
Type 521 card punch	
Type 402 acoounting machine	150 cards/minute
Type 941 auxiliary storage unit (optional)	16 numbers
Type 417 accounting machine (no alphabetic printing)	150 cards/minute

IBM responded to these requests by quickly designing a combination called the IBM Card-Programmed Electronic Calculator, which it announced in May 1949. It was never called that, however; instead it was known simply as CPC. As Table 42 suggests, it consisted of various existing or modified IBM products. The IBM 402 was an updated version of the IBM 405 accounting machine. The IBM 941 could store sixteen signed numbers, each made up of ten decimal digits. The company offered the IBM 417, a variant of the 402 (minus its alphabetic printing feature), which printed 150 lines per minute. All output were numbers. The overall design of the CPC made it possible to run long engineering formulas, provided the memory* required for such evaluations did not exceed about twenty numbers. The IBM 604 was a slave to the IBM 402, providing an early example of electronic storage usage.

IBM began to ship CPCs in the last quarter of 1949, and modifications were made the following year to enhance the operator's control over its functions. More storage was also added. The product was a success, with some 700 having been sold which by the standards of the day was a large volume of machines. It continued to be manufactured for several years and was used by both scientific and commercial users interested in speeding up the processing of jobs that required the usage of multiple types of card punch devices.

CPC helped move IBM closer to the world of computers. First, many of IBM's executives came to realize that a significant number of potential customers existed for expensive computational devices that were more sophisticated than old card punch products. (In the 1940s many managers both within IBM and other firms which built electronic devices or office machines had little confidence that a commercial market existed for electronic computing, let alone computers.) Second, the volume sold made it possible for IBM to hire additional electrical engineers in anticipation of building other computational equipment. That hiring resulted in bringing a generation of technical personnel into the company in the late 1940s and early 1950s who would subsequently develop IBM's computers in the 1950s and 1960s. Finally, the CPC taught IBM and its customers how to use electronic computing as the data processing industry and its technology evolved into the stored-program concept of computers. The advantages of automatic digital calculation became increasingly obvious, so much so that when full-blown electronic digital computers* appeared in the early 1950s, there was a ready market made up of sophisticated users. Digital computers predominated at the expense of analog computers,* giving the data processing industry a distinct signature which is still evident today.

For further information, see: Charles J. Bashe et al., *IBM's Early Computers* (Cambridge, Mass.: MIT Press, 1986); Engineering Research Associates, Inc., *High-Speed Computing Devices* (New York: McGraw-Hill Book Co., 1950, reprinted by Tomash Publishers, Los Angeles, 1983); René Moreau, *The Computer Comes of Age: The People, the Hardware, and the Software* (Cambridge, Mass.: MIT Press, 1984).

IBM SYSTEM 360 (S/360). No set of products had as dramatic an impact on computer technology or on the growth of the data processing industry as this family of processors. In April 1964 International Business Machines Corporation (IBM)† announced the S/360, a series of various-sized computers which, for the first time, were all compatible with each other, that is, programs written for one computer could run on a different-size member of the same family. Peripherals on one machine could be used on another computer as well. The success of the S/360 was spectacular. The sale of the S/360s placed IBM in the forefront of data processing vendors throughout the 1960s and nearly doubled the size of the company. Furthermore, it made IBM one of the most admired companies in American industry while turning its stock in the 1960s into one of the most attractive "go go" issues on Wall Street.

IBM, like many other firms marketing computer technology in the early 1960s, had a variety of different types of processors, none of which were of the same technology. In 1961 the company had fifteen different computer offerings, all of second-generation vintage and categorized by the company into seven separate product lines. In the slang of the industry, they did not "talk to each other"— they were incompatible. Because peripheral equipment was unique to each processor type, there was little compatibility across the entire product set. Thus, for example, a user could not take a printer from one type of computer and attach it to a different processor. The cost of moving to a new computer was not limited to a new processor; it usually involved the whole system, often raising the cost just for hardware by an additional 200 percent. The costs of hardware migration, although high, were less of a disincentive than the total expense of conversion which always required rewriting all the application software* and replacing whatever operating system was available. Such conversion costs usually exceeded the price of the new hardware and entailed enormous frustrations and many long weekends spent testing new software. The complexities of conversion or migration slowed the growth of data processing, particularly of systems that were compatible and could share data. From a vendor's point of view, the consequence was obvious: the sale of new products with larger capacities frequently proved difficult, even when customers wanted more computing power. The technical and financial obstacles required a breakthrough.

Despite its successes with its current product line at the start of the 1960s,

IBM recognized the need to develop new technologies if it was to remain in the computer market. The number of vendors supplying computer-related products was already growing, and the threat of rapid technological advances indicated that the survivors in this industry had to make dramatic and rapid product introductions. Furthermore, IBM's management had concluded that survival also meant all-out commitment to the new industry, a lesson lost on many other companies that would not survive in the 1960s and 1970s. In order to determine the best way to move forward with product development, IBM vice-president, T. Vincent Learson,** appointed one of the most important task forces in the history of data processing in the fall of 1961. The task force, called the SPREAD* Committee, was made up of thirteen people, many of whom went on to make significant contributions to the company, such as John W. Haanstra, chairman of the task force, and Bob O. Evans,** its vice-chairman. Illustrious names in the history of product development, such as Frederick P. Brooks, Jr.,** were also members. The group was charged with defining IBM's product strategies for the 1960s. The task force made its recommendations in December 1961.

First, these managers suggested that IBM no longer produce a proliferation of computer products, but rather a family of compatible computers based on similar technologies. Second, they recommended that all peripherals for one computer be usable on any other member of the family of processors. Furthermore, they stressed that all the processors should use the same or compatible operating systems. They also made a series of suggestions regarding what features and functions should be incorporated in the next generation of computers. The corporation adopted these recommendations and in 1962 started work on what ultimately would become the S/360.

The primary reason for adopting this radical new course of action became evident from the committee's analysis of the current products. Almost all of these machines were destined to have short lives as marketable items owing to the pressure of other manufacturers introducing newer, less expensive, more reliable technologies. Such actions would motivate IBM's customers to return their leased equipment to the company with potential disaster for sales in the immediate three to five years. Thus, for example, the popular 1401 could be expected to end its marketable life in 1965. The SPREAD Committee said that new products had to be announced during the first quarter of 1964 if IBM wanted to maintain its position in the marketplace and avoid potential financial disaster.

The molding of the new product set therefore began with a concern to replace current products in a timely fashion, thereby avoiding inroads into the company's accounts by other firms. In addition, the company saw an opportunity to expand the overall demand for data processing products if systems could be made easier to install and use, and if conversions to larger and newer systems were facilitated. According to the company's management, failure to introduce a new set of products would mean ruin otherwise. So strongly did management believe in the need for new technologies that IBM made a total commitment of its financial and managerial resources to the project proposed. This is what made the company the leading vendor of data processing products and services by the late 1960s and enabled the company to introduce the S/370* and maintain a strong market position throughout the 1970s.

A number of specific recommendations and strategies emerged from design groups: merging of scientific and commercial processing within the same system, better price/performance over second-generation technologies, compatibility up and down the entire family of new computers, a complete set of compatible software offerings, greater compatibility of peripheral equipment than ever before, and modularity of components and equipment making up a system. The name System 360 was chosen to symbolize a total family that encompassed all of a user's computing requirements, or the full 360 degrees of a circle.

The design requirements were met. In the area of price/performance, success outstripped all early projections and targets. For example, the Model 30 S/360 was six times faster than the 1401* for less than a one-third increase in rental cost. The price/performance of the S/360 family was significantly better than with any second-generation technology when compared on a device-by-device basis. No competitive offering at the time of its announcement in April 1964 could match it.

Part of the company's ability to offer radically better price/performance derived from new technologies. The company committed itself to use solid logic technology (SLT) which provided far superior circuitry over earlier systems. IBM management required SLT to be manufactured at half the cost of the then current technology (called Standard Modular System or simply SMS) and to be four times its speed. Such third-generation technology had to be delivered within eighteen months for use in the new family of computers. These goals were met, thereby incorporating more efficient technology in new computers while driving IBM's manufacturing costs down per unit. The introduction of new components and tough cost targets forced IBM to modernize its plants and invest in automation—a process repeated again in the late 1970s and early 1980s with new technologies evident in the 308X computers. The cost advantages to IBM during the middle to late 1960s compelled almost all other competitors to manufacture their equipment outside of the United States to reduce labor-intensive costs or to automate production more than ever before.

When IBM introduced SLT in its computer line in 1964, no other vendor had products based on such sophisticated technology. The decision to base the S/360 on this new technology was perhaps the riskiest ever taken by the company but it paid off. At the time the SPREAD Committee was doing its work, IBM scientists and engineers were dabbling in research on SLT. Yet up to then no serious attempt had been made to apply such technology to products. Also untested were the changes in manufacturing processes that would be required to meet the cost objectives assigned to the various plants. The process of managing the development and then the manufacture of so much new technology, computers, peripherals, new operating systems,* and additional software, not to mention the training of field engineers to install and maintain such equipment and salespersons to sell it, was new to the company and the data processing industry. Although IBM on the one hand and the armed services of the United States on the other had worked on complex systems during the 1950s, the System 360 project dwarfed anything that had been seen by any American company up to that time, even surpassing Ford's decision to make automobiles via the assembly line method earlier in the century.

Less risky, but very important for customers, was IBM's decision to merge scientific and commercial computing within the same processor. Until the early 1960s computers were designed for use primarily in scientific applications or almost exclusively for commercial uses. The emphasis on peculiar design considerations for one or the other was important at a time when the characteristics of a machine's elements were usually determined by its use. Thus, for example, in scientific computing, the need was generally for more computer power than memory,* whereas in commercial applications the opposite was frequently the case. The ability to do both required more memory and computing power within a single system, and that would not come as a practical product to sell until costs could drop and technology could improve sufficiently to justify having both capabilities within a single system.

By the late 1950s users were paying less attention to the distinction between the two sets of applications and were designing them with characteristics of both, forcing the demand for more computing power at reduced costs on all vendors. From IBM's perspective, producing a computer that could be sold to either application user would provide economies of scale, thereby contributing to a downward spiral in the costs of marketing and manufacturing of such systems. Armed with the requirement that the S/360 provide "a full circle of" applications caused the designers to come up with products that contributed significantly to customer acceptance of the new computers.

Compatibility represented a characteristic of the new system with important benefits. Until the S/360, no vendor had offered compatibility up and down a product line. Compatibility of computers alone would have marked the S/360 family as a new generation and a profound improvement over the past. Customers could now protect their investment in earlier programming—an investment that was frequently far larger than any placed in equipment—as they moved into larger processors while allowing them to phase in the acquisition of new applications. The costs of training data processing personnel could be slowed, while the pool of trained people who could be hired would expand. Applications could be moved from one computer to another with far less effort than in the past and, therefore, more quickly at less cost. In retrospect, compatibility motivated many computer buyers to migrate to the S/360 family, leaving behind many older IBM and other vendors' equipment while increasing the overall population of computers in general.

In addition to the obvious benefit from IBM's sales of many new computers were cost reductions. Training expenses for salespersons and systems engineers decreased because they no longer had to be trained on multiple product lines, only one. IBM manufactured fewer operating systems that could be used in more computers. The company streamlined manufacturing since many components could be used interchangeably in various models of the S/360. Similar benefits were evident in peripheral equipment of various models. The exact same model of a particular piece of peripheral equipment could operate attached to various sizes of S/360 computers, such as printers, card input/output gear, and tape and disk drives.* Compatibility therefore dramatically improved price/performance in computing throughout the 1960s.

The greatest proof of the wisdom of the S/360's design characteristics could be seen in its acceptance and survival. Laid down in 1961/1962, the basic precepts of the S/360 remained in IBM's processors even in the mid-1980s, despite the introduction of newer technologies, more reliable equipment and software, and less expensive machines. Programs that ran in a S/360 could, with fewer modifications than it took to go to this first family of computers, operate on the S/370s, 308X, and 4300s as late as the mid-1980s.

For those who did not want to immediately convert their application software to S/360 architecture, IBM provided emulation software which allowed non-S/360 code to function in the new machines. Peripherals were replaced with new announcements across the board, beginning in April 1964 and continuing to the end of that product line in 1970, only to be carried on in an uninterrupted line through the S/370 to the present. At announcement time in 1964, more devices were made public than had ever been done at a single time by any vendor in the history of the industry. Tape, disk, printers, card I/O, operating systems, communications gear, and multiple models of computers were all brought out at the same time.

Buried in these initial announcements was the 1403 N1 printer which, even in the late 1970s, was still considered the most popular system printer in the industry. The 1403 had first appeared with the 1401 computer and had been widely accepted. The 1403 N1, however, operated at nearly twice the speed of earlier models, producing over 1,100 lines of high-quality print per minute. No vendor had either that speed or quality for several years. The significance of such a device could hardly be lost on users who printed paychecks and invoices, managed normal accounting applications, and practiced inventory control.

Other important equipment announced with the S/360 included disk drives which were the best then available in the industry. The 2311 disk drive was new and unique technology, providing IBM with significant marketing exclusives for several years. That circumstance would be repeated again in the early 1980s with IBM's 3380 disk drives which were technological advances over existing 3350 disk drives. The 2311 and, later, the 2314 disk drives were important in the 1960s because IBM assumed, correctly as it turned out, that applications developed in the 1960s would favor online systems as opposed to the more traditional batch applications. Online systems, usually involving the use of cathode ray terminals (CRTs) for real-time transactions, required relatively cheap, reliable, and direct access to information which was stored on disk drives and not on the older tape-oriented data storage systems of the 1940s and 1950s.

IBM's commitment to computers was total and it paid off. Almost the entire data processing staff was involved in the development of this product line. Plants were changed to produce new products, while laboratories devoted all of their energies to the project. Every IBM lab in the world worked on the S/360—from Endicott, New York, where the 360/Model 30 was developed to Mainz, Germany, where various models were also manufactured. SLT was developed in Poughkeepsie and East Fishkill, New York, and manufactured there and in France and Germany. Between 1961 and 1964, research and development expenses rose from $175 million to $275 million per year, and these expenditures continued to rise sharply throughout the 1960s.

For months before the announcement, activity at IBM was feverish while rumors pervaded the industry. Production models of a wide variety of computers and peripherals were being readied, and on March 18, 1964, the company selected April 7 as the day of announcement. On that day IBM unveiled the System 360 throughout the United States. The reception of this product line exceeded everyone's expectations. By October 1966 orders represented a threefold increase over worldwide revenues of all products for IBM prior to announcement. IBM had projected that in 1965 customers would order 589 systems when in fact it was 668. For the following year the forecast was for 2,897 processors when 3,232 were actually ordered for a total gross backlog of 9,013 systems. IBM scrambled to expand plant capacities all over the world to meet the large demand for its systems. Between 1964 and the end of 1967, the company's workforce grew by 50 percent to over 70,000 employees. At peak time in 1966 and 1967, it was not uncommon for field engineering personnel charged with responsibility for installing these systems in customers' data centers to work seventy-hour weeks. Despite this success, there were some initial problems with the operating system (OS/360) which were fixed, with some 1,000 people assigned to this product.

The S/360 represented the single most important event in the history of the company, transforming it dramatically. Some have called the product the "greatest in the history of American industry." IBM's revenues suggest its impact. In 1965, prior to mass shipments of the S/360, IBM's revenues stood at $3.5 billion. By the end of 1970, they totaled $7.5 billion. Put another way, prior to the S/360, IBM had sold a total of 11,000 computers in the United States. At the time of the S/370's announcement in 1970, the figure had grown to 35,000. IBM was in the computer business to stay.

Throughout the 1960s, IBM announced additional models of this S/360 computers, peripherals, and new releases of its software. Competitors, initially stunned, soon reacted. They were forced to quickly reduce prices on their existing product lines to remain competitive in cost/performance. But it was done with the understanding that profit margins would shrink or disappear. They also adopted strategies calling for compatible product lines, and in some cases,

Table 43
S/360: Dates of Announcement and Shipment

Model	Announced	First Shipped
22	April 7, 1971	June 1971
25	January 3, 1968	October 1968
30	April 7, 1964	June 1965
40	April 7, 1964	April 1965
44	August 16, 1965	September 1966
50	April 7, 1964	August 1965
60	April 7, 1964	Not shipped[a]
62	April 7, 1964	Not shipped[a]
65	April 22, 1965	November 1965
67	August 16, 1965	May 1966
70	April 7, 1964	Not shipped[b]
75	April 22, 1965	January 1966
85	January 30, 1968	December 1969
91	November 17, 1964	October 1967
92	August 17, 1964	Not shipped[c]
95	d	February 1968
195	Augst 20, 1969	March 1971

[a]Replaced with Model 65.
[b]Replaced with Model 75.
[c]Renamed Model 91.
[d]Available only on a U.S. government contract.

SOURCE: A. Padegs, "System/360 and Beyond," *IBM Journal of Research and Development* 25, no. 5 (September 1981): 387, Table 2. Copyright 1981 by International Business Machines Corporation; reprinted with permission.

decided to produce equipment and software that could run with S/360 systems. These strategies were implemented to various degrees by Radio Corporation of America (RCA),† General Electric (GE),† Control Data Corporation (CDC),† Sperry Rand†, Burroughs†, Honeywell,† and others. These responses caused IBM to make more S/360 product announcements (see Tables 43 and 44). It brought out the 2314 disk drive on April 22, 1965, and the most popular of all the S/360 computers, the Model 20 on November 18, 1964. By 1970 over 7,400 had been shipped and installed just in the United States.

The S/360 thus contributed to the overall dramatic growth in the use of data processing. Just using the number of computers installed indicates the nature of this enormous growth. In a U.S. government report of 1969 dealing with installations of computers in the United States, the comptroller general noted that some 400 computers were installed in the late 1940s, 6,000 in 1960, and over 67,000 in 1968. Nearly 15 percent of the 1968 install base consisted of S/360s—achieved in less than four and a half years from initial introduction. The hardware market that year approximated $7.2 billion for all vendors and

Table 44
S/360 Model Characteristics

Central Processing Unit (CPU)		Control Storage				Processor Storage		
Model	Cycle Time (ns)	Size (K words)	Word Size (bits)	Type	Cycle Time (ns)	Size (K words)	Bus Width (bytes)	Cycle Time[a] (ns)
22	750	4	50+5	RO	750	24–32	1	1,500
25	900	8	16+2	RW	900	16–48	2	900
30	750	4	50+2	RO	750	16–64	1	1,500
40	625	4	52+2	RO	625	32–256	2	2,500
44	250	None				32–256	4	1,000
50	500	2.75	85+3[b]	RO	500	128–512	4	2,000
65	200	2.75	87+4[c]	RO	200	1,024–8,192	4X(1–2)	8,000
						256–1,024	8X2	750
67	200	2.75	87+4[c]	RO	200	1,024–8,192	8X(1–2)	8,000
						256–1,024	8X2	750
75	195	None				256–1,024	8X(2–4)	750
						1,024–8,192	8X(1–2)	8,000
85	80	2	105+3[d]	RO	80	512–4,096	16X(2–4)	960
		0.5	105+3[d]	RW	80			
91	60	None				2,048–6,144	8X16	780
95	60	None				1,024	8X16	180
195	54	None				1,024–4,096	8X(8–16)	756

[a]Certain registers and paths were 17 or 18 bits wide where a main storage address was processed in one cycle. RO: Read-only; WO: Write-only.
[b]Extended to 90+3 for the IBM 1410 emulator, or 92+3 for the IBM 7070 emulator.
[c]Extended to 94+4 when any emulator was installed.
[d]Extended to 122+4 for part of control storage when any emulator was installed.

SOURCE: A. Padegs, "System/360 and Beyond," *IBM Journal of Research and Development* 25, no. 5 (September 1981): 388–389, Table 1. Copyright 1981 by International Business Machines Corporation; reprinted with permission.

was then on a growth curve of 15 to 20 percent compounded per year for the next five years. If one measured the total revenues of all computer companies in the 1960s, growth of the industry approximated 33.5 percent just for the first three-fourths of the decade.

Such growth was characterized and supported by new technologies, a sharp increase in the number of companies and individuals using computers, the dramatic expansion and development of new and additional applications (particularly in business), the increased reliance on online over batch transactions but with dramatic growth in the amount of software for batch installations as well, and the ease of use of computerized technologies when compared to those of the 1950s. All of these phenomena continued unabated into the 1980s.

Because the economics of the S/360 encouraged greater use of computers, while forcing all of IBM's competitors through another round of price cutting and the introduction of new, third-generation technology by the major vendors, price/performance improved sharply for all. To illustrate the degree of this improvement, one can begin by measuring IBM's 650 computer of 1953 to the S/360. The 650 processed 700 instructions per second. The 360/30 of 1964 (a smaller S/360) performed 30,000 transactions in the same second. Memory from the first machine to the second grew in size 6.5-fold along with the additional forty-fold increase in processing speed. Yet both machines cost approximately the same to rent. Dramatic price/performance improvements were also evident with virtually the entire cluster of peripheral equipment.

For further information, see: J. D. Aron et al., "Discussion of the SPREAD Report, June 23, 1982," *Annals of the History of Computing* 5, no. 1 (January 1983): 27–44; Bob O. Evans, "System/360: A Retrospective View," *Annals of the History of Computing* 8, no. 2 (April 1986): 155–179; Franklin M. Fisher et al., *IBM and the U.S. Data Processing Industry: An Economic History* (New York: Praeger Publishers, 1983); Katharine D. Fishman, *The Computer Establishment* (New York: Harper and Row, 1981); J. W. Haanstra et al., "Final Report of SPREAD Task Group, December 28, 1961," *Annals of the History of Computing* 5, no. 1 (January 1983): 6–26; A. Padegs, "System/360 and Beyond," *IBM Journal of Research and Development* 25, no. 5 (September 1981): 377–390; Emerson W. Pugh, *Memories That Shaped an Industry: Decisions Leading to IBM System/360* (Cambridge, Mass.: MIT Press, 1984); William Rodgers, *Think: A Biography of the Watsons and IBM* (New York: Stein and Day, 1969); Robert Sobel, *IBM: Colossus in Transition* (New York: Times Books, 1981).

IBM SYSTEM 370 (S/370). This family of computers, introduced and marketed by International Business Machines Corporation (IBM)† during the 1970s, was the company's major line of large processors and built on the phenomenal success of the System 360* machines of the 1960s. The S/370 family continued to support IBM's growth throughout the decade, reinforcing its continued position as one of the largest corporations in the world. The S/370 also reflected many of the technological changes evident across the entire data processing industry in the 1970s.

Although the S/360 family had nearly doubled IBM's revenues between 1965 and the end of the decade, and had insured that the company would have an important set of technological competitive edges for several years, that exclusive rapidly waned in the late 1960s and early 1970s. The company was motivated once again, as it had before deciding to develop the S/360s, to make a round of new product announcements to avoid product obsolescence and competitive inroads into its install base of mostly leased S/360s. Competitive pressures were evident in the form of new technological introductions in computers from various other vendors. Competition appeared through the remarketing of S/360s by leasing companies with terms and conditions different from IBM's. They did the same with S/370s as well. Plug-compatible offerings (machines that were duplicates of existing IBM products, especially with peripherals such as tape* and disk drives*) became common during the 1970s and represented about 20 percent of all IBM losses to competition. These various forms of competition therefore contributed to IBM's flat earnings for the four years after 1968, despite the shipment of large numbers of processors in an industry that continued to expand.

Efforts to develop follow-on products to the S/360 had begun by the mid-1960s and, once new products had been developed to replace the 2314 disk drive, the initial S/370 processors were announced on June 30, 1970. Additional models, introduced over the next several years, provided a range of machines to replace the entire product set from the S/360 with attractive price/performance relative to competitive offerings (Table 45). These announcements also evolved as machines relied less on look-alike S/360 technology and compatibility and instead incorporated new features that were not compatible with the old line.

The first two machines announced were the Models 155 and 165 which were compatible with the S/360. Yet they were faster and had larger memories, which meant more information or larger programs could be run. The 155 and 165 ran up to four times the speed of an S/360 Model 50 and could attach up to 400 percent more memory.* Subsequent products increasingly were not as compatible either with older systems or peripherals. Operating systems* evolved away from the S/360 world to offer new functions, additional system automation, and the ability to support new peripherals. The major leaps forward in products included the intermediate S/135 and 145, the S/158 which incorporated virtual storage, the small S/115 and 125 with their integrated adapters, and finally midlife kickers in the form of new models of the S/135 and 145 followed by the S/138 and S/148. Variations of some other machines also appeared.

Equally significant was the introduction of a new disk drive for use with the S/370, code-named Merlin and officially dubbed the 3330 and its later fixed-head version, the 3350 using Winchester technology. The 3330 could move data two and a half times faster than the 2314 and housed three times as much information on a disk pack as its predecessor. It became an extremely reliable piece of equipment which established the standard for performance for all disk drives throughout the 1970s. Following a now established tradition in the data

Table 45
S/370: Dates of Announcement and Shipment

Model	Announced	First Shipped
115	March 13, 1973	March 1974
115-2	November 10, 1975	April 1976
125	October 4, 1972	April 1973
125-2	November 10, 1975	February 1976
135	March 8, 1971	April 1972
135-3	June 30, 1976	February 1977
138	June 30, 1976	November 1976
145	September 23, 1970	June 1971
145-3	June 30, 1976	May 1977
148	June 30, 1976	January 1977
155	June 30, 1970	January 1971
158	August 2, 1972	April 1973
158-3	March 25, 1975	September 1976
165	June 30, 1970	April 1971
168	August 2, 1972	May 1973
168-3	March 25, 1975	June 1976
195	June 24, 1971	August 1973

SOURCE: A. Padegs, "System/360 and Beyond," *IBM Journal of Research and Development* 25, no. 5 (September 1981): 387, Table 20. Copyright 1981 by International Business Machines Corporation; reprinted with permission.

processing industry, the cost of storing data on a 3330 dropped when compared to 2314. For one dollar of rent, a user of a 2314 could store 38,200 bytes of data while, for the same amount of money, one could house 145,600 bytes on a 3330. With the 3350, the amount of data per dollar rose to 470,500 bytes. In addition, the S/370s were enhanced to provide more control over disks. Thus, computers could now conduct some error correction analysis on disk drives and issue commands through system programs. These improvements in function and cost encouraged the continued rapid growth in the use of online systems and the storage of machine-readable data in computers. By the late 1970s, it was not unusual for data centers to increase their amount of online storage by 40 to 50 percent per year.

A new printer to replace the 1403 N1, called the 3211, was introduced. It operated at 2,000 lines per minute versus 1,100 lines per minute for its predecessor. The new device was generally used more frequently with larger processors of the S/370 family, while the 1403 remained very popular with smaller computers despite the introduction of other system printers by IBM (such as the 3203) in the mid- to late-1970s.

Faster tape drives came in the form of a November 1970 product announcement: the 3420 tape subsystem, originally code-named Aspen. Faster, more reliable, and less expensive than its predecessor, the initial models, called the 3, 5, and 7, and later augmented with others, became the tape workhorses for most IBM processor installations. They remained the most popular tape drive in use for over fifteen years. The technology in the 3420 was also replicated by plug-compatible manufacturers who also did well with their versions of these machines. Not until 1985 did IBM or any other vendor begin shipping a newer technology to replace the 3420s. When it came, the new tape drives (IBM 3480s) took the form of smaller devices, using tape cartridges instead of reels of tape, cost less, and had greater capacity to store and move data about.

Other elements necessary for a full S/370 processor configuration also appeared. Cathode ray tubes (CRTs), called the 3270s, appeared, while a minicomputer to handle telecommunications for the S/370 family (3705) represented a significant enhancement to the family of hardware. Old operating systems for the S/360, such as the OS/360, were replaced with a 370 OS, called the OS/VS1, a smaller operating system for the junior-sized members of the S/370 family called DOS (Disk Operating System), and later DOS/VS when virtual storage was announced. Late in the 1970s VM and MVS appeared. Software* enhancements therefore made it possible for S/370 users to operate more and larger programs, using greater amounts of data than was possible with the S/360.

Technological innovations took place in the computers themselves over the decade. Fewer technological introductions were made on the day the S/370 was announced than was the case with the S/360. IBM elected to bring out new technologies with individual product announcements, a strategy that has continued into the late 1980s. Thus, one has to look at almost every computer announcement to identify the evolution of technology, a process that often reveals significant enhancements to computers. For example, beginning with the 158 and 168, computers began to rely on advanced semiconductor memory. Called field effect transistor, or simply FET, this technology was denser and easier to manufacture than its predecessor. This meant computers could have much larger memories at even greater price/performance ratios than in the past. With these two processors, internal speeds also increased. When the 155 was compared to the 158, the increase in speed equated to approximately 20 percent more per dollar of cost. Better memory alone created an enormous increase in the requirement for FET, while the improvements in speed per dollar spurred demand as well. Smaller memories meant that an entire computer system of some size could fit into one data center (also called a machine room) with the expectation that the amount of electricity needed to operate the machine and decreasing amounts of air conditioning would provide substantial cost avoidances as systems grew.

Virtual memory was also introduced in the S/370, beginning with the machines announced in 1972. For example, operating systems went from OS and DOS to OS/VS and DOS/VS. Virtual memory combined hardware and software, effectively allowing a computer user to ''pretend'' that main memory in the processor was much larger than it actually was. This became possible because with virtual memory (also known as virtual storage), the computer would bring into main memory those portions of software housed in a disk drive that were needed for processing in the next immediate period of time. These were brought into main memory in blocks usually of about 4,000 bytes each, with traffic controlled by the operating system (also known as the SCP—System Control Program). Before virtual storage, an entire program had to be loaded into the computer's main memory before it could be run. Thus, no program could be larger than the space available for housing it within a computer. Now only a portion of it had to be resident in the machine and only when it was to be used. This change allowed users to write much larger programs for more complex and sophisticated applications. Hence, a program that required up to 16 million bytes of memory could run in a machine that might have had only 256,000 or fewer bytes of real memory. For a listing of model characteristics, see Table 46.

As Table 44 indicates, faster speeds, more function, and more capabilities were added throughout the early to mid-1970s. Channel attachment of peripheral equipment was standardized in these years by using outboard control units. Thus, disk drives would be attached to a disk control unit (3830) and 3420 tape drives to their control unit (3803), which in turn were hooked to a computer via a standard channel that either could access. Multiple control units could be attached to a particular channel into the computer. Later in the decade, control unit attachment capabilities were built into the computers themselves. For example, on the Model 115 or 125, some peripherals could be attached directly to the computer without a control unit, such as card input/output equipment. This function not only reduced the cost of a system, but also knocked out the possible use of competitive or second-hand IBM control units.

Another innovation involved the use of a cache subsystem. The objective of such a feature is to allow storage of information somewhere between the main memory of a computer and the outboard storage, such as disk drives. That allows data to be brought in faster. The initial offering of a cache subsystem on an IBM computer came with the S/360 Model 85, making it the first such feature made available on a commercial computer by any vendor (Table 47). It typically was housed between the central processing unit and main storage in all subsequent, large S/370 computers. Because systems were using larger amounts of data by the early 1970s, such a subsystem became increasingly important on large computers in order to speed up the movement of data back and forth from one part of a computer to another. In the late 1980s cache systems have become integral parts of any large computer, and their function also exists in sophisticated disk control units as well. These systems therefore reduced the dependence of a computer on the access time to get data, a function that was often slower than the system's capability to process information.

The S/370 family did not contribute proportionately as much to IBM's growth in the 1970s as had the S/360 the decade before when measured as a percentage

Table 46
S/370 Model Characteristics

Model	Central Processing Unit (CPU) Cycle Time (ns)	Control Storage Size (K words)	Control Storage Word Size (bits)	Control Storage Type	Control Storage Cycle Time (ns)	Processor Storage Size (K words)	Processor Storage Bus Width (bytes)	Processor Storage Cycle Time (ns)
115	480	20-28	20+3	RW	480	64-192	2	480
115-2	480	12-20[a]	19+2	RW	480	64-384	2	480
125	480	12-20	19+2	RW	480	96-256	2	480
125-2	320-480[b]	16-24	19+2	RW	320	96-512	2	480
135	275-1,485[b]	12-24	16+2	RW	275	96-512	2	990[c] R / 935[c] W
135-3	275-1,485[b]	64	16+2	RW	275	256-512	2	990[c] R / 935[c] W
138	275-1,430[b]	64	16+2	RW	275	512-1,024	2	935[c]
145	203-315[b]	8-16[d]	32+4	RW	203	160-2,048	8	540 R / 608 W
145-3	180-270[b]	32	32+4	RW	180	192-1,984	8	405 R / 540 W
148	180-270[b]	32	32+4	RW	180	1,024-2,048	8	405 R / 540 W
155	115	6	69+3	RO	115	256-2,048	8	2,070[e]
155-II	115	8	69+3	RO	115	256-2,048	8	2,070[e]
158	115	8	69+3	RW	115	512-6,144	8	1,035 R / 920 W
158-3	115	8	69+3	RW	115	512-6,144	8	920
165	80	2	105+3[f]	RO	80	512-3,072	8X4	2,000[e]
		2	105+3[f]	RW	80			
165-II	80	4	105+3[f]	RO	80	512-3,072	8X4	2,000[e]

Table 46 *(cont.)*

Central Processing Unit (CPU)		Control Storage				Processor Storage		
Model	Cycle Time (ns)	Size (K words)	Word Size (bits)	Type	Cycle Time (ns)	Size (K words)	Bus Width (bytes)	Cycle Time (ns)
168	80	2–3.5	105+3[f]	RO	80	1,024–8,192	8X4	320
		0.5–1	105+3[f]	RW	80			
168–3	80	2–3.5	105+3[f]	RO	80	1,024–8,192	8X4	320
		1–2	105+3[f]	RW	80			
195[g]	54	None				1,024–4,096	8X16	756

[a]The 115–2 contained a separate input/output (I/O) processing unit for some functions that were executed on the CPU in a 115; hence, the smaller CPU control-storage capacity.

[b]Variable, depended on type of operation performed.

[c]Four bytes could be accessed and transferred in this time. R: Access for reading; W: Access for writing; RO: Read-only; WO: Write-only.

[d]Part of this capacity was physically in the processor storage and thus had to be subtracted from the available processor-storage capacity.

[e]The model used magnetic-core technology.

[f]Extended to 122 + 4 for part of control storage when any emulator was installed.

[g]The 195 had certain facilities (e.g., time-of-day clock, control registers, MOVE ALONG) not available on the S/360 Model 195.

SOURCE: A. Padegs, ''System/360 and Beyond,'' *IBM Journal of Research and Development* 25, no. 5 (September 1981): 388–389, Table 1. Copyright 1981 by International Business Machines Corporation; reprinted with permission.

Table 47
Cache Characteristics in S/370 Computers[a]

Model	Size (K bytes)	Line width (bytes)	Cycle Time (ns)	Type	Associativity
155	8	16	115–230	T	2
155-II	8	16	115–230	T	2
158	8	16	115–230	T	2
158-3	16	16X2	115–230	T	4
165	8–16	8X4	80–160	T	4
165-II	8–16	8X4	80–160	T	4
168	8–16	8X4	80–160	T	4–8[b]
168-3	32	8X4	80–160	T	8
195	32	8X8	54–162	T	4

[a]Cache appeared in two S/360 computers first. The S/360 Model 85 had 16–32K byte size and a line width of 8X8, with a cycle time of 8–160. It, too, was a T type and had an associativity of 16. The Model 195 was 32K in size with a line width of 8X8. Its cycle time was 54–162. It had a T type and an associativity of 4. T stood for store-through; on storing, the value was placed in main storage. No values were placed in cache unless a line had been assigned to the main-storage location. K means a number $2^{10} = 1,024$.
[b]Depended on cache size used.

SOURCE: For footnote and data, Table 1 of A. Padegs, "System/360 and Beyond," *IBM Journal of Research and Development* 25, no. 5 (September 1981): 388–389. Copyright 1981 by International Business Machines Corporation; reprinted with permission.

of revenue contributions. In the early 1970s growth in IBM's revenues was flat or slow. In 1976, according to *Datamation*, a leading data processing industry publication, IBM garnered some 44 percent of the industry's revenues with all products, not just the S/370s. That same publication later calculated that in 1979, the figure had gone down to 34 percent. The use of plug-compatible equipment by competitors which were designed to look and function like specific IBM products but at lower costs ate into the company's sales while other firms introduced various new products. Second-hand processors and peripherals also took a share of the market from IBM's new processors. New in the industry was the use of minicomputers to offload individual applications which could be "distributed" to the place where they were most needed rather than be housed in the main data center where the majority of the S/370s were installed.

Despite these competitive pressures and the innovative use of computing technologies, the company's sale of S/370s contributed to the general growth in IBM's revenues throughout the decade, especially after 1975. Revenues increased from over $14 billion in 1975 to over $22 billion in 1979. Furthermore, the S/370 architecture virtually became the industry's standard for processors, operating software, and peripherals.

The actual number of S/370s sold is not yet fully known. However, IBM acknowledged that the actual number of each model exceeded sales of equivalent S/360s by wide margins. Because the data processing industry continued to grow at over 11 percent per year in the 1970s, it is understandable how this family of products could produce more sales than the S/360s and yet account for a smaller portion of IBM's total revenues. That the architecture and technology of the S/370 were relevant in the late 1970s and early 1980s could be demonstrated by the fact that the 4300 products which replaced many models of the S/370s and the 308Xs processors of the early 1980s continued to rely on the fundamental architectural precepts of the S/360 and S/370. At the same time they were incorporating new technologies that made their function easier and broader and their capacity larger and faster, while driving computing costs down in general.

For further information, see: Franklin M. Fisher et al., *Folded, Spindled, and Mutilated: Economic Analysis and U.S. v. IBM* (Cambridge, Mass.: MIT Press, 1983); Franklin M. Fisher et al., *IBM and the U.S. Data Processing Industry: An Economic History* (New York, N.Y.: Praeger Publishers, 1983); A. Padegs, "System/360 and Beyond," *IBM Journal of Research and Development* 25, no. 5 (September 1981): 377–390; Robert Sobel, *IBM: Colossus in Transition* (New York: Times Books, 1981).

ICES. The Integrated Civil Engineering System (ICES) was a programming tool developed in the 1960s by the Department of Civil Engineering at the Massachusetts Institute of Technology (MIT) for use in applications common to civil engineering. It initially ran on an IBM System/360* in 1967. ICES was an instant success and became publicly available in November 1967 with over 300 companies, universities, and government agencies placing orders for it. Thus, ICES quickly became one of the most important software* packages available to civil engineers, maintaining that status throughout the 1970s.

Its collection of functions accounted for its popularity. First, ICES had a series of engineering subsystems that could be used to solve problems in civil engineering. Among these subsystems were structural analysis, design of structures and highways, and soil mechanics. Second, it had special commands to make these subsystems effective. Third, other functions permitted the modification of existing commands and subsystems while offering the ability to develop new languages to operate with existing software within the system. Fourth, ICES allowed a user to integrate the functions of all of the subsystems together; thus, one could get to one subsystem from another. Finally, the software was clean, that is, it was relatively error-free and performed as advertised.

Additional facilities were added throughout the design phase to enhance the language's usefulness for engineers. The most dramatic enhancement was a piece of code called ICETRAN which extended the functions of FORTRAN IV.* Armed with ICETRAN, a programmer could write programs for applications serviced by ICES. ICES became an anthology of functions found in such packages as COGO, STRUDL, ROADS, BRIDGE, TRANSIT, and PROJECT—all software tools to support the work of civil engineers.

ICES was one of many packages that appeared for the first time in the 1960s to help engineers with specific applications. The two other more important tools of the period included Coordinate Geometry (COGO), which was developed around 1960 at MIT to solve problems in plane geometry, and Structural Engineering System Solver (STRESS), developed at MIT in 1962. Its main purpose was to analyze framed structures.

For further information, see: Jean E. Sammet, *Programming Languages: History and Fundamentals* (Englewood Cliffs, N.J.: Prentice-Hall, 1969).

IDS. Also known as the Integrated Data Store, this programming system for business computing was developed at General Electric† and was made available by 1964. It was originally designed as an extension to COBOL,* and its purpose was to take advantage of larger memories on computers by more sophisticated file-handling methods. Thus, one could add fields to a data record, called *chain fields*, that would have the addresses of other data also being managed by IDS. This feature enhanced the file management capabilities of COBOL's Record Description and DATA Division. Additional commands allowed the programmer to move and manipulate information with greater flexibility than in COBOL. IDS was a phenomenon of the 1960s, and subsequent enhancements to COBOL and other programming languages* eliminated the need for IDS.

For further information, see: C. W. Bachman, "Software for Random Access Processing," *Datamation* 11, no. 4 (April 1965): 36–41.

ILLIAC. This series of computers was built at the University of Illinois beginning in the late 1940s and ultimately led to the construction of the ILLIAC IV, the first supercomputer built in the United States. The ILLIAC I was a classic example of a von Neumann* digital vacuum tube computer which was fashionable in the late 1940s. Work on that computer made it possible for the University of Illinois to launch an important program in computer science which produced significant technical results for the data processing industry over the next three decades and a generation of computer scientists.

The University of Illinois worked on its first two computers—ILLIAC I and the ORDVAC*—at the same time. Professors Ralph Meagher and Abraham H. Taub headed the projects through their university's Digital Computer Laboratory. The laboratory eventually emerged as the Department of Computer Science, granting undergraduate and graduate degrees in the technology of data processing. The key founder of the University of Illinois's computer program was the dean of the graduate school following World War II, Louis Ridenour. It was he, for example, who hired the two project leaders for the ORDVAC and ILLIAC in 1948.

Of the various computers built at the University of Illinois, the two most important were the ILLIAC I and the ILLIAC IV. The first reflected trends in the technology of the 1940s, and the second the architecture of large scientific computers in the 1970s. The most significant of the two machines for the early history of digital computing was the ILLIAC I, which was completed in September 1952. To speed the process of design, many of the technical components of this machine were borrowed from one being built by John von Neumann** at the Institute for Advanced Study at Princeton, New Jersey. The fourteen-member task force building the ILLIAC I understood von Neumann's work well and was able to employ his ideas along with currently available technology to design and construct the University of Illinois's own machine.

There were three major components of the system following the von Neumann model: arithmetic unit, input/output, and memory.* The arithmetic logic unit was built with registers and an adder, all controlled asynchronously. It could do addition and subtraction in 72 microseconds (μs), multiplication between 642 and 822 μs, division in 722 μs, and shifts up to sixty-three digital positions in 16 μs per digital position. Input/output followed conventional norms of the day by using five-hole punched paper tape* which would allow the computer to receive data at a rate of some 300 characters per second. The output medium could receive information at the rate of sixty characters per second. Teletypewriters were also used for output. A ten hexadecimal character format was used on tape for every 40-bit computer word. All the devices were controlled primarily by a panel of three telephone switches, a neon indicator, and two push buttons. Memory on this machine relied on Williams tube cathode ray terminal (CRT) technology. Each CRT housed 40-bit words cycling through at 18 μs. This was the exact same memory used, for example, on the ORDVAC and other computers.

The ILLIAC I was packed and shipped to Maryland in late summer, 1952, reassembled, and tested by the end of Labor Day. It was a competent computer, capable of performing 11,000 operations per second for the U.S. government and used primarily for military projects.

The ILLIAC II was completed in 1963; it performed 500,000 operations per second and was based on transistor technology. The ILLIAC III was built in 1966 to conduct automatic scanning of large quantities of data. Then came the ILLIAC IV, designed as a parallel processor capable of having up to sixty-four different computers working on one problem in tandem, resulting in 300 million operations per second.

The ILLIAC IV functions as a large control unit that sends instructions to sixty-four processors, making it an extremely powerful system and, unlike the boxey and wirey ILLIAC I, has miniaturized circuitry and smaller components. It was developed in the 1960s and 1970s and made operational in November 1975. Like its predecessors, it is used primarily for scientific applications in such diverse subjects as fluid analysis, simulation of weather and climates, seismic stress wave propagations, image processing, and astrophysics. Ironically, the motivation leading to the construction of this supercomputer was nearly the same as that which had motivated government agencies and universities to work on computers in the 1920s through the 1950s: ballistics, reactor designs, weather prediction, and large mathematical projects. The only difference was that bigger machines were needed.

The supercomputer was the brainchild of Professor Daniel Slotnick who conceived of the original design and found financial support in the U.S. Defense Advanced Research Projects Agency. Components were brought together at the Burroughs Corporation's* plant at Paoli, Pennsylvania, for assembly before the system was sent to the National Aeronautics and Space Agency's Ames Research Center outside of San Francisco in 1971. Unlike earlier computers that were single processors, and before the concept of the diatic processor became a reality (multiple computers within one and common by the mid-1980s), the fundamental design of ILLIAC IV was different.

One processor instructed the others. Each processor had a logic unit, memory, and, where appropriate, input/output devices attached. With such an arrangement, intermediate results from a calculation could move from processor to another so that multiple calculations could take place, feeding each other in a coordinated manner. When built in the 1970s, main memory on the system was 16 million words—which was large by the standards of the day, but only average by those of large processors in the mid-1980s. The billion bytes of disk memory allowed users to process applications in quantum sizes larger than before, when memories had been 1 megabyte or less. All peripherals were controlled by a Digital Equipment Corporation (DEC)† PDP–10 minicomputer, while a Burroughs Corporation B-6700 compiled all programs into machine-level code for the ILLIAC IV.

After a painful process of development, the device proved to be reliable for specialized scientific computers. Using 1977 as an example, two years after it went into productive use, it was averaging 72.5 hours of work per week, although that figure dropped somewhat the following year.

This computer—which is still in operation today—could be programmed in two languages: CFD, which was like FORTRAN,* and GLYPNIR, which resembled ALGOL.* CFD was named after the Computational Fluid Dynamics Branch of the Ames Research Center. The second language had been in existence since 1969. Because few packages were ever developed to help programmers, working with this computer was difficult. Yet 80 percent of all users were comfortable working with GLYPNIR alone. As a one-of-a-kind processor, it was never fully understood other than by a small community of users and the developers of the system.

Thus, a tradition of computer development at the University of Illinois was established and prospered because of the ILLIAC series which stood near the edge of technology through four different generations of equipment. ILLIAC IV, like ILLIAC I, and more so than ILLIAC II or III, represented exciting contributions to the scientific applications of data processing. Both were built at the time others were constructing similar devices (in the case of ILLIAC IV, the Cray Research Corporation's machines). Yet each supported major scientific projects while training new computer scientists.

For further information, see: on ILLIAC I: James E. Robertson, "The ORDVAC and ILLIAC," in N. Metropolis et al., eds., *A History of Computing in the Twentieth Century* (New York: Academic Press, 1980): 347–364; on ILLIAC IV: R. Michael Hord, *The ILLIAC IV: The First Supercomputer* (Rockville, MD.: Computer Science Press, 1982).

INFORMATION THEORY. See ARTIFICIAL INTELLIGENCE

IPL-V. One of the more important early programming languages* for string and list processing, IPL-V (Information Processing Language) introduced one of the most influential features in programming called list processing. It was the brainchild of Allen Newell** and his colleagues at the RAND Corporation. List processing involved taking a piece of information in a computer's memory and adding to it the address in memory of the next element of data, using a logical sequence. This concept was in contrast to the conventional approach of the late 1950s through which data were simply located in memory* in sequential order. When implemented in a programming language, a list processor could support the creation of applications that needed dynamic allocation of storage, usually in large and complex environments. For example, proofs for theorems in propositional calculus could now be calculated on a computer. Another early use of such languages was in playing chess and, more importantly, in conducting research in artificial intelligence.* IPL-V represented a major contribution to these efforts.

IPL went through various editions, marked IPL-I through V. Its earlier versions were designed as application-oriented assembly languages. IPL-I was never implemented; IPL-II ran on the JOHNNIAC* at RAND in the late 1950s under Newell's supervision and with the help of J. C. Shaw and Herbert Simon; and IPL-III was scarcely used because it required more memory than the technology of the late 1950s permitted. IPL-IV, however, was extensively employed in doing research on artificial intelligence and, although never fully documented, was similar to IPL-V.

IPL-V was first developed at the Carnegie Institute of Technology where Newell did graduate work, beginning in 1957, using an IBM 650* computer. By early 1958 it was operational, and by the end of the summer 1959 a final version of the language ran on an IBM 709. The first full description of IPL-V became available the following year. No further development of the language took place after that because its authors turned their attention to solving the problems that led them to create the language in the first place.

During the 1960s, IPL-V became widely used as a language for research and operated on a variety of computers. It ran on IBM's 650, 704,* 709/7090, and 1620,* UNIVAC's* 1105 and 1107, Control Data Corporation's† 1604 and G-20, Burroughs'† 220, Philco's† 2000, and the AN/FSQ-32 (a military computer). But the features of this language became useful to scientists and computer specialists in the early 1960s.

IPL-V resembled an assembly language more than others available at the time. Theoretically, it was a machine language that just did list processing. It was intended for use only by professional programmers in batch mode, although later a time-sharing version was created and, finally, one for remote job entry. IPL assumed that the machine it ran on was made up of cells that were storage locations in memory, each with a specific address. It had a collection of symbols with which a programmer could create *expressions* (addresses). Finally, IPL also had a repertoire of instructions called *processes*. IPL permitted a programmer to define two types of data addresses: data list structures (they held the data to be processed) and routines (definitions of the processes). IPL was intended to work on lists of information: data lists, data list structures, and list structures.

IPL profoundly influenced the evolution of programming and programming languages in the 1960s. First, it introduced list processing and demonstrated how it could be done in a practical manner. Other list processing languages that emerged as a result of IPL's influence included COMIT,* SNOBOL,* and TRAC.* Second, it offered early examples of pushdown and popup methods of operations that allowed programmers to control the order in which data were presented for processing (like LIFO methods in inventory control). Third, it simplified the use of subroutine hierarchies even beyond those of FORTRAN* (a language contemporaneous with it for scientific computing) and COBOL* (which emerged in the early 1960s for commercial applications). Fourth, it introduced the use of attribute value lists which are now common elements in programming and programming languages. Fifth, its architecture was suitable for research in artificial intelligence and thus contributed enormously to the expansion of such work in the 1960s and 1970s using computers.

Finally, IPL employed a method for dynamically expanding areas of data (size of areas in a computer's memory allocated to a particular program) during the execution of a program. By 1970 various operating systems began to introduce such a feature called virtual storage. With that, not only languages but also operating systems* (also known as system control programs, or SCPs) could judge how much memory in a computer had to be provided to run a program. IPL, and specifically IPL-V, was the first piece of software* to illustrate how to automate programming and the amount of data required in the real memory of a computer. From this small seed would germinate the software to manage databases in the 1970s.

For further information, see: N. Chapin, "An Implementation of IPL-V on a Small Computer," *Proceedings of the ACM 19th National Conference* (1964): D1.2–1—D1.2–6; E. B. Hunt and C. I. Hovland, "Programming a Model of Human Concept Formulation," in E. A. Feigenbaum and J. Feldman, eds., *Computers and Thought* (New York: McGraw-Hill, 1963): 310–325; A. Newell and H. A. Simon, "The Logic Theory Machine—A Complex Information Processing System," *IRE Transactions, Information Theory*, IT–2, no. 3 (September 1956): 61–79; A. Newell and F. M. Tonge, "An Introduction to Information Processing Language-V," *Communications ACM* 3, no. 4 (April 1960): 205–211 and Newell et al., *Information Processing Language-V Manual*, 2d ed. (Englewood Cliffs, N.J.: Prentice-Hall, 1965); Jean E. Sammet, *Programming Languages: History and Fundamentals* (Englewood Cliffs, N.J.: Prentice-Hall, 1969).

IT. Internal Translator (IT) was a programming language created for use on the IBM 650* computer. Its developers included important designers of programming languages* in the 1950s at International Business Machines Corporation (IBM)†: Alan J. Perlis,** J. W. Smith, and H. V. Zoeren. An earlier version of their language was developed at Purdue University for Datatron. This language provided a programming tool for scientists and engineers who had numerical scientific problems. It became available during 1957, and with the demise of the 650, IT also disappeared.

The language was flawed from the beginning, insuring that its usage would be limited. IT conformed to the hardware language of the 650 which was a primitive computer for mathematical and complex scientific computing tasks. Because of IT's scanning technique, the language frequently caused unnatural and incorrect evaluations of expressions described in mathematical terms. Despite these severe handicaps, IT proved that a compiler could be designed to express algebraic notation for use on a small computer such as the 650 which had a maximum memory* of 2,000 words. IT also encouraged the development of successors such as RUNCIBLE, GATE, CORREGATE, and GAT. Each worked more efficiently and on larger computers but drew much of their structure from IT. Work done on the design of compilers at both Case Institute of Technology and the University of Michigan also grew out of their use of IT.

Some functions of IT would become standard in FORTRAN* and other scientific computing idioms. For example, it used floating- and fixed-point variables, subscripts, and mixed mode arithmetic. Subroutines (treated as operands) were a key feature of IT's design. IT was also very difficult to read and work with. Jean Sammet has cited one example of the evaluation of the polynominal

$$y = \sum_{i=o}^{lo} a_r x^i$$

which was expressed by IT taking nonalphanumeric characters and representing them as a single letter:

1	READ	F
2	Y2 Z OJ	F
3	4K 11K 11K M1K 1K	F
4	Y2 Z C11 Y1 X Y2	F
5	H	FF

That was impossible to comprehend. Yet the reference language version of the same statement was intelligible:

1 :	READ
2 :	Y2 ← 0
3 :	4, 11, 11, −1, 1
4 :	Y2 ← C11 + Y1 × Y2
5 :	H

The language, according to the terminology in vogue in the 1980s, was "not user friendly." Despite these handicaps, the language did facilitate scientific computing on a small computer and was popular at Case and Michigan where important research projects on the science of computing were emerging. Eventually, a combination of FORTRAN and IT Appeared called FORTRANSIT* to improve IT's qualities.

For further information, see: A. J. Perlis and J. W. Smith, "A Mathematical Language Compiler," *Automatic Coding, Journal of the Franklin Institute, Monograph No. 3* (Philadelphia: Franklin Institute, April 1957): 87–102; Jean E. Sammet, *Programming Languages: History and Fundamentals* (Englewood Cliffs, N.J.: Prentice-Hall, 1969).

J

JOHNNIAC. Named after John von Neumann,** an important computer designer of the 1940s, this machine was constructed by the RAND Corporation in 1953 and operated until 1966. Often called a Princeton-classed machine, it reflected much of the technology developed after World War II by Neumann at the Institute for Advanced Study at Princeton, New Jersey.

The JOHNNIAC was constructed in response to a need for additional computational power to support projects being done for the U.S. Air Force by the RAND Corporation in 1950. At that time, RAND had six IBM 604* calculators, making it one of the largest single users of computer power in the United States. William Gunning became project leader on the JOHNNIAC. Some of the unique features of this machine included the use of conventionally available card punch equipment for input/output, an octal notation instead of the more common hexadecimal notation, as a means of improving programming, easy access to all tubes by maintenance personnel, and safety features to prevent electrical accidents. The machine used core storage capable of housing 4,096 words. The original memory,* selectron store, was changed to the core store in 1955. The original vacuum tube circuits were partially replaced with transistors in the late 1950s. This machine was highly reliable throughout the 1950s, particularly when compared to other computers. In addition to modifications made to the memory and circuits, peripherals were changed and improved upon. Thus, printers were replaced as better ones appeared on the market.

Software* was constantly under development. In 1958, for example, floating-point subroutines were made available (perhaps even before 1958) along with QUAD, which was an interpretive coding package. The compiler called SMAC was also added to the computer's library of software; both were considered highly reliable pieces of code. The machine was used for a variety of research projects, including the development of chess routines and additional software tools for programmers. The total cost to design and build JOHNNIAC was

$470,000. It was built in less than three years—about the same amount of time taken by other institutions (Princeton, Los Alamos, University of Illinois, Argonne) to make similar machines.

For further information, see: F. J. Gruenberger, "The History of the JOHNNIAC," *Annals of the History of Computing* 1, no. 1 (July 1979): 49–64.

JOSS. Also known as the JOHNNIAC Open Shop System, JOSS was a programming language developed at the RAND Corporation for its own use in the early 1960s. Its software* ran on the installed JOHNNIAC* computer which had been developed at the Institute for Advanced Study at Princeton, New Jersey, by John von Neumann,** the father of the modern computer. Prior to the 1960s, a community of computer scientists at RAND had worked primarily on military systems but had also contributed to the development of data processing technology in general. The creation of JOSS marked the first known development of a language intended for use solely as an interactive tool. Until that time, all programming languages* operated in a batch environment; today they are almost all interactive and online systems. JOSS also encouraged the development of many dialects used in the 1970s and early 1980s which accelerated the trend toward interactive computing.

Discussions at RAND in the late 1950s regarding the future use of their computer focused on its smallness, limited function, and usage at a time when the demand for additional computing capability was growing. At a meeting held on November 8 and 9, 1960, which involved many of the key data processing people at RAND, the father of what would become JOSS, J. C. (Cliff) Shaw,** suggested that a language be developed to allow the computer to support users by way of hard copy terminals and interactively. He had already worked out on paper some of his ideas about how such a system might work and presented these ideas to the people attending the meeting. After it was decided to proceed, two and a half years of work went into the computer and the programming language known as JOSS. Tom Ellis and Mal David made the necessary improvements in the hardware to promote the computer's efficiency, whereas Shaw created almost all the software with some assistance from Mary Lind and Leola Cutler, also employees of RAND.

The result was an interactive, English-like language in syntax, which was extremely easy to use to solve numerical scientific problems. It was designed to allow engineers and scientists to do small numerical calculations quickly, using a tool that required little knowledge of programming. It was easy to write in and relied heavily on a character set of a specially designed typewriter. Later a hard copy terminal was used. Built-in functions could be executed simply by calling them up; no commands took more than one line to describe; all error messages were in English (e.g., Eh? for saying "what did you type?" or "I can't find part 3"). Although the software required that input (typing) be letter perfect, that task was facilitated by conforming to conventional English punctuation, grammar, and usage. The same would later apply to German and French versions of JOSS.

When JOSS I appeared in 1963, it was immediately used at RAND, gaining rapid, wide acceptance. Its few but powerful verbs made it a useful tool, despite even though it ran on a primitive computer. When JOSS II appeared two years later, operating on a PDP-6 as well as on JOHNNIAC, its enhancements took advantage of improved technology. JOSS II, for example, had more data names and a larger library of functions, including the capability for a user to define his or her own functions and store them in a computer. This last feature was particularly convenient for those using a specific function repeatedly, thereby avoiding the need to code that function each time a new program or problem was required. As with JOSS I, JOSS II made it possible for many results to be obtained by using a small list of commands (e.g., using such verbs as FORM, DELETE, GO, CANCEL, TYPE, SET, DO, of which there were thirteen). The users did not have any housekeeping functions associated with how the computer operated; thus, every user felt that the computer was dedicated 100 percent to him or her at that moment, not realizing that others were also communicating back and forth with the computer at the same time.

JOSS I was first run on April 16, 1963, and the final version used at RAND operated on February 11, 1966, a week before the JOHNNIAC was dismantled and given to the Los Angeles County Museum. However, JOSS had been used at RAND between January 8, 1965, when a full and complete version of JOSS was made available, and February 1966. JOSS had an important impact on other scientists and engineers designing programming languages. Hundreds toured RAND's facilities and saw demonstrations of this simple language, and went on to develop their own interactive tools in the next two decades. They witnessed the evolution of the language from one that could service only twelve users (JOSS I) to a final release (of JOSS II) that handled thirty at the same time, many of whom were located at various U.S. Air Force bases and thus communicated interactively over telephone lines with RAND's computer. The language was also used on the PDP-6, 6/10, and 7, on International Business Machines Corporation's (IBM's)† System 360* Model 50, and on the Burroughs† B55000. Some of the languages derived from JOSS included TELCOMP, SON of JOSS, ISIS, CAL, PI L/L, and CITRAN. They were all minor languages of the 1960s and 1970s.

For further information, see: Charles L. Baker, "JOSS—JOHNNIAC Open-Shop System," in Richard L. Wexelblat, ed., *History of Programming Languages* (New York: Academic Press, 1981): 495–508; G. E. Bryan and J. W. Smith, *JOSS Language* (Santa Monica, Calif.: RAND Corporation, August 1967); J. C. Shaw, *JOSS: A Designer's View of an Experimental On-Line Computing System* (Santa Monica, Calif: RAND Corporation, August 1964) and his *JOSS: Experience with an Experimental Computing Service for Users at Remote Typewriter Consoles* (Santa Monica, Calif.: RAND Corporation, May 1965).

JOVIAL. JOVIAL was one of the most widely used programming languages*
for military applications in the United States during the 1960s and 1970s. It was
one of many multipurpose languages first developed in the late 1950s to handle
a large variety of applications that were both massive in size by the standards
of the day and complex, often using large amounts of data. One reason, therefore,
why work began on JOVIAL in 1958/1959 was to develop better methods for
automatic coding.

Work was originally done by the Systems Development Corporation (SDC)
at its facilities in California. The project originally resulted in the design of the
Compiler Language for Information Processing (CLIP), based on ALGOL* 58.
In early 1959 a project similar to CLIP was started by SDC's Strategic Air
Command Control System (SACCS) Division in New Jersey. This East Coast
effort eventually became known as JOVIAL.

Initial design began in February 1959, and one year later an early version of
JOVIAL was running on an IBM 709 using an interpreter. A compiler for
JOVIAL ran on the same machine by Christmas 1960. With these early releases
of JOVIAL, the U.S. Air Force could begin the process of sharing information
by separating data from specific programs and, therefore, allow various pieces
of code to use common data sets called the Communication Pool (COMPOOL).
This process represented an evolution from experiences and functions found with
the earlier SAGE* project.

More specifically, the reasons for having the language were tied to the Air
Force's heavy use of data processing in defense projects throughout the 1950s.
The Air Force had successfully worked on SAGE, its major nationwide defense
system which relied on what was leading edge technology during the 1950s.
The next important project was the SACCS system, which was to be written
and developed with entirely new hardware, software,* and programs. The system
was to rely on a military computer built by International Business Corporation
(IBM)† called the IBM AN/FSQ-31. The need for a more sophisticated, modern
language than earlier existed thus became the prime motivation for developing
JOVIAL.

Jules I. Schwartz one of the key architects of the language, has described how
this language acquired its name. He states that it originated from a report
submitted by SDC in New Jersey during December 1958 entitled "OVIAL—
Our Version of the International Algebraic Language." According to Schwartz,
manager of the project, the development group felt that the name OVIAL had
to be changed because it was too similar to the word *ovary*. Therefore, during
a staff meeting held in January 1959, someone suggested it be called JOVIAL,
using Schwartz's first name—Jules—supposedly because he was project leader.
The name would be "Jules' Own Version of the International Algebraic
Language." Schwartz believed that the title was proposed in jest, but to his
surprise he soon after learned that the new title had been used in a contract drawn
up between his organization and the U.S. government while he was away.

Once this contract was signed, work progressed rapidly, and numerous versions of the language appeared. It ran on both a 709 and the Q-31. By May 1961 a second major version of the language, called JOVIAL 2 or simply J2, was available on the 709, IBM's 7090, and SACCS' Q-31. In time JOVIAL 3 appeared, which became the edition most widely used throughout the 1960s. J3 ran on a variety of machines, including a CDC 1604 and 3600, Philco's† 2000, the SAGE AN/FSQ-7, and the AN/FSQ-32—all major systems of the early to mid-1960s. While the language was being developed, SDC had to decide on which version to use because variations were being developed on both in California and New Jersey. In May 1960 SDC chose to standardize the development work being done in California. Once that happened, versions of the language could be better controlled, and it became a more universal tool for the Air Force. The Pentagon subsequently encouraged other branches of the military to use the language. The U.S. Navy relied on JOVIAL for its NAVCOSSACT project and the U.S. Army for a variety of applications. In June 1967 the U.S. Air Force released specific statements of standardization for JOVIAL based on version 3 which insured even further commonality in the commands and syntax of the language.

The original design objective for JOVIAL called for a language that could be used to program very large systems. It had to be reasonably flexible, when compared to other existing languages, and relatively machine independent so that it could operate on a variety of computers. What emerged was an ALGOL-based language that was relatively easy to read and write. Later, a generation of military programmers argued that it was difficult to learn, yet they used JOVIAL for a large variety of applications ranging from defense systems to data handling, from scientific and numerical calculations to business-oriented uses. JOVIAL also had to work with data defined in COMPOOL. Students of the language have characterized it as procedure- and problem-oriented and also effective for defining problems. JOVIAL was designed at three levels: for reference, publication, and hardware. JOVIAL was batch-oriented and intended for use by professional programmers. A later version, called JTS, was made easier. It could be used in time-sharing systems. Yet another subset, called TINT, operated in an online environment. Both appeared in 1965.

The language was used primarily within the military community. A number of reasons have been offered over the years to suggest why JOVIAL was not more generally adopted. Especially noted has been the computer manufacturers' lack of support for JOVIAL. It was essentially the domain of SDC, especially in the 1960s. The language also had some design deficiencies, including slower compilers than those used in other languages, a lack of elegance, the complexity of learning how to use it, and too many variations. In fact, nearly seven years elapsed before any definitive standards were set (1967). Others have complained that early versions of JOVIAL lacked facilities for input/output—a problem subsequently solved.

JOVIAL had many of the same features as those found in IAL and ALGOL. However, it was unique in relying on very specific and detailed data definitions and on its use of standards for handling information originally developed during the SAGE* project. The language lent itself to accessing a wide variety of logical or arithmetic values. More important, the language could change over the years without affecting the data it had to use.

JOVIAL met its original objectives, even though its earliest compilers were too slow and it was complicated to learn. Compilers were continuously improved during the first half of the 1960s, but the stigma of having poor compilers remained with the language. In terms of innovations, JOVIAL was one of the earliest to have a compiler coded in its own language. It was also one of the first that could do both scientific computing and nontrivial data handling simultaneously. JOVIAL handled general information processing as well. The use of COMPOOL with a compiler once again proved the wisdom of building data sets independently of programming and yet of allowing languages to access information. Put in other words, data definitions were independent of programs—an early step in the direction of database management systems* of the 1970s and 1980s. JOVIAL used its own compiler rather than one written in another, usually machine-level, language. However, one of its most important technical contributions was the great flexibility which it gave the programmer to control the allocation of storage in a system while not insisting that it be done with each program.

JOVIAL became a widely used language by the military. As late as the 1970s the U.S. Air Force was continuing to enhance the language. One of the sets of specifications from the 1970s, called J73, resulted in new versions of JOVIAL. Various government agencies in addition to the military adopted the language. The Federal Aviation Administration (FAA), for example, built systems based on JOVIAL. In the 1960s although other languages emerged that were more efficient, and some already existing ones were more effective (e.g., FORTRAN* and COBOL*), JOVIAL, through military, particularly U.S. Air Force support, gained acceptance. Indeed, the Air Force was probably the most extensive military user of data processing in the 1950s and 1960s, and its influence on other government agencies was extraordinary. JOVIAL was a case in point.

For further information, see: U.S. Department of the Air Force, *Standard Computer Language for Air Force Command and Control Systems*, Air Force Manual AFM 100–24 (Washington, D.C.: U.S. Government Printing Office, 1967); M. H. Perstein, *Grammar and Lexicon for Basic JOVIAL* (Santa Monica, Calif.: Systems Development Corporation, 1968); Jean E. Sammet, *Programming Languages: History and Fundamentals* (Englewood Cliffs, N.J.: Prentice-Hall, 1969); Jules I. Schwartz, "The Development of JOVIAL," in Richard L. Wexelblat, ed., *History of Programming Languages* (New York: Academic Press, 1981): 369–397.

K

KIEV. This early Soviet computer was built to support scientific applications at Dubna, a nuclear research center. The machine was designed and constructed at the Computing Center of the Ukrainian Academy of Sciences at Kiev and was completed in 1959. It did not have transistors, the common building block of computers by the end of the 1950s, nor did it have an internal clock. Its memory* consisted of 1,000 words, each made up of 41 bits. Memory's access time was 10 microseconds per word. KIEV was equipped with additional secondary memory in the form of three drums, each capable of storing up to 3,000 words. The drums were reported to revolve at 1,500 revolutions per minute. It also had a read-only memory of 512 characters, and the vacuum tubes in the KIEV reached 2,300. Input was performed with paper tape,* and output went to paper tape or a tabulator.

Soviet progress with computers always trailed behind that of the West largely because the USSR did not take computers seriously until nearly twenty years after the United States and Great Britain had. The KIEV, built for the same kind of applications and in the same period as the LARC,* provides a benchmark of the differences in development. In a quick survey of LARC, one notes that the American machine had 20,000 words of memory which could be expanded to 97,500 words. Access time was 0.5 microseconds. Up to twenty-four drums, each capable of housing 250,000 words, could be attached, and it also used tape drives* and had a printer.

For further information, see: Andrei P. Ershov and Mikhail R. Shuri-Bura, "The Early Development of Programming Languages in the USSR," in N. Metropolis et al., eds., *A History of Computing in the Twentieth Century* (New York: Academic Press, 1980): 137–196; René Moreau, *The Computer Comes of Age: The People, the Hardware, and the Software* (Cambridge, Mass.: MIT Press, 1984).

KLERER-MAY SYSTEM. This programming language, designed to solve numerical scientific problems online using a GE 225 or 235 computer, was developed at the Hudson Laboratory of Columbia University in the early 1960s. This particular language, unlike many others of the period, reflected the interest of its designers in experimenting with solving problems through a careful balancing of optimal use of hardware and software.* The system taught users how to program by having the computer print out how it intended to interpret particular programming statements. This language may also have had the smallest reference manual in history: a single 8½- by 11-inch laminated card with instructions on both sides. Later, a one-page addendum was issued to provide instructions on how to use the language online. Although there were no other manuals, articles describing the language eventually appeared in data processing journals.

The Klerer-May System initially relied on a modified Frieden Flexowriter for input. All subscripts and superscripts were constructed with the keyboard, a paper tape* reader, or under the control of the computer. Like many other programming languages* of the early 1960s, it had eighty-eight symbols which could be typed consisting of twenty-six capital letters, ten digits, fourteen lower-case letters, eighteen Greek letters, and a variety of other characters (such as $+$, $-$, $\times$, $/$, $=$, $.$, $()$, etc.). However, six special characters allowed a user to create new symbols. Like many other systems, the design objectives included relying as naturally as possible on normal mathematical notation. It also had a library of routine mathematical functions.

Yet unlike many other languages of the period, this one provided more information to users than might otherwise have appeared. The most important of this kind of data was the system's description of how it intended to interpret a line of code, a useful feature since the language could be ambiguous. Such information allowed a programmer to correct or modify statements to do exactly what he or she intended. The system appeared to be a compromise between improving the efficiency of a compiler and making programming easier for an individual.

Despite its limited use—essentially to portions of Columbia University—the Klerer-May system made one important contribution, namely, it showed a way of making two-dimensional input/output general. Its method of dealing with ambiguous source statements from a programmer was unique, a feature that appeared later in the 1970s and 1980s as HELP functions in many applications. It eliminated the requirement for a programmer to write dimensional declarations.

For further information, see: M. Klerer and J. May, "A User-Oriented Programming Language," *Computer Journal* 8, no. 2 (July 1965): 103–109 and their short laminated card, *Reference Manual* (Dobbs Ferry, N.Y.: Columbia University, Hudson Labs, revised edition, July 1965); Jean E. Sammet, *Programming Languages: History and Fundamentals* (Englewood Cliffs, N.J.: Prentice-Hall, 1969).

L

L⁶. This list programming language* was known more formally as the Bell Telephone Laboratories Low-Level Linked List Language. It was the creation of Kenneth C. Knowlton at the laboratory, and it ran on an IBM 7094. L⁶ was one of the better designed list programs of the 1960s and provided significant improvements over earlier languages of this type, such as IPL-V,* in certain quantitative applications.

In the 1960s it was debated whether L⁶ was a language or rather a piece of software* that enabled one to develop languages. In large part it had the appearance of an assembly program, but it gave the operator the ability to define how to use storage and to structure lists and pointers to such lists. The L⁶ program appeared as a series of commands in highly symbolic form, line by line. Jean E. Sammet gives one example of these commands:

$$IF(XA,E,O)THEN(R,F,C,X)DONE$$

which describes a test and its resultant operation. Whether or not a language, it was a useful tool employed at Bell Labs† where it ran on an IBM 7040, later an S/360,* and a MOBIDIC B, PDP 6, and an SDS 940—all computers from the 1960s.

The most elementary piece of data in such a program was a block of memory* or, to use the correct data processing terminology, a memory block defined by the programmer containing up to 128 words. Thirty-six fields could exist in a block; obviously, then, fields could vary in size. Interestingly, a block might not only be up to the size of a word but might also overlap with or contain pointers to other blocks (as one would later see in such access methods as International Business Machines Corporation's popular VSAM). The program handled both digits and letters in its files. Thus, like other list processing languages, considerable attention was paid to provide flexibility in the definition and manipulation of files of data.

Operations, better known as instructions, were of two types. *Tests* and *operations* provided all the usual commands for managing and moving data, and performing calculations or actions with them. Unlike earlier list processors, L^6 gave programmers an enormous amount of control over arrangements of lists, fields, and size of data files. According to Sammet, this language proved extremely useful for "problems requiring maximum efficiency at object time."

For further information, see: K. C. Knowlton, "A Programmer's Description of L^6," *Communications, ACM* 9, no. 8 (August 1966): 616–625; Jean E. Sammet, *Programming Languages: History and Fundamentals* (Englewood Cliffs, N.J.: Prentice-Hall, 1969).

LANING AND ZIERLER SYSTEM. This system was one of the first interpretive algebraic languages ever developed to do numerical scientific computing. Often called a system instead of a language, it was developed at the Massachusetts Institute of Technology (MIT) by J. H. Laning and W. Zierler between 1952 and 1953 as part of the effort to build the WHIRLWIND* computer. This particular piece of software* was the first in the United States that permitted mathematical expressions to be written in notation that appeared in normal format. Like many other programming languages* of the 1950s, researchers working on this language wanted to develop a system for the mathematician and engineer that would be as similar to arithmetic, algebraic, and mathematical notation as possible. Variables were expressed as single letters, and both paper tape* and Flexowriters were used.

Some of the features of future languages appeared in this system. It had both conditional and unconditional control transfers. There were floating point, numerical, and subscripts, yet it did not have loop controls; to handle loops one had to use a series of sequences of values. One could also simply increment fixed values. The system had twenty mathematical functions, could print and stop, did differential equations, and relied heavily on the design characteristics of the WHIRLWIND.

Compared to other early scientific programming languages, the Laning and Zierler system had a particularly rich set of functions. It also worked well.

For further information, see: C. W. Adams and J. H. Laning, Jr., "The M.I.T. Systems of Automatic Coding: Comprehensive, Summer Session, and Algebraic," *Symposium on Automatic Programming for Digital Computers* (Washington, D.C.: Office of Naval Research, Department of the Navy, 1954): 40–68.

LARC. This computer, also called the Livermore Atomic Energy Research Computer (LARC), was Sperry Rand's* supercomputer of the 1950s and served the same kind of scientific user as International Business Machines Corporation's (IBM's)† STRETCH.* It was developed in response to pressure from the Lawrence Radiation Laboratory in Livermore, California, which needed a very large processor to conduct research primarily in atomic energy. This computer is considered the last of the large systems built with decimal-only memory.* It had banks of memory, each of which could store 2,500 words made up of twelve-decimal digits each. LARC had eight banks, providing the capability of storing 20,000 words—a very large memory system for the 1950s. The system was so designed that additional banks could be added to bring the total potential memory up to 97,500 words. With a configuration of eight banks, the system could access a word each in 0.5 microseconds.

The LARC consisted of two processors. The first managed input and output functions, and the second did the actual arithmetic. The system was so designed that another arithmetic processor could be attached to provide enhanced performance. Floating-point addition took 4 microseconds for each instruction and for multiplication, 8 microseconds. By the standards of the day it was faster than commercial machines, although somewhat slower than STRETCH, its only counterpart during that era. It could handle up to twenty-four drums of memory, which was impressive when one considers that each drum could store up to 250,000 words. The system required only two drums to function as designed.

The UNIVAC division of Sperry Rand treated LARC as a special-purpose machine. Its construction costs were too high to justify manufacturing a commercial version; there simply were not enough potential customers to warrant taking that action. Thus, only two of them were ever built. The Lawrence Radiation Labs took possession of its system in 1960, and the second went to the U.S. Navy Research and Development Center soon after near Washington, D.C. They were both shut down in 1969.

This advanced machine is frequently compared to STRETCH as representing projects that taught their manufacturers many things about large computers reflected in subsequent products. It is overlooked, however, that both companies also recognized and, for decades believed, that the demand for supercomputers was too limited to justify major investments. LARC was built with components that existed at the time of its initial construction (1956), whereas STRETCH depended on newly emerging technologies. IBM managed to take greater advantage of the technology in STRETCH, pushing STRETCH into subsequent commercially available machines. Yet both devices were important. The LARC was also the largest decimal computer ever built.

For further information, see: M. M. Maynard, "Livermore Automatic Research Computer (LARC)," in Anthony Ralston and Edwin D. Reilly, Jr., *Encyclopedia of Computer Science and Engineering* (New York: Van Nostrand Reinhold, 1983): 872–873; Michael R. Williams, *A History of Computing Technology* (Englewood Cliffs, N.J.: Prentice-Hall, 1985).

LEO COMPUTERS. These British digital computers* were built and sold during the 1950s and represented the first significant venture by the British into commercial data processing. The history of these computers and the organizations that constructed them provides a window into the development of the data processing industry in Great Britain during its early days. The machines also reflect research performed at British universities in the 1940s and 1950s on electronic digital computing.

The first important digital computer in Great Britain was the EDSAC* built by Maurice V. Wilkes.** Although not the only such device, it was important because it provided a direct link to the future LEO computers. J. Lyons and Company, a catering firm, sought to enter the field of data processing, believing that such devices might be useful in office applications. In the autumn of 1947 the firm contacted Wilkes to propose a joint venture, and, beginning in 1949, T. R. Thompson and J.M.M. Pinkerton of Lyons managed their firm's part of the project to build their first computer. It was named the Lyons Electronic Office (LEO), and the plan called for building a new edition of the EDSAC with some enhancements. Peripheral equipment was handled as a subcontracting operation with various British firms building them. By the spring of 1951 the first machine could run programs in a test environment. More important, by November commercial applications were running on the machine for the J. Lyons & Company, perhaps the first such commercial use of a British digital computer.

This machine had improved input/output equipment, including magnetic tape storage. Additional improvements made the machine and its peripheral equipment fully operational for a variety of business applications by the start of 1954. It was also operating as a service bureau computer in that the company ran scientific computing jobs on it for other organizations. These included both the British Ministry of Supply and the de Havilland Propellers Ltd.

Within six months, the company considered the possibility of constructing yet other devices and came up with the design of what became known as LEO II. To facilitate its construction and eventual marketing, on November 4, 1954, LEO Computers Ltd., came into existence. This was a bold step, but it made sense. The Lyons organization's positive experience with a computer made the step toward manufacturing and selling them an easy and logical one. This was a startling event in British industry, for the data processing industry as we know it today did not exist as yet. The driving force behind this effort was Raymond Thompson who was enthusiastic about the project even before the EDSAC had even been completed. With the help of Wilkes's team at Cambridge University the technical risks involved were lessened.

The final product was impressive. It had some 7,000 tubes and 2,048 18-bit words of storage, two line printers, and card punches and readers made by the British Tabulating Machine Company. The lessons learned in this project would be carried over to the commercial versions of the machine. The original device operated at Lyons until January 1965, making it one of the longest operating electronic computers in the history of digital equipment.

The creation of LEO Computers was a practical venture with a simple objective: to build and sell the LEO II. When completed, that computer boasted a mercury delay-line store of 1,024 39-bit words, along with four magnetic drums and magnetic tape drives for additional storage. As Table 48 suggests, it was an impressive early computer compared to the first machine built for Lyons. The LEO II was first shipped in 1957, and by the time it was withdrawn from

Table 48
Technical Features of LEO II

Technology	Valve
Word length (bits)	39
Store size, fast	1,038
Store type, fast	Delay
Store size, backing	Up to 65,536
Store type, backing	4 drums
FXPT add time	340 microseconds
FXPT multiplication time	0.6–3.5 ms
Date first delivered	1956
Approximate basic cost	90 British pounds

SOURCE: Reprinted with permission from *Early British Computers*, Simon Lavington, First Edition.
Copyright © Digital Press/Digital Equipment Corporation (Bedford, MA), 1980.

marketing, thirteen copies had been sold. It was followed by LEO III, announced in 1962, of which about 100 were constructed.

In 1963, with business prospects continually expanding and the record of success good, the English Electric Ltd. and LEO Computers Ltd. merged together to form the English Electric LEO Computers Ltd. Simon Lavington noted, after English Electric bought out Lyons' share of LEO, "a fascinating connection between teashops and computers" ended. Although the history of that firm was not relevant to the development of LEO computers, it did market them during the 1950s.

For further information, see: Simon Lavington, *Early British Computers* (Bedford, Mass.: Digital Press, 1980); J. M. M. Pinkerton and E. J. Kaye, "LEO—Lyons Electronic Office (Part 1)," *Electronic Engineering* 29 (1954): 284–291, Part 2 by E. H. Lanaerts, "Operation and Maintenance," ibid.: 335–341, Part 3 by E. J. Kaye and G. R. Gibbs, "A Checking Device for Punched Data Tapes," ibid.: 386–392.

LINCOLN RECKONER. This online language, developed in the mid-1960s to solve numerical scientific problems, was designed and used at Lincoln Laboratory† on an experimental computer called the TX–2. It was the first language to introduce the concept, later called coherent programming, by which a person could write a program that was later used by another individual in his or her own program. The designers of this language intended it to be a user of routines rather than simply a facility for creating them.

A large part of the programming language* comprised a library of routines for scientific computing. Its greatest applicability was to the computation of arrays of data, primarily by engineers and scientists. Later *Junior*, a compiler with FORTRAN*-like functions, was added. Yet the intent was always to provide a language designed to handle only small- to medium-sized computations. The idea of coherence made a large number of small computations work together.

Thus, for example, the answers gained from one program could become inputs to other programs. Apparently, the language was used by only a small number of people, primarily at the Lincoln Laboratory during the 1960s and 1970s.

For further information, see: Jean E. Sammet, *Programming Languages: History and Fundamentals* (Englewood Cliffs, N.J.: Prentice-Hall, 1969).

LISP. This programming language was developed in the late 1950s and is still in use today, primarily by scientists and programmers working in the field of artificial intelligence.* LISP was a product of research done at the Massachusetts Institute of Technology (MIT) and represented an attempt to advance the science of programming languages.* Ultimately, however, it contributed to a general understanding of the theory of computability—a subject that is still evolving and a subject of investigation.

Initial work on LISP began in 1956 when a mathematician, John McCarthy,** attempted to develop an algebraic list-processing programming language to use in research concerning artificial intelligence at a summer program at Dartmouth College. He wanted the language to operate on an IBM 704* computer which International Business Machines Corporation (IBM) was to give to MIT and which Dartmouth would also be able to use. During 1957 and 1958 McCarthy worked with FORTRAN,* which at the time was the leading algebraic language. In the fall of 1958 he joined the faculty at MIT where he and Marvin L. Minsky** (later considered one of the great pioneers in research on artificial intelligence) began their work. Specifications for what would become a list processing language (hence the name LISP) began with FORTRAN-like notations. From the earliest design work came two basic ideas involving the use of words: *car* (contents of an address register) and *cdr* (contents of the decrement register)—acronyms that would be critical to the design of LISP as powerful functions.

The language underwent changes over the years. The earliest release was known as LISP 1. The first significant version, however, was LISP 1.5, and the most important and enduring was LISP 2. LISP 1.5 was historically closer to FORTRAN, whereas LISP 2 was based on ALGOL.* The languages were readily accepted by highly qualified programmers and researchers. LISP 1.5, for example, was run on the IBM 1620,* PDP-1, and Scientific Data Systems' 940. LISP ran on newer machines beginning in the mid-1960s and continuing through the early 1980s.

Although users represented a small, highly technical community, they found the language practical for its symbolic integration, in addition to other various algebraic formulations, in proving theorems, in solving problems in geometric analogy, and even in analyzing bidding in contract bridge. Therefore, LISP 1.5 could hardly be characterized as a general-purpose language. It was best suited for the manipulation of symbols and in processing lists. Users found it heavily prone to errors because of its excessive use of parentheses that had to be properly matched. Despite such handicaps, researchers found this early version useful for processing lists and for doing recursive operations. The language was also intended for batch processing, although by 1961 an online version was available at MIT.

By the early 1960s the limitations of the language were preventing its wide acceptance. It was very slow, taking much more time than other languages to do computations. In addition, its management of memory* was too limited. It could not represent data by using blocks of registers, and its input/output functions were very restricted. Therefore, with the support of the Systems Development Corporation and Information International, Inc.—two data processing consultants located in Cambridge, Massachusetts—MIT decided to develop a new version of LISP. McCarthy and his associates and students relied on recent work done to define an ideal, universal language called ALGOL.* Discussions concerning the design of a new language began in the fall of 1963, and by 1966 an initial release of LISP 2 was available.

The language failed to attract broad support, however; it was still too specialized to be considered a general-purpose language. No formal standards were ever established by the data processing community, and, unlike the case with many other languages, no formal group was charged with responsibility for maintaining and improving it. For the most part, those who had developed the language in the first place provided the changes that had appeared in 1.5 and 2, and later other modifications that came in the 1970s. Yet most of the recent changes were minor once LISP 2 had initially been made available.

For further information, see: John McCarthy, "History of LISP," in Richard L. Wexelblat, ed., *History of Programming Languages* (New York: Academic Press, 1981): 173–191; Jean E. Sammet, *Programming Languages: History and Fundamentals* (Englewood Cliffs, N.J.: Prentice-Hall, 1969).

LOLITA. Language for the On Line Investigation and Transformation of Abstractions, a programming tool for list and string processing, was a variation of the Culler-Fried System.* LOLITA was an attempt to simplify the processing of strings of symbols. Its design was relatively uncomplicated in that a user could employ nine lists in memory* of a computer and additional lists on a computer's drum memory (auxiliary memory in the early 1960s), using the Culler-Fried console. The console comprised two keyboards especially designed for the system and a screen (oscilloscope in reality) to display data.

For further information, see: F. W. Blackwell, "An On-Line Symbol Manipulation System," *Proceedings of the ACM 22nd National Conference* (1967): 203–209.

M

MACARONI BOX MODEL. This machine was a comptometer* invented by Dorr E. Felt** in the mid-1880s in the United States. The commercial version of the machine, marketed primarily in the 1890s, launched the business machine industry. The rival to this machine was the Burroughs Adding and Listing Machine which was developed by William S. Burroughs.** The two devices were the most widely used adding machines in the world before World War I.

The Macaroni Box was the pilot model built by Felt and was in fact housed in a wooden macaroni box—hence the name. He used meat skewers for keys and staples for key guides. His springs were constructed out of rubber or elastic bands. The commercial version of the machine, called the comptometer, was manufactured by Felt's manager, Robert Tarrant,** in Chicago beginning in the late 1880s.

For further information, see: J.A.V. Turck, *Origin of Modern Calculating Machines* (Chicago: Western Society of Engineers, 1921).

MAD. This programming language for solving numerical scientific problems was developed at the University of Michigan in the early 1960s. Also known as the Michigan Algorithm Decoder, it was an example of a language designed to provide very fast processing in an academic environment where large numbers of programs had to be run on computers that, by the standards of the 1980s, were relatively small. It was yet another attempt to improve the efficiency of a system through design of compilers and translators in programming languages.*

Work on the language began at the University of Michigan in the spring of 1959. Computer scientists there wanted a language that could run on an IBM 704* and later on the 709/90/94. They completed the first version in February 1960 and subsequently offered it to other institutions through SHARE.† What they produced met their initial expectations for a fast translator that was easy to use, had few rules to learn, and could be enhanced conveniently. Most important, however, it was a language that could be taught quickly to students. Subsequently, additional enhancements were made to MAD at Michigan throughout the 1960s. By the end of the decade the language was also being used extensively at MIT and at other large universities.

The developers originally wanted a faster, more efficient ALGOL* language, but they ultimately developed not ALGOL but a different language. The software* also caused other packages to be developed. For example, some people wanted the speed of a MAD compiler but the facilities of a scientific language such as FORTRAN.* The result was a package called MADTRAN. MADTRAN took correctly written FORTRAN programs (that is, programs free of errors) and translated them into MAD and then compiled these. MADTRAN correctly translated such programs and was used in several universities during the 1960s. MAD itself ran on a variety of computers in the 1960s, including the IBM 7040, the UNIVAC* 1107, and even the Philco 210-211. MAD's compiler was fast by the standards of the 1960s because of its design. All key words had to have more than six letters, which told the compiler when a data name was being processed. Abbreviations were allowed, making it faster for a programmer to write in MAD. After running, the software expanded abbreviations into full listings. Finally, the compiler used different algorithms than had been employed in compilers, making it more efficient.

The language was used successfully to teach students how to do problem-solving with software, particularly numerical scientific problems. A large number of programs also ran in a short period of time. The language, however, was little used in either government or business. The primary reason may have been the lack of support for it by major vendors at a time when other languages were widely used, such as FORTRAN for scientific and mathematical computing and COBOL* for commercial applications.

For further information, see: B. W. Arden et al., "MAD at Michigan," *Datamation* 7, no. 12 (December 1961): 27–28; Jean E. Sammet, *Programming Languages: History and Fundamentals* (Englewood Cliffs, N.J.: Prentice-Hall, 1969).

MADCAP. This programming language was developed at the Los Alamos Laboratory of the University of California for use on the MANIAC II* computer. Along with COLASL,* also developed at Los Alamos, the purpose of this language was to do scientific computing on an existing computer. Considered more experimental than COLASL, MADCAP nonetheless was used throughout the 1960s until the MANIAC was dismantled. Furthermore, unlike COLASL which appeared in only one version, MADCAP underwent constant evolution while in use.

MADCAP had good notational facilities for a programming language of the 1960s. Like COLASL, MADCAP was an easy language to read and apparently easier than the first to program. It was best suited for numerical scientific computing, although some engineers and scientists also found it convenient for combinatorial and set theoretic exercises. It was both a hardware and a publication language. Initially, it relied on a specially built Frieden Flexowriter for input, recording information on paper tape* and on normal paper. As the sample program suggests, it was relatively free form, with instructions sometimes exceeding one line. Yet all procedures were treated as independent subprograms (one could call them blocks). A variety of data types made MADCAP a useful programming language for scientific problems.

Arithmetic variables or constants were characterized as *real*, and constants were even allowed to be defined in hexadecimal. Boolean variables per se were not used, yet strings of characters were employed either in patterns described alphanumerically or in bits. Both single- and double-precision arithmetic was available. The language permitted complex number arithmetic, while Boolean was handled as operations in set theory. In this early scientific programming language, no distinction was made between arithmetic involving integers or floating point. It had a stock list of mathematical functions in a library but no debugging statements.

MADCAP had a powerful looping capability compared to those of other contemporary languages. Furthermore, loops could be defined in various ways, making them even more flexible than those in other languages. In addition, the language had powerful input/output and formatting control systems that were flexible and numerous in function, and that allowed users to define the role of various peripheral equipment attached to the computer.

Scientists used the language extensively for solving numerical exercises; yet, because of the features inherent in this programming tool, they also employed it in resolving combinatorial problems. Its ability to handle such problems made MADCAP one of the technologically most interesting scientific languages of the early 1960s. It would be some years before that kind of function would be available as standard features in scientific programming languages* in general.

For further information, see: D. H. Bradford and M. B. Wells, "MADCAP II," in R. Goodman, ed., *Annual Review in Automatic Programming* (New York: Pergamon Press, 1961) 2: 115–140; Jean E. Sammet, *Programming Languages: History and Fundamentals* (Englewood Cliffs, N.J.: Prentice-Hall, 1969).

MADM. See MANCHESTER MARK I

MAGIC PAPER. Magic Paper was designed as a programming language* to do formal algebraic manipulation online in the early 1960s; however, it was never completely implemented. It was designed to operate on a PDP-1* complete with typewriter, cathode ray terminal (known as a scope in the early 1960s), paper tape* reader and punch, normal magnetic tape, and a light pen for both input and output. Because a user was expected to use a typewriter as the main form of input, the program was equipped with a collection of symbols for algebraic expression that might otherwise not have been possible to type. The cathode ray terminal was also designed to be used heavily as an input/output device. Results on the screen could be photographed, punched out on paper tape, or recorded on magnetic tape.

This software* allowed numeric evaluations to be performed along with other mathematical operations. A different and more recent version of the late 1960s was created at the Computer Research Corporation for use on a modified PDP-1 computer and specially designed terminals. This particular version of the language permitted users to define new operators, that is, to define them to be prefices, infices, or suffices. Other features included the ability to choose whether or not to have parathensices on arguments (values or addresses passed to procedures or functions at time of call) and how many arguments a program should have, and allowed the user to decide whether the software should be associative and commutative.

The software was apparently not widely used, for it was only a minor language of the 1960s. However, it was an early attempt to make algebraic manipulation an online activity.

For further information, see: L. C. Clapp and R. Y. Kain, "A Computer Aid for Symbolic Mathematics," *Proceedings, FJCC,* 24 (November 1963): 509–517.

MANCHESTER MARK I. Known also as the Manchester Automatic Digital Machine (MADM) and as the Manchester University Computer (MUC), this was one of the first stored-program computers to operate in Great Britain. It was the result of work done by a team at the University of Manchester headed by Frederick C. Williams** and Tom Kilburn in the late 1940s.

The original motivation for constructing this computer grew from Williams' work on what would later be known as the Williams tube. This was a form of computer memory* using cathode ray tubes (CRTs). Both Williams and Kilburn arrived at the University of Manchester in December 1946 in order to develop this form of memory. They were so successful in developing the Williams tube into the most reliable computer memory available in the late 1940s that almost all computers built at the end of the decade and in the early 1950s used their technology. Their tube stored bits of data on a portion of a screen that was electronically charged, all controlled by an electronic beam. Because the charge on the screen had to be refreshed about every fifth of a second, even on older technology, the breakthrough involved the way data were kept fresh on a CRT. Once that problem was solved by a simple form of regeneration, then the developers put together their memory from inexpensive components. But the most important feature of this memory was that data could be accessed randomly rather than in earlier memories which could only be done sequentially. Thus, one could get to a piece of information directly without having to read all the data in front of the desired one. That meant computers could be faster and have larger memories. The Williams tube was used in the Manchester Mark I.

From the beginning, therefore, the developers of the computer rejected the

Table 49
Technical Features of Mark I (Manchester Mark I)

Word length (in bits)	40
Instruction length	20
Instruction format	1-address
Instruction set	26 operations
Store size, fast	128 full words
Store type, fast	Williams tube storage (CRT)
Store size, backing	Drum
Average add time	1.8 ms
Average multiply time	10 ms
Input medium	Paper tape or hard copy
Output medium	5-track paper tape reader
Digit period	8.5 microseconds
Main valve type	EF 50
Approximate number of valves	1300
Approximate number of GE diodes	None

SOURCE: Reprinted with permission from *Early British Computers*, Simon Lavington, First Edition. Copyright © Digital Press/Digital Equipment Corporation (Bedford, MA), 1980.

use of delay-line storage systems common to other devices in England and in the United States. The initial version of the machine could house 128 40-bit words and an additional 1,024 words on slow magnetic drum storage. An addition could be performed in approximately 1.8 milliseconds—a very respectable speed for the 1940s—while input/output relied on 5-bit paper tape* and a teleprinter. For a listing of the computer's features, see Table 49.

The Williams tube was developed in 1947, and the computer was constructed in 1948 out of surplus supplies of thermionic tubes intended for use in World War II. The machine ran its first program on June 21, 1948. The system was expanded and enhanced with more function into 1949; in April of 1949 it ran its first complex program to solve a problem involving Mersenne prime numbers. During June 16–17, 1949, the machine ran error-free for nine hours, which was a good record of reliability for computers of that era. The machine lived a short life, however. It was taken apart in August 1950 so that a more advanced computer could be constructed in the same facility.

This technology incorporated in the Mark I was so well received that the British government contracted with Ferranti, Ltd.† to build copies of the computer. This led to a long-standing relationship between the University of Manchester and the computer industry that involved five major projects between the early 1950s and the end of the 1970s. Ferranti manufactured nine Mark Is (1951–1957). The first one was installed at the University in February 1951. That particular machine was used by professors and technicians in companies that required large computing power. Modifications to the Manchester computer continued. Mark II went into use in May 1954; it was twenty times faster than the Mark I. The Mark II, sometimes called the Megacyle engine (MEG), had many improvements. It used floating-point arithmetic, consumed half the electricity required by the Mark I, and took up less space.

For further information, see: Simon Lavington, *Early British Computers* (Bedford, Mass.: Digital Press, 1980); F. C. Williams, "Early Computers at Manchester University," *Radio Electrical Engineering* 45, no. 7 (1975): 327–331, and with T. Kilburn, "The University of Manchester Computing Machine," *Review of Electronic Digital Computers* (February 1952): 57–61.

MANIAC. This American computer was built at the Los Alamos Scientific Laboratory under the direction of Nicholas C. Metropolis.** It was constructed between 1948 and 1952, primarily for use in the development of atomic energy applications, including a bomb. It originated with work done on the ENIAC* at the Moore School of Electrical Engineering,† University of Pennsylvania, where engineers from Los Alamos successfully ran programs. That experience inspired them to develop their own machine. Their device contributed to the development of the hydrogen bomb and to other government projects of the mid-1950s while expanding the number of computer scientists in the same era.

The machine they designed and built consisted of an arithmetic unit to do calculations and memory* to hold data, and had input and output devices attached that were commonly available. Four panels of electronics made up the logic unit. MANIAC's high-speed electrostatic memory was built in forty units, each capable of housing 1,024 bits of data. Half the memory units could operate in parallel (more than one at a time could thus send or receive data from MANIAC). It also had a 10,000 word magnetic drum memory unit for additional information storage. A photoelectric paper tape* reader/punch was used for input/output along with a -inch magnetic tape medium. A typewriter served as an output device along with an Analex Model 1 line printer. MANIAC was completed in March 1952 and soon after became the inspiration for additional models built at Los Alamos (MANIAC II begun in 1955) and at the University of Chicago (MANIAC III, late 1950s).

For further information, see: "The MANIAC: A Great Big Toy," *Datamation* 24, no. 8 (1978): 80; N. Metropolis, "The MANIAC," in N. Metropolis et al., eds., *A History of Computing in the Twentieth Century* (New York: Academic Press, 1980): 457–464.

MARK I. See HARVARD MARK I, II, III, IV

MATHLAB. This early formal algebraic manipulation programming language emerged in 1964 from the MITRE Corporation where it was the result of work done by a handful of employees, most of whom were summer employees. Little is known about the language since it was developed for experimental use within the company and underwent numerous revisions throughout the 1960s. Some of the early computers it ran on included the AN/FSQ-32 and MITRE's STRETCH* (also known as the 7030) from International Business Machines Corporation (IBM). The most significant version of the language was a later release called MATHLAB 68 which became available in the fourth quarter of 1967.

This relatively simple tool relied heavily on English-like language in output (e.g., THANKS FOR THE EXPRESSION) but was intended for mathematics. Its designers developed it for use online to do such tasks as additions of expressions or equations, and to perform substitutions, differentiations, integrations, and Laplace transforms. Users keyed in their programs on a typewriter, and output came on the same device. MATHLAB was both problem-solving and a hardware language. Users were assumed to know little or no programming, and, to help make it easy for such people, it was written in LISP.*

MATHLAB contributed to the development of programming languages* in the late 1960s. It was the first fully developed programming language that could handle algebraic expressions online; earlier, all efforts had been batch. MATHLAB was not effective as an online system because of its dependence on a typewriter for input/output, but nonetheless it pointed the way to future uses of terminals with screens. It was also the first to use very high-level functions (e.g., solve, integrate) that would become common in languages of the 1970s. A byproduct of MATHLAB was a series of routines that had to be written in order to offer such high-level commands. MATHLAB 68 in particular allowed a system to use ALGOL's* type of notation in a language. With this later version, one could either store a program or introduce it to the computer line by line. It also had a series of subexpressions for mathematics.

For further information, see: C. Engelman, "MATHLAB—A Program for On-Line Machine Assistance in Symbolic Computations," *Proceedings, FJCC* 27, Part 2 (November 1965): 117–126; Jean E. Sammet, *Programming Languages: History and Fundamentals* (Englewood Cliffs, N.J.: Prentice-Hall, 1969).

MATH-MATIC. By today's standards within the data processing industry, MATH-MATIC may be considered an ancient programming language for solving numerical scientific problems. It was developed between 1955 and 1957 to satisfy the need for a scientific computing language that would soon be satisfied by FORTRAN* (produced at the International Business Machines Corporation—IBM†), the most widely used language of the late 1950s. MATH-MATIC (Sperry Rand's answer) did not compete effectively, and so it was rapidly relegated to the dust heap of outmoded computer languages. Had Remington Rand† and the UNIVAC* series of computers competed well against IBM at the end of the 1950s, MATH-MATIC might have been the important language that FORTRAN became.

MATH-MATIC was first called AT-3 and was developed at Remington Rand† under the direction of Grace B. M. Hopper** whose name is also associated with the emergence of COBOL* as the most widely used programming language for commercial applications of the 1960s and 1970s. The specific purpose of

MATH-MATIC was to increase the speed and ease with which scientists and engineers could solve numerical scientific problems using the UNIVAC I computer. Although limited by the lack of floating point on the computer and the hardware's memory* of 1,000 words, the language nonetheless contained a repertoire of basic mathematical functions. It used a Unityper to prepare input tape, and the keyboard associated with the system allowed users to put together numerical subscripts for use as exponents. The language also allowed subscription of data names, while floating-point types of calculations were conducted via subroutines.

The language was very limited in capability largely owing to the smallness of its computer. Yet its IF command was very flexible by the standards of the mid-1950s, if for no other reason than there were no limits on the number of IF clauses one could define. In addition, loop commands were also very powerful. Both sets of commands would later be standard in such languages as ALGOL* and COBOL. MATH-MATIC could operate with statements prepared in the computer's machine code or in A-3, which was simply an intermediate-level language with its own compiler. The ability to accept and execute commands written in either format was unusual for any language and remained uncommon in those developed in subsequent decades.

Jean Sammet, a noted historian of programming languages, found MATH-MATIC's handling of automatic segmentation important. This was the capability of the language to take a program that was too big to fit into the computer's small memory and to break it up into pieces, bringing these into memory and taking them out as necessary. Thus, the compiler always worked to insure that only that piece of programming that was about to be executed in the computer was in its little memory. This concept was hotly debated and analyzed by scientists designing computer languages in the 1960s. The concept would later by employed first for operating systems* by the end of the decade and soon after for programming languages* and data management in general. In the 1970s this function would be called virtual memory. The most unusual characteristic of this function of segmenting programs was MATH-MATIC's ability to find an entire loop and to keep all of its commands together in a single segment of memory, thereby improving the efficiency of the language. It would make segments of computer larger or smaller (within limits) so that an entire loop could be brought into main memory as a segment unless it was too big. Users liked this feature because it meant their programs ran faster.

Yet the language failed to take off, primarily because the computer it relied on was too small and sales of the machine were too limited. IBM's computers and its FORTRAN surpassed MATH-MATIC by the late 1950s as the dominant, and even standard, scientific programming language in the data processing industry. The acceptance of the IBM 704* and 705* computers in particular, and IBM's more effective marketing, caused MATH-MATIC to pass into history by the end of the 1950s. It died not because of its lack of features but because of the failure of its manufacturer to sell enough of the right kinds of computers.

The experience in the 1950s with MATH-MATIC and other defunct languages that were directly dependent on a particular type of computer led many language designers in the 1960s to advocate "machine independent" programming tools. This meant that languages should not rely on a particular machine but should be designed to run on various types. This call for universality led to such languages as ALGOL,* COBOL, and BASIC* which could run on various types of computers. The experience gained with such languages as MATH-MATIC propelled computer scientists into designing computers and associated technologies differently in the 1960s than they had in the 1950s.

For further information, see: Jean E. Sammet, *Programming Languages: History and Fundamentals* (Englewood Cliffs, N.J.: Prentice-Hall, 1969); A. Taylor, "The FLOW-MATIC and MATH-MATIC Automatic Programming Systems," in R. Goodman, ed., *Annual Review in Automatic Programming* (New York: Pergamon Press, 1960) 1: 196–206.

MEMORY, COMPUTER. All electronic computers consist of two parts: the central processor which performs all calculations and transactions within a computer; and the machine's memory which houses information and programs used by the processor. This memory may also be a system of various memory types, operating at different speeds and based on fundamentally varied technologies. Memory should not be confused with other forms of electronic storage that can make up a computer system. There are other auxiliary storage devices, such as tape and disk drives*, which also store information but outside of the computer. When their data are needed, the computer reads into its memory such information and when no longer required writes out to disk and tape drives. That process has remained much the same since the 1960s. Before then and down into the 1970s, some configurations included card input/output equipment. In some of the earliest machines (particularly from the 1930s and early 1940s), tape was used as part of a computer's memory system. These distinctions are important because with all these various devices and technologies at play it became easy for historians to confuse their use of the term *memory*. For our purposes, memory consists of storage capability within a computer.

No history of the evolution of computer technology would be complete without an appreciation of the development and changes in memory over the past half-century. It is a history marked by increasing reliability from one generation of computers to another and by rapid expansion in capacity. A description of memory will elucidate the development of electronics. In short, it was an important cause for the expansion in the use of data processing. As its capabilities improved and costs dropped, data center managers had the necessary incentive to use this technology. More memory, for example, meant that greater and more complex applications could be written to do more with computers. The necessary cost justification for these uses became easier when the price of memory began to drop. As organizations became increasingly dependent on computers to conduct business, the need for reliable computers that did not break increased.

In the early decades of computers, memory systems were very unreliable; by the early 1970s they performed better than the processors themselves.

Ever since the first computers were constructed, the nature of memory determined the overall architecture of the processor. The lack of large memories in early machines posed one of the greatest problems in expanding the capabilities and use of such technology. This obstacle was serious and existed over twenty years, especially after scientists began designing equipment that used the concept of the stored program, which is to say, of having a program and appropriate data in the computer at the same time that they were to be acted on by the machine. Computer scientists sought memories that were reusable (like an electronic blackboard), that moved data in and out of the computer very quickly (at least as rapidly as the processor could take, use, and return information), but that could also store information for long periods of time and in sufficient quantities, and that functioned reliably. That agenda has not changed materially in over forty years; its execution and form has, however.

From the earliest days (for practical purposes since the 1940s, although work had begun in the 1930s), magnetic memory systems were seen as a logical path to follow, and in effect the history of memories involved the development of magnetic, electronic units. In its least complicated concept, if a surface could be magnetized or be charged electrically in some positive or negative manner or in some pattern of charges, these physical states of being could represent binary ones and zeros. Any combination of these two numbers could represent other numbers, letters, and, hence, words and information. That idea gave birth to the concept that computers should be binary machines and that collections of ones and zeros could be the machine-level representation of data (in a system called hexadecimal numbers).

In the 1940s and early 1950s various paths were pursued, information was shared among research groups, and results were compared. Important progress was made. By the mid-1950s, when the construction of memories had been taken over in the vast majority by computer companies competing against each other, the evolution of memories became more secretive, characterized by multiple and often concurrent changes appearing in the industry. Given the lack of electronic parts during World War II for nonmilitary projects, various exotic yet creative approaches were often explored. For the purpose of relevant history, the early days of memory began with the use of relays, probably first with the Harvard Mark II* processor and with relay calculators, the machines built at the Bell Telephone Laboratories.† Relays were slow devices compared to the operating speeds of vacuum tubes and, therefore, were not employed for long. Scientists were also discouraged by the hugeness of memory systems built from relays and could ill afford the vast quantities of electricity which such contraptions consumed when compared to vacuum tubes.

The first important breakthrough in the evolution of computers concerned thermal devices. Immediately after World War II in Great Britain, A. D. Booth, at Birkbeck College in London, tinkered with such a device, although he never built one completely. Yet the concept was well understood by those working in the field and gave inspiration to other projects. His project consisted of one small drum heated by wires, warming a portion of the surface. As the drum rotated, hot surfaces would pass by detectors, causing hot spots to be recycled back through wires, which in turn copied to cool parts of the drum. A fan cooled the back of the drum, thereby performing an erasing function, allowing that surface to be able to receive more information through heating by wires.

In Germany Konrad Zuse** also studied how to build mechanical memories and was the first to do so on a relatively large scale for his Z* series of computers. The Zuse memory suffered from the same problem of slow speed as did other early mechanical and electromechanical memories. Its access time was many times slower than the internal processing speed of the computer. For decades (probably beginning in the mid-1930s), scientists had understood that memories and internal processing speeds of computers had to be synchronized so that a slower component did not keep the entire machine from waiting for work or from operating at optimum speed. Booth also worked with such memories, building one system made up of rotating disks. Each disk had holes close to its outer edges, and each also had a pin that went in and out. Thus, as a disk rotated, pins went in and out thanks to a solenoid while a brush established electrical contact with protruding pins. The brush thus "read" the binary signals reflected by the pins. A number of disks on a shaft made up a storage unit with one bit of information per disk. A modified version used wires protruding from the edge of the disk. Booth used this newer memory in the ARC system, an early British computer.

Yet the most important early system (because it became widely used and thus spurred the construction of useful computers) was delay-line memories in the late 1940s and early 1950s. This type of memory appeared in the EDSAC,* EDVAC,* UNIVAC I* (all descendants of the ENIAC*-class machine), the British Pilot ACE,* the SEAC,* and LEO* I. These machines were built for all kinds of work: military, experimental, and commercial. The memory attempted to delay series of electronic pulses for a few milliseconds, pulses that represented binary numbers. Such delays amounted to storing information before sending it back to the machine's delay-line system, which again stored the impulses for another period of time. These short delays repeated were in effect long-term storage.

Throughout the late 1940s improvements were made to this kind of memory, particularly by Booth, working with the Ferranti Ltd.† By 1951 he had developed an acoustic delay line known as the magnetostrictive delay line. In this system a thin wire, and sometimes a tube, along with two coils—one at either end of the wire—represented the heart of the mechanism. A pulse going through the coil on the left side caused a magnetic field to exist, which in turn made a longitudinal compression wave to go down through the wire. As the wave passed through the second coil, it created a current in it. Next multiple coils were packaged together to increase capacity. Information was the magnetized portion of the nickel wires used. This concept for memory systems still existed in some computers until the mid-1970s. After that time, large-scale integrated circuits overwhelmingly dominated memory technology.

Because information in a delay-line system of the late 1940s and early 1950s could become available only after reaching the opposite end of the memory device, scientists sought better methods of moving data about and storing information. The next answer came in the form of a new random-access memory developed by Frederick C. Williams** and Tom Kilburn, both of whom were working at the University of Manchester. They called it the electrostatic storage mechanism. In practice, they used cathode ray tubes (CRTs) which looked like large television sets in which to store information. Their first attempt came in 1946, and by the end of the next year they had one that housed 1,000 bits of data. A beam of electrons would strike the phosphor surface of the tube, charging it for about 0.2 seconds. By refreshing the charge five times per second, one could store a piece of information for long periods. Patterns of charges (dots and dashes) symbolized information that could be read by a computer.

The Williams tube, as it commonly became known, was very popular. It relied on known technology and thus could be made out of existing components. It was used with WHIRLWIND* and even on commercially available computers such as the IBM 701.* The earliest devices were serial, housing a complete binary word on one single-sweep line on the surface of the tube. John von Neumann** at the Institute for Advanced Study (IAS) in Princeton, New Jersey, took it one step further and developed a system whereby an entire word could be read at one sweep rather than one bit at a time. Thus, the IAS computer* used a CRT for each word and had a configuration consisting of forty such devices. Jan A. Rajchman** at Radio Corporation of America (RCA)† also developed a CRT system which was strictly a digital electrostatic storage tube known as the selectron. It became only an historical curiosity because RCA did not develop an efficient way to manufacture this memory and so it was little used. Only about 2,000 tubes were ever made, each of which could store 256 bits of data and were eventually used on the JOHNNIAC* computer at the RAND Corporation. Other scientists who worked on similar systems included Louis Couffignal in France, who experimented with neon lamps for his memory.

Yet another early line of development involved rotating magnetic memories. Magnetic tapes were too slow, whereas other memory systems had too little capacity. Scientists therefore experimented with the idea of coating nickel or ferrite on a rotating drum. This expedient proved such an outstanding success that, with the Williams tubes and delay lines, it became one of the most popular of the early memory systems. This technology remained in use into the late 1970s. It was slow and thus had to be used with a faster system such as the Williams tube, but it was effective. The versatile Booth also worked on magnetic drums and in 1947 made one that was used in early 1948 in the Automatic Relay Computer (ARC) constructed for the British Rubber Producers' Research

Association Laboratories. Booth continued to build such memories until the early 1950s.

At the IAS in Princeton, work continued with magnetic drums. As part of that exercise, scientists invented what became known as the bicycle magnetic wire memory. It used two old wheels from a bicycle covered with a magnetic recording wire that ran from one to the other wheel and passed over a read/write mechanism. A small motor, along with differential gears, drove the wheels. This system did not work effectively since the read/write mechanism kept cutting the magnetic wire. Similar projects elsewhere also failed, most notably the one at the National Bureau of Standards† which was building the SEAC.*

Despite these frustrations, magnetic memory systems continued to evolve because they were the first to provide reliable memories at reasonable costs. The most attractive of these were the static magnetic memory systems of the early 1950s. First created at the Massachusetts Institute of Technology (MIT) in the late 1940s and early 1950s as part of the WHIRLWIND* computer system, by the mid-1950s they were operational. Early experimenters included Jay Wright Forrester** at MIT, both Dudley Buch and Jan Rajchman at RCA, An Wang,** who later established Wang Laboratories (a major manufacturer of word processing equipment in the late 1970s), and Frederick Viehe.

Viehe's story is an important part of the history of memory. It is also an odd one. An inspector of streets and sidewalks in Los Angeles, he would tinker at home with various electronic projects. In 1947 he filed for a patent which included a description of a magnetic core storage unit. In 1956 International Business Machines Corporation (IBM)† purchased the rights to his idea. Viehe retired from the Los Angeles Department of Public Works and died in 1960 while on an expedition in a desert.

By the early 1950s IBM scientists and general management had become very interested in computing technology. The company conducted research on the subject but also purchased patent rights for other components from MIT concerning memory and access to transistors from American Telephone and Telegraph.† With computer companies now building core memories, quality improved rapidly. With the arrival of such memories, capacities grew, costs dropped, and by the late 1950s the number of computers built worldwide began to increase sharply (Table 50). The impact was great. For example, the use of core memory on WHIRLWIND, as opposed to just electrostatic memory (e.g., the Williams tube), doubled the speed of the computer's operation while increasing the rate at which data came into it by a factor of four. The amount of time necessary to maintain this memory was slashed in half, and mean time between memory failures increased from two hours to two weeks. By the mid-1950s builders of computers had recognized that with this kind of memory they now had a breakthrough in technology.

With the start of a second generation in digital computers,* beginning in about 1959 additional enhancements to memory systems appeared. In second-generation equipment the use of transistors (precursor of the chip*) led to

Table 50
Early Computer Memory Systems, 1942–1986

Era[a]	Type	Sample Computers Used On
1941–1944	Relays	Harvard Mark series
1945–1947	Thermal	Never implemented
1934–1950	Mechanical	Z series, ARC
1946–1953	Delay-line	EDSAC, EDVAC, UNIVAC I, Pilot ACE, SEAC, LEO I
	Electrostatic	WHIRLWIND, IBM 701, IAS
1947–1950s	Rotating magnetic	ARC, IAS
1952–1959	Static magnetic	WHIRLWIND, IBM 704, IBM 7090
1964–1967	Core memory	IBM S/360 Models 50, 65
1968–1971	Thin film	IBM S/360 Model 95
1971–1973	Bipolar	IBM S/370, UNIVAC
1973–1979	MOSFET	IBM S/370 Models 158, 168
1979–1986	SAMOS	IBM 4300s

[a]Dates are approximate and subject to debate since not all computer manufacturers converted to new technologies at the same time. The equipment listed is representative but does include some of the most widely used systems of their day.

miniaturization, the demise of the vacuum tube, and the introduction of modern microelectronics in data processing. Core memories became the norm, and with the advent of third-generation equipment in 1964, most symbolized by IBM's S/360* family of computers, chips became the building blocks of such systems. Costs and sizes continued to shrink as reliability and capacity rose. Memories were now measured not in terms of 2,000- to 4,000-bit systems but amounts as large as 16,000 (known as 16K). Disk drives* came into their own in the 1960s, providing additional online direct access storage that theoretically was limited only by the size of the computer room and by how much a user wanted to buy in the way of disk drives. Manufacturers obviously had to limit the number of disks both because of the availability of channels from the drives to the computers and because of the capabilities of operating systems to manage large amounts of data. But memory sizes continued to grow.

A summary of the capacities of second- and third-generation machines suggests the changes in technologies. The IBM 1401* used ferrite core, and early models had storage capacity for 1,400 to 4,000 characters. By the end of that machine's life in the early 1960s, memory could be increased to 16,000. It took 11.5 microseconds to transfer a bit of data from memory to the processor. The computer was a huge success; over 10,000 were sold. The PDP-4, built by the Digital Equipment Corporation (DEC)† and first shipped in July 1962, had a memory capacity of 32,768 words. Memory cycle speed dropped to 8 microseconds, while on the PDP-1 (1960 vintage) it was actually 5 microseconds. Some of the larger models of the IBM S/360 had memory speeds of 750

Table 51

Increases in Memory Cycle Speeds, 1950–1975, Selected Years (Microseconds)

Era[a]	Cycle Time[a]	Representative Computers
1950	220	UNIVAC I
1955	15	IBM 704, 705
1960	5	IBM 1401, 7090
1965	1	IBM S/360, 6600
1970	0.50	IBM S/3, S/370
1975	0.10	Cyber/76, IBM 303X

[a]Dates and cycle times are approximate. Memory cycle times varied by machine and are thus estimated. For specific times, consult the dictionary entries on each computer. For a general discussion of the above data, see Montgomery Phister, Jr., *Data Processing Technology and Economics* (Bedford, Mass.: Digital Press, 1979): 58–61, 538–542.

SOURCE: Business Week April 2, 1984, pp. 70–71.

nanoseconds and storage of up to 1 million bytes in core memory. By about 1966 thin film memory units appeared in these machines, which had operating speeds of about 120 nanoseconds. For a listing of memory cycle speeds, see Table 51.

While capacities increased during second- and third-generation equipment, costs fell. Memory dropped by a factor of ten for magnetic core technologies between 1960 and 1972. The IBM 1401 incremental memory cost $1.50 to $5.00 per byte. On an IBM 370 Model 155 (vintage very early 1970s), that was $0.40 to $0.50 per byte. Prices continued to drop during the 1970s. Thus, 1 megabyte of memory on an IBM 4341 in 1979 (like all computers of the 1970s built with chips) cost $110,000, or one-fourth what it did a decade earlier. A similar drop was seen in systems of the 1980s. As of 1986, one could buy an IBM PC with 256K of memory for less than $2,000; that price included the processor and memory.

Chips began to be widespread in computers in the mid-1960s and dominated by the early 1970s. Their capacity has continued to grow while their physical form and price have kept on shrinking. Measured in thousands of positions of storage, chips went from 4K to 8K units in the 1960s to 16K, 32K, and 64K during the 1970s to 256K and 500K in the 1980s. That meant more memory capacities, which by the 1980s were available in sizes ranging from several K of memory to billions.

The establishment of chip manufacturing companies beginning in the early 1960s, which encouraged the creation of Silicon Valley in California, made it possible for commercial rivalry to establish a subindustry within data processing just to make chips (semiconductor industry). Packaging chips together in increasing amounts (such as with bipolar technologies) made it possible to enhance computer speed and capacity. Per bit prices in the 1970s dropped by about 35 percent annually, causing worldwide demand to grow at nearly 36 percent each year. By the end of 1983 that demand had become a $1.8 billion enterprise worldwide.

The implications of better and more capacity have not yet been fully analyzed, but several trends are evident. First, more computers were used and sold. Second, as more were needed and built, manufacturing costs dropped, forcing prices down while preserving appropriate profit levels for computer companies. Third, beginning realistically in the early 1960s, computer users began to write online applications instead of just batch jobs. Online systems required more storage space in memory for ever larger programs. Such applications also necessitated even faster computers which, of course, meant memory systems that could move data in and out of themselves quickly. As failures declined and as confidence in systems rose, organizations relied on computers more than ever. In the early 1950s, hardly any employee of any U.S. organization was required to know computing in order to perform his or her job. That circumstance did not begin to change until the late 1960s when about 10 percent did rely on computers. By the mid-1980s over 50 percent of the workforce needed computers to perform their duties. During the mid-1980s the impact of computers for future years was vigorously debated, with many believing that "computer literacy" had become as important as the ability to read and write.

Another aspect of memories involved who manufactured them. The invention and early manufacture of chips had been an exclusively American phenomenon. In 1973 Japanese companies began to export their products to the United States, taking advantage of the American recession which had led chip manufacturers to cut back production and halt plant expansion. The Japanese quickly garnered some 40 percent of the total market for 16K chips. By 1980 they had introduced additional products and had nearly 70 percent of the market. By "market" is meant all manufacturers of chips bought from the manufacturers, and not chips made for internal consumption. Hence, all such statistics did not take into account the very large number of chips which IBM made exclusively for its own use either in internal equipment or for products sold to the general public. Although IBM's production during the 1970s, for example, remained unclear, its internal production of 256K chips in the 1980s rivaled or exceeded worldwide production (Table 52). In the case of 256K chips, IBM and the Japanese dominated production. In 1984, some industry analysts thought that some 40 million of these might have been manufactured by all sources worldwide. The most conservative estimates never dropped below 20 million. By that same year American vendors had pushed back Japan's dominance of the U.S. market to 50 percent and the worldwide market down to 60 percent. In terms of raw volumes of main memories sold worldwide, in 1979 U.S. shipments amounted to 41 gigabytes, and by the end of 1984 they had risen to 872. Industry pundits estimated that by 1989 that volume might rise to 6,800 gigabytes. No longer could capacities be measured in megabytes.

For further information, see: René Moreau, *The Computer Comes of Age: The People, the Hardware, and the Software* (Cambridge, Mass.: MIT Press, 1984); Montgomery Phister, Jr., *Data Processing Technology and Economics* (Santa Monica, Calif.: Santa

Table 52

Changes in Costs of Random Access Memories, 1973–1988, Selected Years

Year	Largest RAM Chip[b]	Price/Million Characters	Annual Sales ($)	Annual Volume/ (Billion Bits)
1973	1K	50,000	0.06	8
1978	16K	5,000	0.03	650
1983	64K	500	1.8	28,000
1988[a]	256K	100	8	800,000

[a]Estimated data.

[b]K represented 1,024 bits of storage. The source for these data estimates that in 1990 1 megabyte chips (equal to 1,024K) would be readily available and probably at a cost of $60, with annual sales approaching $10 billion and volume at 1,400,000 billions of bits.

SOURCE: Business Week, April 2, 1984, pp. 70–71.

Monica Publishing Co., 1976); E. W. Pugh, "Solid State Memory Development in IBM," *IBM Journal of Research and Development* 25, no. 5 (September 1981): 585–602; K. C. Smith and A. S. Sedra, "Memory," in Anthony Ralston and Edwin D. Reilly, Jr., eds., *Encyclopedia of Computer Science and Engineering* (New York: Van Nostrand Reinhold, 1983): 942–955; An Wang, *Lessons: An Autobiography* (Reading, Mass.: Addison-Wesley Publishing Company, 1986); Michael R. Williams, *A History of Computing Technology* (Englewood Cliffs, N.J.: Prentice-Hall, 1985).

METROPOLIS. *Metropolis*, a movie released in 1927, made a strong case for the adverse effects of machines on humans. It was made by Fritz Lang who built the plot around workers held in bondage by a giant industrial machine that had rebelled against the workers. One of the earliest films on robotic devices, *Metropolis* showed the central dynamo of "Metropolis" to be a device that became transformed into Moloch, the god of sacrifice. The device was anthropomorphic. Its creator, a scientist named Rotwang, declared that his robotic worker was created in the image of the heroine Maria. He stated that this machine was built "in the image of man, that never tires or makes a mistake." Therefore, "now we have no further use for living workers."

The movie was one of many artistic expressions of concern in the late 1920s about automation. Machines in factories were feared because they might take jobs away from people, while during World War I technology had been used to kill millions of people. Many writers, movie makers, and artists were suspicious of machines in general, fearing their power, their potential for dominance, and their obvious misuse by people.

For further information, see: Charles Eames and Ray Eames, *A Computer Perspective* (Cambridge, Mass.: Harvard University Press, 1973).

"MILLIONAIRE" CALCULATING MACHINE. The "Millionaire" was the most popular calculator within the scientific community by the early 1900s and a great commercial success. Between 1893, when it first appeared, and 1935, 4,655 of these machines were sold, over 2,000 of them by 1912. The device was the invention of Otto Steiger who, like other inventors working on calculators, sought to improve on existing designs while making commercially viable products. The machine was relatively simple to operate: it relied on the concept of direct multiplication which, according to the promotional literature of the day, meant "one turn of the crank for each figure in the multiplier."

This popular machine was initially sold to government agencies involved in scientific research or in gathering statistics. By the early 1900s, however, it was standard fare in most American university laboratories, and many commercial enterprises bought them as well. The greatest business users were railroads, banks, and insurance companies, business sectors that led the American economy in the collection of statistics during most of the first half of the twentieth century. Many computer scientists of the 1940s had spent long hours using the "Millionaire" and other rival machines. Their desire to make these devices even faster and more efficient led to the development of electronic calculators in the late 1930s and 1940s. These were the precursors of the electronic digital computers* of the mid- to late 1940s.

The "Millionaire" was a relatively good machine. Besides the ease with which it could be used, the manufacturer claimed—correctly—that it was fast. A multiplication of six digits by three could be done in 2 to 3 seconds. A very large calculation involving three digits to the second power times three other digits to the second power would take 8 to 9 seconds. One involving eight factors with four sets of multiplications, resulting in an answer with nine digits, could take between 30 and 35 seconds. Each of the entrants to a calculation involved setting levers quickly in a flat box that sat on a table. Calculations could not be done that rapidly by hand or by relying on previous technologies available from other companies, the most common of which was that of the Colmar Arithmometer.*

For further information, see: Charles Eames and Ray Eames, *A Computer Perspective* (Cambridge, Mass.: Harvard University Press, 1973); E. M. Horsburgh, ed., *Handbook of the Napier Tercentenary Celebration* (Edinburgh: Royal Society of Edinburgh, 1914), reprinted, Los Angeles: Tomash Publishers, 1982); J.A.V. Turk, *Origin of Modern Calculating Machines* (Chicago: Western Society of Engineers, 1921).

MINSK. The MINSK series of Soviet computers, developed in the 1950s and 1960s, began with a machine constructed at the Moscow Institute of Energy by I. S. Bruk. He built the first machine, called the M1, in 1951 and, in quick succession, the M2 in 1952 and the M3 in 1956. Further work led to the MINSK22 series. The first member appeared in 1964, although it had been preceded by the MINSK1 (1960) and the MINSK2 (1963). Prototypes of the last two machines were apparently available in 1959 and 1962, respectively. The MINSK22 series was constructed in Byelorussia for the Ministry of Radio. Both the MINSK2 and the MINSK22 could be characterized as general-purpose computers and, in fact, proved extremely popular. Estimates placed their number at some 2,000 copies constructed between 1951 (with the early M series) through 1976.

These computers were very similar technically. They had one set of 107 two-address instructions, and they used a word length of 37 bits. Computing took place at the rate of 5,000 instructions per second unless floating-point addition was being processed; that consumed 72 microseconds. They used ferrite-core main memory ranging in size from 4,000 to 8,000 words. These systems used magnetic tapes for secondary storage. Apparently, some MINSK22 computers also had drum storage attached. These machines were considered general-purpose computers because they could process letters and not just numbers.

These machines were used primarily for scientific and engineering applications, and innovations were usually the result of those uses. Thus, for example, in 1969 multiprocessing was attempted on a MINSK2 connected to a MINSK22. That combination was called the MINSK222. There were many variants of the MINSK series, not all of which were compatible with each other. Yet as a group, they represented one of the first series of relatively widely used Soviet computers.

For further information, see: Andrei P. Ershov and Mikhail R. Shuri-Bura, "The Early Development of Programming Languages in the USSR," in N. Metropolis et al., eds., *A History of Computing in the Twentieth Century* (New York: Academic Press, 1980): 137–196.

MIRFAC. This simple programming language* was designed to solve numerical scientific problems. It was developed in Great Britain in the early 1960s and used throughout most of the decade. It initially ran on a Dura Mach 10 computer which used a sphere of eighty-eight symbols and a collection of editing operations ranging from recording information on paper to printing subscripts. Variables were noted in English or Greek letters, and all statements in MIRFAC were numbered consecutively beginning with 1 upwards. Data variables and constants could be described as numeric, alphabetic, or bit arrays. This language did not appear to be widely used other than by its developers. However, the work actually done with this little known language appeared to involve the reduction of data, automation of design with a plotter, and retrieval of information.

For further information, see: H. J. Gawlik, "MIRFAC: A Compiler Based on Standard Mathematical Notation and Plain English," *Communications, ACM* 6, no. 9 (September 1963): 545–547.

MODERN TIMES. This movie starring Charlie Chaplin, released in 1936, centers on a mechanized being who becomes insane and turns a factory into bedlam. Workers on a modern assembly line are converted into mechanized automata. *Modern Times*, like René Clair's *À Nous la Liberté*, questions the role of mechanized utopias. Such a world was portrayed many times in the 1930s in both the cinema and literature. On the one hand, machines were seen as liberating people from manual drudgery, while on the other they posed a threat to their humanity. The obvious metaphor in this movie is the human being as a cog in an industrial machine. The most dramatic expression of this theme during this period was Aldous Huxley's *Brave New World* (1932), a book that describes life in a world ruled by technology.

For further information, see: Charles Eames and Ray Eames, *A Computer Perspective* (Cambridge, Mass.: Harvard University Press, 1973).

N

NAVAL ORDNANCE RESEARCH CALCULATOR (NORC). This device, constructed by International Business Machines Corporation (IBM)† for the U.S. Navy, was one of the most advanced computational systems in the early 1950s. It was as much a research project undertaken to advance the technology of computation as it was an effort to satisfy a need. Through NORC IBM engineers in particular learned many technical lessons which were applied to the 701,* one of the most popular of the early commercial computers.

The NORC project was the outcome of a practical need. The U.S. Navy Bureau of Ordnance conducted research and development in the general field of weaponry for ships (and later rockets) and for its aircraft. The heart of that work was ballistics, a traditional field of military research requiring an enormous amount of calculations. Some of the earliest computers had been constructed precisely to calculate firing tables. For example, the ENIAC*—the first electronic digital computer*—was the product of such a need during World War II. By the late 1940s research had continued to grow in complexity. The Navy also needed a high-speed computer to solve other large and complex mathematical problems such as partial differential equations in three space dimensions and time.

Research at a university often created the need for government funding. In this particular case, Byron L. Havens, working at the Watson Scientific Computing Laboratory at Columbia University immediately after World War II, had developed a new circuit. Called a fundamental circuit, his microsecond delay unit could operate very reliably at one million steps per second. His creation offered the opportunity to build a very fast computer. Havens went to IBM with the results of his work and in turn the firm negotiated a contract with the U.S. Navy for the construction of a high-speed calculator employing the results of the new research. The device was built at the laboratory beginning in 1953 and was turned over to the U.S. Navy after its capabilities were demonstrated on

December 2, 1954. During the following summer, NORC was moved to Dahlgren, Virginia, where engineers at the Naval Proving Grounds used it as their main processor until 1958 when it was replaced by the IBM STRETCH.* Other computers were subsequently installed at this facility; yet the NORC continued to operate there until it was finally shut down in 1968.

By the standards of the day, NORC had a very large memory* made up of 264 Williams' tubes. These could store 3,600 words, each of which was sixteen decimal digits in length. All digits appeared on the Williams' tube in binary notation. One of the interesting features of this memory was a seventh digit of each word which allowed the system to detect errors when information was passed from one part of the configuration to another. NORC's tubes provided yet another check digit for each column of data in a tube and, when used with the other check digits, could be used to identify the location of an error and correct any that were single bits. Thus, the reliability of the system's management of data flow rose sharply in comparison to other computers of the day. This memory contributed enormously to NORC having an average "uptime" of 92 percent. Other systems of the day were up between half and three-quarters of the time. Ninety-two percent availability did not become the norm for computers until another decade had passed.

The capability of processing ever larger amounts of data also became a unique feature of NORC. It handled up to 15,000 operations per second by using Havens' delay circuitry, along with a high-speed arithmetic unit that could execute a simple addition in 15 microseconds or a multiplication in 31 microseconds. All calculations were done electronically, which allowed parallel processing, particularly for multiplication, and speed. It had a pipeline of twelve adders which accepted answers from multiplications; the system then turned the multiplication into a single digit result per microsecond. Other circuits checked the results of arithmetic calculations, or a "casting out nines" process on each operand. This was yet another means of increasing the reliability of the operations performed.

NORC's other technological introductions were of benefit to the design of future IBM computers. It had a memory control system, for example, which for the first time freed programmers from having to make frequent checks on the location in memory of their programs while operating. Another advance involved equipping NORC with eight fast tape drives* which individually could read or write 70,000 digits per second. That speed represented a fivefold increase over that of existing technologies. That quality of technology soon after appeared with the IBM 701 computer and gave that company so much early experience with tape drives that its tape products were able to set industry standards of speeds and performance for two decades afterwards.

IBM had always needed manufacturing and packaging advances to justify its introduction of new products. This was also the case with NORC, even though profit and costs played a minor role in comparison to other projects. One of the company's design objectives was to build a system from parts and units constructed at various locations and then lashed together—a process that continues to this day. Thus, subcontractors built components, and the computer was then fabricated and assembled at the Watson Laboratory at Columbia University. Peripheral devices were drawn from various IBM plants, including Poughkeepsie and Endicott, New York. The computer had over 9,000 vacuum tubes, some 25,000 diodes, and several hundred thousand other parts. The configuration included two printers with a rated speed each of 150 lines per minute. Interestingly, the system buffered input, which meant that the computer could be calculating at the same time that the printers operated. It also had card-to-tape-to-card input/output.

The system proved highly successful for scientific operations. In addition to studies in ballistics, NORC was used to calculate the precise positions of the earth, moon, and other planets at all times to the year 2000. This exercise was part of the research that went into Project Vanguard (an early U.S. space program). These calculations, done to complete research conducted by Dr. Paul Herget, director of the Cincinnati Observatory, took only ten hours. NORC also simulated the motions of neutrons within a nuclear reactor. A considerable number of calculations were performed on the orbits of satellites and on the design of space equipment in general. It evaluated the design of rocket systems and the effects of reentry of satellites and rockets into the earth's atmosphere. One large astronomical calculation required NORC to operate for sixty-five hours to do over 75 billion operations, which it did without a single error.

For further information, see: J. F. Brennan, *The IBM Watson Laboratory at Columbia University: A History* (Armonk, N.Y.: IBM, 1971); J. C. McPherson, "Naval Ordnance Research Calculator (NORC)," in Anthony Ralston and Edwin D. Reilly, Jr., eds., *Encyclopedia of Computer Science and Engineering* (New York: Van Nostrand Reinhold, 1983): 1016–1017.

NELIAC. The Navy Electronics Laboratory International ALGOL Compiler (NELIAC) was a programming language designed to solve numerical scientific problems. It was one of the first ALGOL*-based scientific languages and the first that could create its own compilers usable on a variety of computers. NELIAC was used by the U.S. Navy for numerous projects during the early 1960s.

During the summer of 1958 the Navy Electronics Laboratory, located at San Diego, began work on this compiler, eager to develop a language that could be used on new computers being delivered to that facility later that year. Harry Huskey and Maurice Halstead, its two primary architects, had decided to use ALGOL on their systems, but because specifications for that language were not going to be ready in time they began to create an ALGOL-like compiler. By the end of 1958 they had their first compiler, and only four years later eighteen different computer centers were using various releases of NELIAC. Additional locations acquired variations of it after 1962, including data centers that had an

IBM 7094, UNIVAC 1107/1108s, CDC 3100, 3600 and 3800, Burroughs D825, and later the IBM System/360.*

Because the language was created with little organization, no standard version ever existed. Each location developed its own variation, and, despite an attempt in January 1963 to standardize the language as had been done for others, the effort failed. NELIAC was too easy to write in and document, making modifications tempting. Many users rushed forward to install NELIAC at the time that the Navy was deciding whether or not to use this or JOVIAL* as its standard language in hopes of improving NELIAC's chances of being selected. However, the Navy elected to use JOVIAL for many command and control applications.

NELIAC was a formal and succinct language that was easy to learn and use, but it tended to create errors because many of its rules of punctuation were easily violated. Despite this problem, it was designed to support scientific computing and problems in engineering. It was a procedure and problem-oriented language and at the same time, a hardware, reference, and publication idiom. Like other programming languages* of that era, it was batch only and used typewriters or IBM 029 key punches for data entry. Although designed to be machine independent, and in fact it ran on many types of computers, in reality it was not independent because of its storage-handling functions. That might have accounted in part for the variety of NELIACs that existed. It looked a great deal like ALGOL, and transliteration to that language was relatively easy to do. It was most efficient for small programs. Those with 100,000 words of object code or more were not effectively handled by NELIAC, although one could write large programs with it. Users reported that its compilers ran very fast and that it was easy to use its character set of fifty-two upper- and lower-case letters. It had ten digits and twenty-six special characters (such as (), $+$, $\times$, $/$, $=$, etc.). The language allowed users to exercise fixed and floating-point numbers. All the variables in a program were treated as integers. The exception was if they were on a list of nouns with a floating-point number setting the mode. It could not deal with alphanumeric information.

To handle input/output, NELIAC had subroutines. Some versions of the language used other packages. It had no libraries, built-in functions, debugging or error messages, or statements to allocate storage in most releases. Yet it was easy to write a compiler in its own language. Some of the more interesting applications involving NELIAC included a Blood Bank Program, programming at the University of California at Berkeley, and programming for the Command Ship Data System used by the U.S. Navy. All of these projects were established in the 1960s. The Navy program had over 100,000 instructions.

Little is known about the language, primarily because of lack of documentation. Its instructions for users were always very limited. By 1970 it became less popular, particularly in the U.S. Navy, where other languages dominated, especially JOVIAL.

For further information, see: M. H. Halstead, *Machine-Independent Computer Programming* (Washington, D.C.: Spartan Books, 1962); H. D. Huskey, et al., "NELIAC—A Dialect of ALGOL," *Communications, ACM* 3, no. 8 (August 1960): 463–468; K. S. Masterson, Jr., "Compilation for Two Computers with NELIAC," ibid., no. 11 (November 1960): 607–611; Jean E. Sammet, *Programming Languages: History and Fundamentals* (Englewood Cliffs, N.J.: Prentice-Hall, 1969).

NORDEN BOMBSIGHT. This device, one of the first to employ analog computing, was used extensively during World War II in military aircraft. One part of the bombsight was a stabilizer that controlled the aircraft during the period of 10 to 30 seconds when a bombing run was being made. The second part, the actual bombsight, consisted of a computer and a telescope. Both used a gyroscope.

The history of the bombsight may be traced back to the early 1900s when the concept of "feedback," in which work performed by a machine controlled work it was about to do, first became a modern reality. Previously, feedback mechanisms had been most widely used with steam engines and in furnaces controlling temperature. In 1908 Dr. Herman Anschutz-Kaempfe patented a guidance system—the Anschutz Gyroscope—for use in turning turrets in steel battleships. Such devices allowed moving machines (such as torpedoes) to change direction because gyroscopes maintained a fixed position in space. Navies became very interested in such technology. One of the earliest of such equipment was created by Elmer Sperry and Hannibal Ford just before World War I. On torpedoes they controlled rudders, causing these bombs to head in the correct direction set by a crew. By 1910 these devices were used on airplanes and ships. Sperry and Ford had first mounted their invention on an airplane as early as 1909.

In airplanes these machines had to account for altitude, speed, and changes in air currents. Without such devices a change in speed to maintain a particular altitude would cause airplanes to crash into each other. Sperry connected his gadget to a wind speed indicator which then automatically modified the force applied to ailerons and elevators. The feedback on an airplane's speed provided for continuous adjustments.

The next logical step was to stabilize the aircraft for bombings. In 1918 Ford developed a device for the U.S. Army that was a combination of a bombsight and an electrical computer with mechanical parts. His gadget mechanically recorded groundspeed and the direction of the airplane's drift in comparison to the location of the target. It would then automatically calculate when the bomb had to be dropped in order to hit its target. The Norden bombsight of World War II fame was based on this early work on feedback mechanisms.

The bombsight was also part of the military's broader use of computing technologies in World War II to control firing of cannon, use of bombs, and calculation and controls necessary for the movement of aircraft and ships. Many wartime projects in ballistics, for example, grew out of this experience and resulted in the construction of the Mark I* at Harvard University and the ENIAC* at the University of Pennsylvania's Moore School of Electrical Engineering.† It also led to the study of feedback mechanisms following the war, including Norbert Wiener's** study of gun director systems during the war and what eventually became known as artificial intelligence.*

For further information, see: Charles Eames and Ray Eames, *A Computer Perspective* (Cambridge, Mass.: Harvard University Press, 1973).

O

OMNITAB. This software* tool was developed at the National Bureau of Standards† in the early 1960s. Some debate still exists as to whether or not it was a language or simply a software tool to facilitate the resolution of numerical scientific problems. It ran on an IBM 7090/94, acting as a simulator of a desk calculator. However, it did contain a library of mathematical functions along with a series of input/output commands. Commands to perform functions were relatively easy and simplistic. For example: "ADD 3.257 AND STORE IN COL 4."

For further information, see: "OMNITAB on the 90," *Datamation* 9, no. 3 (March 1963): 54. Of some use is J. Hilsenrath et al., *OMNITAB—A Computer Program for Statistical and Numerical Analysis*, National Bureau of Standards Handbook 101 (Washington, D.C.: U.S. Government Printing Office, 1966).

OPERATING SYSTEMS. This class of software* made possible the operation of electronic digital computers* as we know them today. Also known as the system control program (SCP), such software allowed computer designers to automate many complex operations calling for the acceptance, processing, and deliverance of data through computational devices. In very large data centers such configurations involved dozens to hundreds of peripheral devices and multiple computers connected to thousands of users via terminals, all of which had to be coordinated and managed. The creation of operating systems in the 1940s and 1950s contributed significantly to a series of rapidly evolving processes which resulted in ever larger and more powerful computers in the next two decades. Without more sophisticated operating systems, scientists would not have been able to take full advantage of the power of the transistor and later of the chip* in building the computers known today. Yet their contribution to the overall evolution of digital computers has been overlooked by historians of data processing.

Briefly defined, an operating system is a collection of software that controls the execution of all application programs in a computer. It handles such common tasks as the allocation of resources (for example, peripheral equipment or memory* in the computer), scheduling of such resources, controlling input and output, and managing the flow of data. In the majority, operating systems have been software. However, in the past ten years (since 1976) some of the household functions commonly found in an SCP have been buried in chips that made up part of a computer or the control units that managed telecommunications and data storage devices. When these functions are incorporated into hardware, they are said to be in firmware. In summary, operating systems are analogous to traffic police officers at intersections managing the flow of trucks, buses, cars, and people through an area safely and efficiently. Much of the history of operating systems is the story of how those functions came into being and their impact on productivity, in making computers easier to use. Along with programming languages* and database management* packages, the story of operating systems defines the role of software within a computing system.

In terms of the content of such programs, an operating system traditionally consisted of a library of software usually written by the vendor selling the computer in which such code operated. This library might have included compilers, assemblers, resource managers, and so on, and had as its mission to reduce the amount of work a computer operator (that is, a person) had to do to have all the components of a system (peripheral equipment, computer, and people's application software) work together properly. The history of operating systems reveals how additional functions came into being, further automating the performance of computers in ever increasing modes of efficiencies. These environments changed from small batch processing in the 1950s and 1960s to batch and online by the mid-1970s with more hardware and software to manage. In addition, by the late 1970s these systems had to manage acceptance of data from remote terminals in small and large quantities through the process of "distributed processing." The number of end users grew sharply during the 1970s, often forcing data centers to increase the amount of machine-readable data stored and used by 45 percent annually. Many intermediate-sized data centers in the 1960s might have had two to three terminals attached to a system, two dozen by the mid-1970s, and over a hundred by the mid-1980s. Large shops often had over 1,000 terminals "hanging off" a system by the early 1980s, all managed by subsystems and then an SCP.

Operating systems developed in four stages. The first involved the use of International Business Machines Corporation's (IBM's)† operating systems based primarily on the architecture of its S/360* and S/370.* These were the most important operating systems because they became the most widely used within data processing by the late 1960s and remain so today. A second aspect involved the operating systems that either came before IBM's 360/370 programs or that ran on computers that were incompatible with these (e.g., Burroughs,† Honeywell†). A third set of programs concerned Unix systems which were developed within American Telephone and Telegraph's (AT&T's)† Bell Laboratories† in the 1960s and which, by the early 1980s, ran on computers made outside of the Bell network. The fourth stage, a phenomenon of the late 1970s and 1980s, involved programs designed and used in connection with minis and micros.

Because IBM's operating systems from the 1960s to the present represented the lion's share of all those used, their historical value to data processing has been the greatest. They reveal why computers were easily used by the end of the 1960s and beyond. Already in the 1930s and 1940s scientists were seeking ways of automating computer operations while extending hardware functions. By the late 1940s a user still had to personally wire equipment together to perform tasks or "program" by wiring boards. Years later these functions would be managed silently by software, including IBM's operating systems. Jobs in the 1940s and 1950s were submitted to computers one at a time and then ran sequentially; in contrast, by the 1970s multiple programs operated concurrently, often interweaving among themselves in their use of hardware and software resources, all under the control of the SCP.

The first significant extension of the functions of hardware occurred not with the operating systems themselves but with higher level programming languages* that began to appear in the late 1940s. These bloomed fully by 1960. Yet during the 1940s and 1950s—and to a large extent even today—SCPs slowly acquired greater ability to mask users from the mechanics of input/output (I/O) functions and the translation of higher level languages into code intelligible to computers. Finally, common functions that first appeared in programming languages (such as routines for basic mathematical functions) shifted to operating systems. As functions became more widely used in applications, scientists embedded them into SCPs, particularly tasks to handle data files. Eventually, such transactions evolved into separate database management subsystems.

The great breakthroughs in the design of operating systems, in particular at IBM, came in the 1950s. From that point forward SCPs grew in size and function to support rapidly growing amounts of complex computing. Thus, for instance, in 1957 computing power in the United States was approximately 10 million instructions per second, and by 1980 that had grown by a magnitude of four. The lines of software in an operating system also expanded. Hence, for example, the FORTRAN Monitor System (FMS), with its appropriate programming libraries, used on the IBM 704/7090 in the 1950s, could be carried around on one reel of magnetic tape. It took up 10,000 binary cards. IBM's MVS SCP in 1981 occupied seventeen reels of tape, or 13 million cards, and took up 520 million bytes of storage. Some minis in the mid-1980s had as many as a half million lines of code in their operating systems as opposed to one million bytes of storage in FMS and measurably fewer lines of programming.

IBM's operating systems dominated data processing because of the success of the System 360 and 370 families of computers. They created a de facto standardization of technology in SCPs. Over a thirty-year period, the company developed twenty-five different families of operating systems (Table 53). The majority of these have been 360/370 in architecture. Many have survived for more than two decades, with modifications and enhancements simply incorporated into new versions to allow such software to manage new devices or to provide additional functions. For example, the operating system for intermediate-sized computers, called DOS/VS during the 1970s, went through thirty-four releases and then was repackaged and renamed DOS/VSE in 1979. That code has continued to undergo modifications down to the present. Yet it was originally introduced in the 1960s. Emulation of even older operating systems and programs designed and written for use on older equipment but now operating in newer ones also survived. The classic example was the IBM 1401* computer of the late 1950s and early 1960s. The ability to run programs of the 1401 existed in IBM's operating systems of the early 1980s. The same is true of other SCPs from non-IBM compatible machines. Such capabilities were allowed to exist so long in order to protect the investment people made in application programs as they moved from one generation of equipment and operating systems to another. To have forced the rewriting of all application programs might have eliminated the justification for moving to newer products because the costs of conversion could be high enough to offset any anticipated benefits of using less expensive, newer computers.

One of the first attempts to produce an operating system was done for the IBM 701 computer but not by IBM. Users of that class of machine met during the annual Eastern Joint Computer Conference held in Washington, D.C., in 1953 to discuss their need for an SCP. At the General Motors (GM) Research Laboratories one was written for use on the 701 at the same time. In 1955 GM and others joined to develop another for the IBM 704.* Users of that particular type of computer formed SHARE,† an organization which would work with IBM to define their requirements and which led the effort that resulted in the Share Operating System (SOS) in the late 1950s. IBM assumed responsibility for its maintenance and further evolution. In turn, this led to the creation of IBSYS in the early 1960s, derived from SOS as the operating system for the 7090/7094 computers. These early SCPs (1955–1962) were single-stream batch monitors that provided primitive job-to-job connections and could only operate on one particular type of computer. Thus, once loaded into a system, the only additional loads were application programs. Before 1955 every time an application program was loaded into a computer, so were instructions on how to run it. This was the equivalent of an operating system's software. Since an operating system was already in the computer, the amount of time taken to load programs into a computer declined. Hence, a computer could be used to process more frequently since less time was needed to load software into it. In addition, programmers and end users were being weaned from the burden of operations, allowing them to focus on their applications.

By 1960 IBM's operating system had I/O routines, job-to-job linkages, and some error-recovery routines. Job Control Language (JCL) appeared for the

Table 53
Capabilities of MVS and Its Predecessors

Generation	Operations	Extensions of function	
		Hardware functions	Application functions
Pre-operating system (early 1950s) with, e.g., the 701	Manual (e.g., each job step required manual intervention) No multiple-application environment support	Symbolic assembler Linking loader	Subroutine libraries in card tubs, manual retrieval
First generation (late 1950s and early 1960s) with, e.g., FMS, IBSYS on the IBM 704, 709, and 7094	Automatic job batching Manual device allocation, setup, work load scheduling No multiple-application environment support Off-line peripheral operations	Higher level languages— FORTRAN, COBOL Primitive data access services with error recovery	Subroutine libraries on tape, automatic retrieval Primitive program overlay support
Second generation (late 1960s) with, e.g., OS/360 on System/360	Multiprogramming Primitive work load management Primitive tuning (device, core allocation) Spooling, remote job entry Operator begins to be driven by the system Primitive application protection Initial multiprocessing (loosely and tightly coupled)	More higher level languages— PL/I, ALGOL, APL, BASIC Device independence in data access First random access data organizations Primitive software error recovery, full hardware ERP's Array of hardware function extensions Supervisor call routines	DASD subroutine libraries Full facilities for programmed overlays Interactive program development support Primitive automatic debugging aids First application subsystems Checkpoint/restart
Third generation with, e.g., MVS OS/VS on System/370	Integrated multiprocessing (loosely and tightly coupled) Work load management extensions More self-tuning, integrated measurement facilities Less operator decision making, fewer manual operations Full interapplication protection, data and program authorization Primitive storage hierarchies for data	Virtual storage Device independence extended Hardware error recovery extended to CPU, channels Operating system functions begin to migrate to hardware	Growing libraries Overlay techniques obsoleted by virtual storage Symbolic debugging aids Primitive data independence Integration of application subsystems Software error recovery for system and applications

SOURCE: M. A. Auslander et al., "The Evolution of the MVS Operating System," *IBM Journal of Research and Development* 25, no. 5 (September 1981): 473. Copyright 1981 by International Business Machines Corporation; reprinted with permission.

purpose of instructing computers on managing specific application programs. Every developer, whether in IBM or at other companies and universities, sought to extract the greatest amount of work out of the computers of the day. Work was done to manage overlapping of peripheral equipment (both reading and printing on cards for example) and multiprocessing. During the early 1960s IBM, working with American Airlines, brought to full fruition an operating system as part of the SABRE* project. It had one of the first operating systems that handled online transactions. Earlier, SCPs could deal only with batch operations. Second-generation operating systems (those of 1960–1968 but within IBM, 1960–1964), were characterized as batch, real-time, or with some time-sharing. Limited multiprogramming became available, along with device independence (not having to tell a computer which peripheral machine to use with which application program) and the ability to manage data.

But the great watershed in the history of operating systems at IBM—as well as the entire data processing industry—came with the announcement of the S/360 family of computers in April 1964. For the first time, a series of computers was brought out which had components that were compatible with each other. That meant programs which ran on one machine size could also run on others of different sizes and that most peripheral equipment attached to one computer model could be used on other models within the same family of computers. Facilitating this change were operating systems that were relatively independent of any particular machine. In short, IBM created an architecture for all 360/370 SCPs. Generally, these SCPs also had more functions than earlier operating systems. In combination with the new hardware—based on contemporary technologies and considerable innovations—these SCPs revolutionized data processing while the S/360 became the most successful product in the history of American business. By 1970 IBM had doubled in size because of it, and the de facto industry standard for all computer technology had become the S/360. The new operating systems introduced more integrated complex systems that could do multiprogramming and multiprocessing, batch and online operations, and were compatible. The compatibility feature proved critical because before the S/360 many shops were forced to rewrite all of their programs when moving to new, bigger, or different computers. The conversion was very expensive, always required learning how to use a new operating system, and often took years. The entire process was full of risk and failures. The promise of the new SCPs was that this process of conversion would be enormously shortened and someday eliminated.

The base software, called Operating System (OS/360), was announced in 1964 and first shipped in 1965. Within several years it appeared in versions to satisfy specific machine limitations and needs; yet its core was in all forms compatible: OS/PCP, OS/MVT, BOS/360, and DOS/360. OS/360 was the parent software for all future IBM 360/370 SCPs. Besides being usable on multiple machines, it was the first to provide complete sharing of hardware (processor, disk, drums, tape, card I/O, printers, etc.) in a system and across various-sized machines. Furthermore, it used the idea of a ''task'' (which today is called a process) in

Table 54
Computing Services in IBM's DOS/VSE (circa 1979–1980)

```
Initial program load
Resource management
Job control
Library services
Linkage editing of programs
Data management
Systems-to-operator communications
Systems utilities
System serviceability and debugging aids
Assembling programs
Virtual storage
VSAM (access method)
POWER (decreased execution time of I/O)
RJE
VTAM (accessing terminals)
DL/I
Languages: RPG, PL/I, COBOL, FORTRAN IV
```

Typical functions available first in DOS/VS and then in DOS/VSE. For details, see IBM Corporation, *Introduction to DOS/VSE* (White Plains, N.Y., 1979).

which the SCP allocated its resources to a task rather than to a whole program. Later, time-sharing capabilities were added which were still known as TSO in the mid-1980s—a subsystem to manage just those types of jobs.

The evolution of IBM's operating systems has continued unabated to the present. Although many variations of the software were produced, key milestones included the introduction of DOS (Disk Operating System) for smaller 360s in various releases between 1967 and 1969 and, during the same period, of MFT (Multiprogramming with a Fixed number of Tasks) for larger systems. MVT (Multiprogramming with Variable number of Tasks) came out, the predecessor of OS/VS2 (Operating System/Virtual Storage 2) (1973) and MVS (Multiple Virtual Storage) (1974). MVS became the operating system for IBM's largest computers. By the early 1980s it was called MVS/XA (for extended architecture using 31-bit architecture). DOS/VS grew out of DOS/360 in 1973 while DOS/VSE (Table 54) emerged in 1979. The mainstream, however, was OS/360 to OS/MFT (1968), to OS/MFT II (1970), and then to OS/VS1 (1972), which then terminated this line of development. Another descendant from OS/360 was OS/MVT (1969) and to OS/VS2 (for very large systems in 1972/1973), to OS/SVS (1974) or OS/MVS. Thus, over the years these operating systems underwent three lines of development: for the smallest 360 and 370 computers the DOS group, for larger intermediate-sized members the OS/VS1 group, and for the very largest, OS/VS2 to MVS/XA class operating systems (Table 55).

While the names changed, the trend toward continued compatibility did not. One saw more automation of an operator's functions, additional intelligence to

Table 55
IBM Operating Systems, 1955–1985

First Generation (1955–1959)

General Motors: IBM 701	1955
GM and NAA: IBM 704	1956
SAGE: IBM AN/FSQ7	1959

Second Generation (1959–1965)

IOCS: IBM 709/7090	1960
NAA: IBM 709	1960/1961
SOS: IBM 709	1960/1961
IBSYS: 709X	1962
CTSS: IBM 709/7094	1961–1962
SABRE: IBM 7090	1963–1964
OS/360	1964–1965

Third Generation (1965–1976)

OS/360: S/360 Computers	1965
BPS/360	1965–1966
BOS/360	1965–1966
CP-40/CMS: S/360 Model 40	1967
TOS/360	1967
OS/PCP	1968
TSS: S/360 Model 67	1968
OS/MFT	1968
DOS/360	1967–1968
OS/MFT	1968
CP-67/CMS: S/360 Model 67	1969
OS/MVT: Large CPUs	1969
OS/MFT II: Intermediate-sized CPUs	1970–1971
OS/VS1: Medium-sized S/370s	1971–1973
OS/VS2: Large S/370s	1972–1973
VM/370	1973
DOS/VS	1973
OS/SVS	1973–1974
OS/MVS	1974–1975

Fourth Generation (1976–1986)

OS/MVS 3.7	1977–1978
DOS/VSE	1980
DPPX and DPCX: IBM 8100	1980
CPF: S/38	1980
MVS/XA	1981–1986

The dates indicate roughly when these operating systems were widely used; announcement dates frequently were six months to one year before. Third Generation was S/360; Fourth Generation was all S/370.

handle ever larger amounts of resources and transactions more efficiently, greater error detection and recovery procedures, and data security. The ability to handle larger numbers of transactions (jobs) and data with more multiple jobs running concurrently on newer technologies encouraged greater use of 360/370 hardware. With these operating systems, a programmer no longer had to interact extensively with an SCP and so could focus on writing applications.

Another trend in operating systems within IBM influenced the family of 360 and 370 computers which involved yet a different SCP called VM/370. Virtual Machine Facility was an amalgam of three different sets of software, each of which constituted part of an operating system: the Control Program (CP)—a true operating system; the Conversational Monitor System (CMS); and the Remote Spooling and Communications Subsystem (RSCS). The first two were experimental operating systems under development in the mid-1960s, and the third appeared later. CP was an SCP that could use a computer to simulate multiple copies of itself. Hence, with a large enough computer, more than one "system" (each with its own operating system) could function in that computer and share resources.

The intent of CP and CMS was to provide interactive computing and to support programming within IBM on the newly announced S/360s in 1964. In 1966 these two pieces of software ran for the first time on an S/360 Model 40 at the IBM Systems Research and Development Center (today called the IBM Scientific Center) in Cambridge, Massachusetts. In 1972 VM became available to IBM's customers on the S/370.

VM took advantage of virtual memory,* a procedure whereby an operating system could bring into the real memory of a computer out of disk storage only those portions of a program that were going to be run next and do that for multiple programs running at the same time. Using virtual storage (as it was also known) meant a computer's memory could be expanded by the amount of disk storage available to it. Disk storage cost less than a computer's memory. Virtual storage (VS), as it was known when it first appeared in the late 1960s in IBM's operating systems, increased the amount of work a computer could do. The ability to use VS made it possible to develop VM fully which gave end users the impression that they had their own private computer system, even though it was actually shared with other people. CMS provided the mechanics of computing online while VM orchestrated all of the computer's resources, coordinating availability of hardware and software among the humans communicating with the entire system. It was a powerful tool, particularly as CMS's functions expanded. RSCS handled telecommunications between computer and terminal and file management (data transfer). By the mid-1970s, VM also managed "guest" operating systems such as DOS/VS and OS/VS1. Hence, more than one system could function within a computer.

VM was at first used only within IBM. By 1981 more than 50,000 users in over 400 systems within the company were using VM, and by 1986 over 500 systems and thousands more users worked in this environment. As early as the mid-1970s IBM began to market VM. Because it was compatible with S/370s and their operating systems, customers found it useful if they needed to support large-scale interactive computing within their own organizations. By 1985 VM was a primary vehicle for delivering electronic mail systems (such as IBM's PROFS) and supported a large number of the company's scientific and engineering software application packages. The number of major users rose sharply during the 1980s. Typical was Northern Telecom,† manufacturer of telephone switching equipment, which employed VM as the standard operating system for computers it used for engineering and scientific computing. Commercial applications were run on machines using MVS.

IBM had other operating systems which it developed for various minis, particularly DPPX and DPCX (announced in 1978) for the 8100 minicomputers, but these pieces of software were not as significant as the OS/360-370 operating systems. Operating systems for minis, however, mirrored trends and priorities that were evident within that segment of the data processing community involved in the development of operating systems. Because more computers used S/360-S/370 operating systems than any other, the absolute number of dominance of these SCPs insured their place in history. However, other important operating systems were developed by other organizations as well.

One example already mentioned was the early work done at GM in writing an operating system for the IBM 701. Yet other firms (such as Radio Corporation of America—RCA†) worked on systems, while many researchers at various universities conducted important research on the operating system's architecture. Because an operating system could take 20 to 40 percent of a computer's resources, its development was always a major technical concern for those who invented, modified, or designed new computerized technology. By the late 1950s every important computer manufacturer was developing operating systems for its machines. Thus, a user on one found that his or her programs had to be rewritten entirely in order to move to another vendor's devices. The process was made even more chaotic because the user could not go from one machine to another within a particular vendor's product line without the exact same conversion. Hence, the need existed for what came with the S/360. Nonetheless, technical innovations came from other organizations too, many of which were later incorporated into IBM's operating systems. By the same token, functions in OS appeared in other non-IBM operating systems.

The leading developers of operating systems in the late 1950s and early 1960s included IBM, Burroughs, RCA, General Electric (GE),† Philco,† Honeywell,† Bendix, Control Data Corporation (CDC),† and Univac within the United States. In Europe others embarked on similar courses of action. But in addition to the work done primarily at IBM (in particular with IBSYS which made possible asymmetric multiprocessing where two computers could work together) and with SABRE, other companies wrote sophisticated systems.

Burroughs, with its Master Control Program (MCP) for the B5000 computer, was a source of numerous ideas and technical innovations in the early 1960s.

MCP made its appearance in 1963 with such features as multiprogramming, multiprocessing (it could use two computers in a slave/master relationship similar to IBSYS), and high-level language development, and provided source-level language facilities for debugging problems. Most important, it was the first operating system to use virtual storage as a standard feature. These characteristics became standard in all major operating systems by the early 1970s and were incorporated into IBM's by 1970.

Also of importance were developments at the Massachusetts Institute of Technology (MIT) during 1958–1962 which influenced IBM's operating systems, particularly VM. MIT developed a time-sharing system called the Compatible Time-Sharing System (CTSS) which was used across the entire campus. Users could talk to a computer via terminals, enter data and programs simultaneously, while the computer concurrently ran programs already submitted in batch mode, all using this operating system. The developers of VM acknowledged that CTSS was the direct descendant of VM and of similar functions that appeared in TSO, OS, and MVS. The development of CTSS (also called Project MAC*) influenced others outside of IBM, including the developers of RCA's Spectra Series of computers in the late 1960s which were direct competitors to IBM's S/360s.

By the 1970s there was considerable focus on enhanced time-sharing capabilities (often known as third-generation operating systems). The same group at MIT developed MULTICS (for a GE 645 computer) as a further refinement of its earlier work. IBM had its CP-67/CMS, which became VM/370, while RCA announced TSOS, later renamed VMOS. Control Data produced similar systems for smaller computers, including KRONOS for the CDC 6000 which it developed jointly with United Computing Systems. Work on all of these systems took place in the 1960s, with wide usage in the 1970s. VMOS eventually evolved into VS/9 for use on the Univac Series 90 and BS2000 for the Siemens line of computers of the 1970s. Honeywell marketed MULTICS in the late 1970s for its 6180 computer. By the early 1970s the 1100 Series of computers from Univac used for their SCP the 1100 Executive which had online computing capabilities.

Almost independent of these developments was research done at Bell Laboratories, specifically on compilers, high-level languages, and data management software. Bell was also the source of Unix. Two scientists there, Ken Thompson and Dennis M. Ritchie, developed the operating system in the 1960s, completing the first release in 1969. This particular SCP differed from all others in its structure. Unix consisted of many programs that could be put together, much like building blocks, to create an operating system with only those functions desired; other SCPs came prepackaged, with many functions interlocked. In addition, Unix was written so that it could run on a variety of computers from micros to mainframes. It first ran on a mini, the Digital Equipment Corporation (DEC)† PDP-7. Bell Laboratories, AT&T, and other data centers in universities and government agencies were impressed by the features of this operating system. By the 1980s for example, it could run on equipment as varied as the IBM Series 1, DEC PDP minis, on CDC's minis, and on all the major mainframes. It was first marketed by Western Electric, but in the 1970s a commercial version was sold by Interactive Systems Corporation. By 1982 over 3,000 computers worldwide used Unix (1,000 within the Bell community, nearly 2,000 at universities, and at least 600 in government agencies and in private businesses.)

In addition to the software comprising Unix, the two developers also created a programming language called C to help communicate with the SCP. An earlier version of the language was called B, and as of this writing work had continued on a follow-on (possibly D?). In October 1983 the two men received one of the most prestigious forms of recognition within the data processing industry—the A.M. Turing Award—as an acknowledgment of their contribution to the development of operating systems in general and more specifically for their creation of Unix.

The last important arena of activity to involve operating systems concerns microcomputers. These devices first appeared in 1975, and one of the first was the Osborne† 1. Apple Computer† also had a highly successful product line in the late 1970s, and IBM profoundly influenced the market with its widely used PC family of micros during the 1980s. In the beginning these machines sold only hundreds of copies by vendor, but went to tens of thousands by 1980. As of this writing, over 100 vendors were selling over 2 million micros each year. Micros represented a change in data processing that was almost as important as the introduction of the S/360 and for similar reasons: it expanded the use of computing. In the case of micros, for the first time a large segment of the population in the industrialized world could do computing inexpensively in homes, schools, and offices without considerable technical or programming backgrounds. A solid set of skills in data processing was not needed primarily because of the use of efficient, user-friendly operating systems.

The most frequently employed operating system in early micros, the Control Program for Microcomputers (CP/M), was written in late 1973 by Gary Kildall, then a professor at the Naval Postgraduate School at Pacific Grove, California. He did the work on CP/M while involved on another project at Itel,† but he was allowed to market the software as his own. His CP/M was written in two months, initially to link a disk drive* to a microcomputer. By 1977 it had become the standard operating system for such computers and made Kildall a wealthy man.

CP/M initially handled data flow from the micro to disk. Later, Kildall added other, normal functions found in most operating systems, including a debugger and an assembler (both of which were used to write programs), an editor, a BASIC* interpreter, and the functions necessary to print data on a printer or flash on a screen of a CRT. By late 1982 CP/M was serving as the SCP in over 500,000 micros.

The other widely used operating system for microcomputers was MS-DOS—the one IBM chose for its PCs. This software was developed and maintained by Microsoft. It also had several other names and variations. The software was originally called SCP-DOS (after the first developer's name—Seattle Computer Products) but was renamed MS-DOS by Microsoft and, for a while, PC-DOS by IBM; users simply called it DOS. A comparison of functions between this operating system and various releases of CP/M suggests that the one mimicked the other. By 1982 MS-DOS had become the industry standard for all 16-bit micros, and not just for IBM's PCs. Within two years versions of Unix were also available for microcomputers.

The operating systems developed for microcomputers evolved as did others for mainframes. First releases frequently had few but essential functions and little or inadequate documentation. Within several years, they supported tape, printers, diskettes, disk, cathode ray tubes, other devices, BASIC, PL/I*, Pascal, and a half dozen other languages. They took up more lines of code and greater amounts of storage than did operating systems of the 1950s and early 1960s. Yet DOS, for example, remained on one diskette which a user simply inserted into the machine and which in turn loaded itself into the processor. In that sense, it was more advanced than the complicated procedures required to load MVS or Unix into a large computer because they required not only copying large amounts of data into the system but also performing a series of tasks called Initial Program Load (IPL). That last procedure alone could take as little as one hour to complete, or, if installing the operating system for the first time (called generating or "gening"), it could take a couple of days.

Operating systems during the 1950s and 1960s thus evolved rapidly into rich collections of functions, while in the 1970s they became easier to work with on all sizes of machines. During the first half of the 1980s they continued along the previous lines of further automating human tasks, providing additional capabilities for security, data handling, and ease of use by end users oblivious to the household tasks necessary to run increasingly complicated configurations of hardware and software. (See Table 56.) Then functions in the 1970s began to be embedded in chips within computers, the instructions from an operating system being called firmware.

For further information, see: Ruth Ashley and Judi N. Fernandez, *CP/M for the IBM Using CP/M-86* (New York: John Wiley & Sons, 1983) and their *PC-DOS: Using the IBM PC Operating System* (New York: John Wiley & Sons, 1983); M. A. Auslander et al., "The Evolution of the MVS Operating System," *IBM Journal of Research and Development* 25, no. 5 (September 1981): 471–482; L. A. Belady et al., "The IBM History of Memory Management Technology," ibid.: 491–503; G. Bender et al., "Function and Design of DOS/360 and TOS/360," *IBM Systems Journal* 6 (1967): 2–21; W. A. Clark, "The Functional Structure of OS/360, Part III: Data Management," ibid., 5 (1966): 30–51; F. J. Corbato et al., *The Compatible Time-Sharing System, A Programmer's Guide* (Cambridge, Mass.: MIT Press, 1963); R. J. Creasy, "The Origin of the VM/370 Time-Sharing System," *IBM Journal of Research and Development* 25, no. 5 (September 1981): 483–490; Paul Freiberger and Michael Swaine, *Fire in the Valley: The Making of the Personal Computer* (Berkeley, Calif.: Osborne/McGraw-Hill, 1984); S. C. Kiely, "An Operating System for Distributed Processing—DPPX," *IBM Systems Journal* 18 (1979): 507–525; G. H. Mealy, "The Functional Structure of OS/360, Part I: Introductory Survey," ibid., 5 (1966): 3–11; R. P. Parmelee et al., "Virtual Storage and Virtual Machine Concepts," ibid., 11 (1972): 99–130; S. Rosen, ed., *Programming Systems*

Table 56

Features of Operating Systems, 1951–1980

First Generation (1951-1955)

Basic assemblers
Utilities
Sorts
Card I/O controls
Some data handling

Second Generation (1955-1964)

Macro assembler
FORTRAN compiler
COBOL compiler
RPG
Additional utilities
Faster, larger sorts
Basic operating systems
Batch processing

Third Generation (1963-1968)

Additional language compilers
Transaction processing
Multiprogramming (limited)
Multiprocessing (usually limited)
Real-time processing
Time sharing
Virtual storage
Single-mode operating systems

Fourth Generation (1965-Present)

Virtual machines
Data communications
Database management
Management information systems
Interactive query languages
Programmable I/O devices (e.g., disk, mass storage)

Fifth Generation (1980-   )

Interactive application development systems
Advanced high-level languages
Integrated packages, graphics
Decision support systems
Advanced distributed processing
Artificial intelligence (e.g., robotics)

The dates in this table differ from those in Table 55 because fundamental changes across all operating systems, not just IBM's, are reflected here. The typology presented is subject to controversy (e.g., are we really into a fifth generation?) but was chosen for convenience.

SOURCE: Most of the above material was drawn from Marilyn Bohl, *Information Processing* (Chicago: Science Research Associates, Inc., 1984): 404–405, but for a difference of opinion, see also, Norman Weizer, "A History of Operating Systems," *Datamation* (January 1981): 119–120, passim.

and Languages (New York: McGraw-Hill, 1967); Norman Weizer, "A History of Operating Systems," *Datamation* (January 1981): 119–120, 122, 125–126; B. I. Witt, "The Functional Structure of OS/360, Part II: Job and Task Management," *IBM Systems Journal* 5 (1966): 12–29.

ORDVAC. This computer was built at the University of Illinois under contract with the Ballistic Research Laboratories of the Aberdeen Proving Grounds, Maryland, between April 1949 and March 1952. It was constructed at the same time as the ILLIAC* and was the first major product of the University of Illinois' computer research in the 1950s and 1960s. The University of Illinois was one of the first American universities to install a computer for normal administrative applications. The ORDVAC was also one of the first computers of the type described by John von Neumann** in the mid-1940s, employing many of the design criteria established for the computer von Neumann built at the Institute for Advanced Study (IAS) in the late 1940s.

Several future pioneers in the development of the modern computer worked on ORDVAC, including James E. Robertson of the University of Illinois and then a graduate student, Ralph Meagher, chief engineer on the project, and A. H. Taub, who had also been associated with the Instutute for Advanced Study at Princeton.

ORDVAC was an electronic digital computer* and incorporated the concept of the stored program. Many of its design features were copied from the IAS computer,* including that for registers, and some of its circuits design. The

Table 57
Technical Features of ORDVAC

Technology	Parallel
Word length (bits)	40
Instruction length	20
Instruction format	1-address
Main store size	1,024
Main store type	CRT
Backing store type	None
Average add time	72 micro-second
Average multiply time	732 micro-second
Basic clock frequency	Asynch.
Approximate number of valves	2,178
Approximate number of GE diodes	None

fourteen engineers associated with the project also used the then popular electrostatic Williams tube memory which could house 1,024 40-bit words (Table 57). Cycle time was 18 microseconds, which was normal for that era. Cathode ray tubes stored data in increments of 1,024 bits arranged in parallel function.

For further information, see: H. H. Goldstine, *The Computer from Pascal to von Neumann* (Princeton, N.J.: Princeton University Press, 1972); James E. Robertson, "The ORDVAC and the ILLIAC," in N. Metropolis et al., eds., *A History of Computing in the Twentieth Century* (New York: Academic Press, 1980): 347–364; Staff of Digital Computer Laboratory, *ORDVAC Manual* (Urbana: Digital Computer Laboratory, University of Illinois, 1952).

P

PAPER TAPE. Paper tape, one of the oldest media for passing data to and from machines, has been used for submitting and storing information, and ultimately as output. By the end of the nineteenth century, it was used in a variety of noncomputerized equipment such as telegraphs, cash registers, and adding machines. During the early days of computer-related work in the twentieth century, almost every configuration relied on paper tape either as input or as a combination of input/output. Although the notion of using paper tape with punched holes to represent information had been discussed or planned for several centuries, it was not applied effectively until 1857 when Sir Charles Wheatstone developed a punched paper tape for use in telegraphy. That event came only twenty-one years after the invention of the telegraph itself. Yet paper tape quickly became a practical vehicle for trapping information off a telegraph. In 1858 a Morse tape reader-transmitter existed which had a speed of 100 words per minute.

Advances in the quality of paper tape and associated equipment followed in the second half of the nineteenth century, primarily for telegraphic communications. By the early twentieth century, almost a decade before the end of World War I, they had advanced to the point where five-track tape keyboard punches were in wide use. Advances in communications were also evident with paper tape. For example, in 1925 a five-track reader that operated at four letters could be acquired. In data processing terms, such a device functioned at 20 bits per second. That made it fast enough to be used in conjunction with experimental equipment, such as relay machines in telephone systems, for advanced telegraphic and stock quoting equipment, and, in the 1930s, as input/output devices for electromechanical computational devices. By the early 1920s data could also be multiplexed in a communications network, increasing the line speed of a tape device to 80 bits per second. By the late 1940s such technology was actively employed in data processing, forming a major input/output vehicle. Almost every major computer project of the 1930s and 1940s used paper tape.

In subsequent decades the medium became more versatile. The number of tracks one could read or write on grew to eight, and the speed at which such tape could be used increased as well. But the application remained essentially unchanged. Sprocketed holes along the edge caught onto a toothed wheel that rotated the tape over a surface, which in turn either had spikes to punch holes in the paper or sensors to read and then interpret these holes. In devices appearing by the 1960s these holes were also being read by photoelectric heads.

Within data processing, five-track tape was common through first-generation computers (1940s). By the late 1950s vendors were offering six-track tape which raised the combination of codes possible on tape to sixty-four. A seven-track tape quickly became available offering parity checking, such as in International Business Machines Corporation's (IBM's)† products of the period. Tape was typically three-quarters of an inch wide until the early 1960s when eight-track tape led to one-inch wide paper, also increasing the number of combinations of code usable to sixty-five. By then, IBM's tape had become the de facto industry standard.

Historians noted that paper tape stabilized at eight-track as a result of the large character sets used with third-generation computers beginning in the mid-1960s. Such computers handled alphabetic data in upper and lower cases and a wide variety of special symbols. These were in addition to the more traditional numbers. Other codes were used to transmit control instructions. Formal standards were established, the most famous and current being ISO Standard No. 646 (1973) which dictated formats in the 1970s and 1980s. The format of tape also influenced ASCII's form, a standard method of laying out bits and bytes (letters and words) in machine-readable data.

Traditionally, tape came in rolls of 1,000 feet and, over the decades, varied in width from a fat 2 to 3 inches down to less than 7/8 of an inch; 11/16 of an inch was relatively standard for telegraphy, and 7/8 or 1 inch for computers. Paper has more often than not been used but so have other substances. Konrad Zuse** and his series of Z computers* in the late 1930s and early 1940s used exposed film, whereas plastic appeared on some experimental computers of the early 1950s. Polyester plastic tape, a suitable, tough tape, was in much evidence by the 1960s. In addition to being used by telegraph operators and computer centers, they remained in vogue down to the present for adding machines and cash registers, making tape one of the oldest output media in the Western world.

For further information, see: Charles J. Bashe et al., *IBM's Early Computers* (Cambridge, Mass.: MIT Press, 1986); Douglas R. Hartree, *Calculating Machines: Recent and Prospective Developments and Their Impact on Mathematical Physics and Calculating Instruments and Machines* (Cambridge, Mass.: MIT Press and Tomash Publishers, Los Angeles, Calif., 1984).

PDP SERIES. This series of widely used minicomputers was constructed by the Digital Equipment Corporation (DEC)† during the 1960s and 1970s. This series was officially known as the Programmed Data Processor but was better known simply as PDPs. They consisted of four different families of processors, complete with peripheral equipment and software.* Membership was determined by the word lengths employed by each processor. They were extremely popular with engineers, particularly those employed in American manufacturing companies. Students of engineering were also important users of these machines.

Various lines acquired unique characteristics. Thus, the 12-bit family reflected decreasing costs for processing power over time and included the PDP-5, 8, and 12. The 18-bit devices focused on more performance at the same price and included the PDP-1, 4, 7, 9, and 15. The 16-bit family offered a wide range of growth and size. This particular group of machines was called the PDP-11 and was made up of various models, all labeled PDP-11/ then the model number, for example, 11/34 or 11/04. The 36-bit units were very large and often were not even called minicomputers. The PDP-6 and the PDP-10 made up the two primary members of that family.

The PDP-1, the company's first machine, appeared in December 1959. Its cost at approximately $120,000 proved very attractive when compared to the cost of competitive systems, which typically sold for more than $1 million each. Although never formally announced, the PDP-3 was constructed privately by the Scientific Engineering Institute in Waltham, Massachusetts, as a 36-bit machine in 1960; it continued to run into the 1970s. The PDP-4, like the PDP-1, was an 18-bit device, had slower memory* and more efficient architecture, and sold for $65,000. The PDP-5 (introduced in 1963) was the first commercially manufactured minicomputer. Earlier machines were little more than tailor-made devices, assembled almost to order. The PDP-6 came out in 1964 as the company's first large processor. It had nearly one megabyte of memory and used a 36-bit word. It was introduced with a time-sharing software package, the first such offering by a commercial computer vendor. The PDP-7, announced in 1964, was a smaller version of the PDP-4. A variant called the PDP-7A was introduced a year later.

The PDP-8 series was particularly significant for the company. These appeared in 1965 and sold in larger quantities than earlier machines. At a price of less than $20,000 for a simple configuration, they made it possible for minicomputers to be widely used, particularly in American industry during 1965–1973. Over time they developed in size and function. The PDP-8/S began with 4K of memory and sold for less than $10,000, while the PDP-8/I and PDP-8/L introduced integrated circuits and therefore qualified as third-generation computers. Additional models also incorporated integrated memory, program-controlled I/O, and access to memory into an omnibus, thereby increasing performance while continuing the downward spiral in costs. By 1977 the PDP-8 computer's processor lived on one CMOS chip.* By that time, a PDP-8 could be acquired for about $2,500. That price made it convenient to use PDPs for such applications as word processing, limited time-sharing, and in process control.

The PDP-7 was replaced with the PDP-9 (1968) and the PDP-9/L (1970), both of which were less expensive models with more memory. Technically, it was distinguished by the first use of microprogramming in a DEC machine. The

PDP-10 replaced the large PDP-6 during the 1970s with four models (KA10, K110, K110, and K120), each of which represented improvements in performance. Its operating system was TOPS-10, which was replaced in 1976 with TOPS-20 as an alternative enhancement of mixed batch and time-sharing processing with real-time computing. That series of machines was particularly attractive to universities. This family, first introduced in 1963, remained essentially the same until the 1980s; over 1,500 were sold.

The PDP-11, a 16-bit machine, was introduced in 1970 as the PDP-11/20. By 1980 there were more than a dozen models, and they dominated DEC's product line. The PDP-11 was also the most widely used minicomputer in the world. This family of processors was particularly attractive because it combined improved performance and cost simultaneously made available over a wide range of models during the course of its life in the 1970s. Its operating systems* evolved from simple paper tape* and disk-based systems to a variety of software tools. Hardware remained compatible across all models but not always so with software. Some 50,000 of these machines were sold between 1970 and the end of 1978.

Owing to the limitations imposed by 16-bit words, DEC chose to announce a follow-on called the VAX (Virtual Address Extension) series in 1978. The first model was the VAX-11/780 and employed a 32-bit virtual addressing scheme. By the mid-1980s a wide variety of models had been introduced, dominating the product line.

Other members of the PDP family are also of interest. The PDP-12, for example, was a processor that could run programs originally written for the PDP-8 or the Lincoln Laboratory† Instrument Computer, otherwise known as LINC in the 1960s. It was first shipped in 1969, and in time some 1,000 were sold. The PDP-14 and two subsequent models, the PDP-14/30 and the PDP-14/35, were constructed to offer a relay logic control processor. Thus, it could frequently evaluate Boolean equations in industrial applications. The PDP-15 replaced the PDP-9 in 1970. It had a few technical improvements, used integrated circuitry, was less expensive than its predecessor, and was one-third the size of the PDP-9. DEC designed the PDP-16 for applications whose cost could not be justified on a general-purpose computer. It was simply printed circuit modules. It was introduced in 1971 and later renamed Register Transfer Modules (RTMs). It faded quickly into oblivion as the cost of computing dropped in the 1970s to the point where it became relatively easy to justify the cost of many commercially built computers for applications which a few years earlier could not have been rationalized.

The entire PDP family proved to be an important set of computers and the most significant minicomputer. By 1982 over 250,000 had been sold worldwide. Their architectures influenced those of other minicomputers. The acceptance of PDPs made distributed processing and stand-alone computing (usually dedicated to specific applications) a reality by 1970. These uses were a major feature of computing throughout the 1970s and into the 1980s, operating side by side with more traditional large, mainframe computing in data centers.

For further information, see: C. Gordon Bell et al., *Computer Engineering: A DEC View of Hardware Systems Design* (Bedford, Mass.: Digital Press, 1978); H. M. Levy and R. H. Eckhouse, Jr., *Computer Programming and Architecture: The VAX–11* (Bedford, Mass.: Digital Press, 1980).

PEGASUS. This stored-program digital computer* was built and sold in Great Britain in the late 1950s. It was one of the most popular computers marketed by the Ferranti Company, Ltd.†

There were two commercial models. The PEGASUS 1 appeared in 1956, and in time the company sold twenty-six, three outside of Great Britain. The second device, called PEGASUS 2, was first shipped in 1959. Twelve of these were sold, one outside Great Britain. In conjunction with support from the National Research Development Corporation (NRDC),† which was the vehicle used by the British government to support many computer projects, it had been decided by the mid-1950s to attempt a large computer using virtually "off-the shelf" components and modified existing technologies. As early as 1953, a researcher named Christopher Strachey had been developing concepts that would appear in the PEGASUS. He proposed a machine that would not optimize programming (a departure from the past). He argued that programmers spent more time than it was worth writing tightly written, short programs all in an effort to preserve precious hardware resource. He believed that components were sufficiently large (in terms of capacity and speed) and inexpensive to eliminate the need to be so careful with resource. Yet he wanted the computer to be designed to facilitate programming in general. The entire machine had to be built relatively inexpensively and therefore be cheap to buy. In 1954 Ferranti Ltd. began work on what ultimately would be the PEGASUS, although at first it was called the Ferranti Packaged Computer (FPC1). The NRDC agreed to acquire ten copies of the machine when completed.

The construction of the machine involved new packaging techniques in which subassemblies of components were built and then lashed together to construct the computer. That approach to manufacturing would soon become standard throughout the computer industry but was new at the time when computers were built as "one-of-a-kind" machines. Modularity did allow for standard, low-cost quantity production. Malfunctioning parts were simply exchanged for similar, new components, reducing the amount of maintenance and time involved. New enhancements could be introduced in a modular fashion, thereby allowing engineering changes to be introduced into already built machines. Initially, there were technical problems, particularly in connecting subassemblies, and accounted for minor differences with packages of components. Redundancy of parts reached 10 percent of the computer, but the ability to swap subassemblies

Table 58
Technical Features of PEGASUS

Technology	Valve
Word length	39
Store size, fast	56
Store type, fast	Nickel
Store size, backup	5,120
Store type, backup	Drum
FXPT[a] add time	300 micro-seconds
FLPT[b] add time	Unknown
FXPT multiplication time	2 ms
FLPT multiplication time	Unknown
Date first delivered	1956
Approximate basic cost	50,000 British pounds

[a]FXPT means fixed-point arithmetic.
[b]FLPT means floating-point arithmetic.
SOURCE: Reprinted with permission from *Early British Computers*, Simon Lavington, First Edition. Copyright © Digital Press/Digital Equipment Corporation (Bedford, MA), 1980.

as part of the process of maintenance made "down time" drop, justifying the additional cost of components. By the 1960s redundancy of parts was standard in almost every line of computers throughout the world. That strategy was arrived at independently in every country, but in the case of manufacturers in Great Britain, and possibly in the United States, PEGASUS suggested a new way.

The PEGASUS used a fifty-six-word fast store (memory*) made out of nickel delay lines (Table 58). A magnetic drum added an additional capacity of 5,120 words. It had a full complement of registers to handle input/output. Words had a length of 39 bits, while two additional 19-bit instructions also were packaged into each word. An extra bit was used to help in the manual debugging of programs. The computer also used a parity-bit in each word for checking. The PEGASUS was designed to use a large number of instructions, which made programming easier. It had eight accumulators, all except one of which could be employed as index registers. The entire processor was so attractive that many of its features appeared in subsequent computers built by Ferranti, including the ARGUS, ORION, and FP6000, and even in the ICL 1900 series of the 1960s.

The PEGASUS had a relatively large memory when compared to its peers but operated at the same speed as, for example, the Elliott 402. The machine performed well in terms of both function provided and physical durability. The first PEGASUS shipped (in 1956) remained in service for thirteen years—nearly twice the average time for a first-generation processor. The PEGASUS 2 added refinements and additional capacity to the series. Users considered both models reliable and held fond memories for these not evidenced with other British computers. For Ferranti, the PEGASUS maintained the company as a serious vendor in the young data processing industry in Great Britain.

For further information, see: Simon Lavington, *Early British Computers* (Bedford, Mass.: Digital Press, 1980).

PILOT ACE. See ACE

PL/I. PL/I was the second most widely used programming language by International Business Machines Corporation's (IBM's)† customers for commercial applications and for scientific and engineering work. COBOL* remained first in the area of commercial applications, but by the early 1980s PL/I had surpassed FORTRAN* in the scientific area. This software* tool was developed in the early 1960s and was used extensively by companies in the United States beginning in the mid-1970s and in American universities, particularly engineering departments, by the early 1970s. PL/I was considered a third-generation multipurpose programming language. It was the product of a joint effort by IBM and two groups of customers—SHARE† and GUIDE†— who wanted to develop a language that improved on the features of FORTRAN, which was the most extensively used language in the late 1950s. The leadership necessary to launch the language came from IBM's study team and members of SHARE.

FORTRAN, though recognized as a good tool for scientific and mathematical applications, was unable to handle character and alphanumeric data (needed for commercial applications), features more evident in COBOL. Furthermore, with the introduction of new data processing hardware, it also worked less efficiently than with older computers and peripheral equipment. Just as disconcerting was FORTRAN's failure to take full advantage of functions appearing in new operating systems.* Therefore, in September 1963 IBM (the developer of FORTRAN) and SHARE agreed to form a committee charged with improving the language. Known as the Advanced Language Development Committee (Table 59), it represented a continuation of a long-standing effort known as the SHARE FORTRAN project. Forming an industry committee to attack the problem was a common strategy for programming languages.* Already ALGOL* and COBOL had been created that way, and each major language was monitored by such committees.

The new group was asked to make recommendations that would allow a larger community of users to apply FORTRAN to their programming needs without compromising the language's effectiveness for mathematicians, scientists, and especially engineers. The initial intent was to enhance FORTRAN, but the members of the committee were not necessarily committed to that idea. Yet both IBM and many users expected FORTRAN to be improved. The committee, however, quickly recognized that the kinds of improvements necessary for the language would make compatibility with FORTRAN impossible. There were too many applications which FORTRAN could not support, including transactions such as using blanks or establishing card formats. After considerable

Table 59
Members of the SHARE Advanced Language Development Committee

Member[a]	Company
Representing IBM	
George Radin	IBM
C. W. Medlock	IBM
Bernice Weitzenhoffer	IBM
Representing SHARE	
Hans Berg	Lockheed
James Cox	Union Carbide
Bruce Rosenblatt[b]	Standard Oil of California

[a]Additional help was provided by Tom Martin of Westinghouse, H. Paul Rogoway of Hughes Dynamics, Larry Brown of IBM, and Ray Larner of IBM.
[b]Served as chairman of the committee.

deliberations, the group decided to recommend that a new language be developed and so reported to SHARE on March 1, 1964.

FORTRAN was a well-established language with many users who had invested large sums of money in it. The suggestion that SHARE and IBM develop a new language was therefore a bold one and was initially received with mixed feelings by both sides. Within IBM this new, still unnamed language was designated the New Programming Language (NPL), but because the initials also represented the title of the National Physical Laboratory in Great Britian, it had to be changed. The final designation, PL/I (pronounced PL/one), was not an acronym but the actual name of the language.

Based on the reports generated by SHARE during 1964 (March, April, June, and December), IBM established a department to design and develop the language, giving prime responsibility for the project to its laboratory at Hursley in Great Britain. IBM issued its first manual on PL/I in the first half of 1965 along with a compiler, and in August 1966, another set of documentation and compilers appeared that worked with the S/360* family of computers and operating systems. In subsequent years new releases were developed, and variations of the language appeared from others. Byproducts included NICOL I from the Massachusetts Computer Association and EPL at both Bell Laboratories† and the Massachusetts Institute of Technology (MIT).

The language was well received, and today it is one of the most widely used languages in the United States and Western Europe. IBM marketed PL/I aggressively, increasing its acceptance within the data processing community. But PL/I was successful largely because it helped address some of the problems programmers faced in the early 1960s. By the start of 1963, for example, programming had separated into two camps: scientific/engineering and commercial/business. Each had its own needs, languages, and equipment; yet it was becoming increasingly apparent that the functions available on one side should be shared with the other. Thus, for example, commercial applications increasingly required functions existing with FORTRAN such as floating-point arithmetic, fast computation, and the ability to work with matrices. Scientific users also wanted capabilities found in COBOL such as the use of decimal arithmetic, rapid input/output (I/O), and sort programs. Data centers objected to maintaining two sets of hardware and software* because of the cost involved and the complexity and redundancy of managing two staffs. Meanwhile, with its announcement of the S/360* in 1964, IBM was rapidly reaching the point where one product line of computers, operating systems, and peripherals could be used by both scientific and commercial users. It was the effort to bridge these two worlds of data processing users that created the conditions for accepting PL/I.

Quality also helped. The Hursley lab produced clearly described software, focused on making it usable, and thoroughly documented it. The lab satisfied the requirements of providing functions originally evident in COBOL and FORTRAN, often making these features more efficient than in older languages. As a result, by the early 1970s, in addition to scientific users still relying heavily on FORTRAN, many were also considering PL/I as an option. COBOL users continued to prefer their business-oriented language, but those working with engineering applications increasingly were acquiring PL/I. George Radin, who was heavily involved in launching PL/I and represented IBM, noted in 1981 that IBM had sold more licenses of the new language than it had for FORTRAN to its users of S/370* computers.

The design criteria offered yet another explanation for the success of the language. The original committee, and later IBM's staffs, wanted to design a language that had considerable freedom of expression while taking full advantage of modern operating systems that no longer required programmers to code in assembly language. By using functions in modern operating systems, many household chores required in earlier programming were no longer necessary, leaving programmers free to focus on the applications themselves. Modularity allowed programmers to ignore functions within the language without worrying that their programs would not compile and run. An attempt was also made to keep the language machine independent so that it could operate in a variety of computers. Finally, the developers wanted to make PL/I easy for new and inexperienced programmers to learn and use. Radin noted years later that "convenience was the key word rather than simplicity." It was to be "a language designed more in the spirit of English than of Mathematics."

Much of the language's structure (such as external and internal procedures and block-structured formats) was similar to that found in ALGOL. One of the greatest benefits of PL/I was its ability to handle a large range of data types: numeric, character/bit strings, pointers, files, entry/labels, mixed type expressions, and implicit or explicit declarations, while incorporating such storage classes as static, automatic, based, and controlled. Numbers could be binary; decimal, fixed and floating point became available; and complex or real descriptions could be used. Characters could be in fixed or variable length. These features allowed the programmer greater flexibility in mixing and matching data and using it in programs than in older languages. In addition, the developers of the language added other features which made the language more convenient, namely, including extensive I/O support, the ability to handle multi-tasks, additional compile-time facilities, and ON-conditions.

PL/I was considered one of the most general-purpose languages available, with a language rich in function and a succinct rather than English-like notation. Its intellectual ancestors were ALGOL and FORTRAN rather than COBOL. It was less error-prone than older languages and was considered easier to read and write. In the final analysis, it culminated an era in the development of procedure-oriented languages by encompassing all the best features that had been discovered, implemented, or developed in earlier languages.

Yet PL/I did not remain frozen in time. Throughout the 1960s and 1970s IBM and SHARE worked on functional enhancements, new releases reflecting changes in operating systems and the availability of newer equipment, and expanded the documentation available on it. During the 1970s many vendors began marketing application packages, some of which were written in PL/I. By the early 1980s the language had become almost universally available on all of IBM's computer systems, those of IBM-compatible vendors, and some non-IBM systems.

PL/I was the culmination of a generation of language development that saw the creation of ALGOL, COBOL, FORTRAN, JOVIAL,* and many other languages, using their best features. It became one of the earliest languages to interact with operating systems to coordinate such tasks as the allocation of storage, tasking, and interrupt handling. Its ability to handle default conditions, although not new to languages in the mid-1960s, became very efficient within PL/I. The use of macro facilities in which clusters of commands could be represented by one command became a major feature of almost all languages that came after PL/I. This feature greatly enhanced productivity because it allowed programmers to write fewer lines of code to do a task, thereby speeding up the process of writing programs and at a lower cost.

The original committee that established PL/I proved effective. Its members also turned out to be very wise in their selection of functions from existing languages. From FORTRAN they borrowed such ideas as independently compiled subroutines each sharing common data; from COBOL they took the concepts of data structures and its facilities for handling I/O and its manner of generating reports; and from ALGOL, which for many represented the ideal of a universal programming language, they copied methods for handling block structures and recursions. The net result was a language that even in its earliest releases could be used for large and complex applications, both scientific and commercial. It became a language through which programmers could even develop systems programs. The programmer never had to write any assembly language (lower level, very complex programming), thereby facilitating the task of programming. Proof of the wisdom of this committee's work and of the efficiency of IBM's development exercises was the wide acceptance of the language and its use, particularly during the 1970s.

This language developed much as other idioms of the late 1950s and early 1960s. Already mentioned was the reliance of an industrywide committee. As with other languages, original ideas concerning a possible language were frequently the work of a handful of people, often fewer than a dozen. Typically, specifications were reported out of committee and modified within the first eighteen months and often in less than a year. Several releases of compilers, interpreters, and documentation appeared within the first year and a half. Successful languages always had the support of a particular vendor or organization. PL/I had both SHARE and IBM to encourage its use. Like other languages, PL/I was quickly standardized and policed by the industry to reduce the number of variations. When this practice was not followed, a language ran the real risk of not being used. Over the years, new releases kept PL/I compatible with newer operating systems and hardware, while incorporating functions requested primarily by users, such as members of SHARE who sat on committees charged with monitoring the language.

For further information, see: R. W. Conway and D. Gries, *A Primer on Structured Programming Using PL/I, PL/C, and PL/CT* (Cambridge, Mass.: Winthrop Publishing Co., 1976); B. W. Kernigham and D. M. Ritchie, *The C Programming Language* (Englewood Cliffs, N.J.: Prentice-Hall, 1978); George Radin, "The Early History and Characteristics of PL/I," in Richard L. Wexelblat, ed., *History of Programming Languages* (New York: Academic Press, 1981): 551–589; Jean E. Sammet, *Programming Languages: History and Fundamentals* (Englewood Cliffs, N.J.: Prentice-Hall, 1969).

PRINT. PR-edited Interpretive system (PRINT) was an early programming language* designed to perform numerical scientific problem-solving using specifically an IBM 705* computer. Its intent was to provide simulations of floating-point instructions, and hence it was an interpretive programming system. Development of the language began in February 1956, and it was first delivered to a customer in July 1956. PRINT appeared before International Business Machines Corporation's (IBM's)† widely used FORTRAN.* PRINT had a series of operation codes, variable fields, arithmetic functions, some mathematical commands, and a series to handle input/output, testing, and other housekeeping functions. It was of the same generation as Remington Rand's† A-2* and A-3.* PRINT was one of several efforts made in the mid-1950s to allow scientific computing on specific digital computers* intended for the general data processing market.

For further information, see: R. W. Bemer, "PRINT 1—An Automatic Coding System for the IBM 705," *Automatic Coding, Journal of the Franklin Institute*, Monograph No. 3 (Philadelphia: Franklin Institute, April 1957): 29–36.

PRODUCT INTEGRAPH. This advanced adding machine was built by Vannevar Bush** in 1920 to aid engineers. He constructed variations during the first half of the 1920s at the Massachusetts Institute of Technology (MIT). When completed, the machine could solve problems in electrical theory in hours as opposed to months. He and his staff would plot equations to be integrated, after which they were passed under pointers in the machine. Operators insured that the pointers stayed on the curves that had been hand-drawn earlier. As the pointers went back and forth (also described as up and down), the flow of electricity changed as measured by an ordinary electricity meter. That meter in turn caused a motor controlling a pencil to pass over moving paper. The curve drawn represented the results of the problem.

Like many of Bush's machines, the integraph was a long contraption on a table-like form. The machine gave him considerable and early experience with electrical components and electronic calculations. From product integraphs he moved to solving equations related to electrical power failures. Within a few years Bush had learned to solve differential equations using what by 1930 came to be known as the differential analyzer.*

For further information, see: Charles Eames and Ray Eames, *A Computer Perspective* (Cambridge, Mass.: Harvard University Press, 1973).

PROGRAMMING LANGUAGES. Although the entire field of data processing languages has not evolved as rapidly as computers themselves, the availability of programming languages has permitted the efficient use of complex hardware. Programming languages are part of the entire subject of software.* Other components of the topic include operating systems,* database managers, application programs, and productivity tools such as report generators, spreadsheets, and graphics.

In its simplist form, a program is a collection of instructions or actions that cause a computer to interpret human commands and that it can then carry out. The International Standards Organization (ISO), which has provided definitions for many terms in data processing, suggests that a programming language is "an artificial language established for expressing computer programs." That is, it is a set of syntax (rules of grammar) and characters (like words) used to write computer programs. Part of the confusion in an otherwise straightforward definition of programming languages is that they have been characterized as either machine-level code or higher level languages. Code (another word synonymous with a program) is slang for what by the end of the 1940s appeared as a rapidly growing idiom of phrases to describe technical features of data processing.

Machine-level code is that set of instructions which a computer recognizes directly and acts on. Typically, these non-English-looking instructions were some of the earliest programming done for computers in the 1930s and 1940s. A programmer wrote programs directly in the language recognized by a computer.

Some of the more formal sets of such languages, which became common during the 1950s and even 1960s, included assembly languages. A widely used report generator called RPG was frequently classified with this type of language. Of greater concern to historians of computing, however, are higher level programming languages. The ISO refers to high-level languages as any programming idiom "that does not reflect the structure of any one given computer or that of any given class of computers." Common examples include COBOL,* FORTRAN,* PL/I,* and ALGOL.*

Jean Sammet** has constructed a taxonomy of programming idioms that has helped decide what constitutes such a language versus other programming tools. She argues, first, that a high-level programming language does not require a person to know machine-level code (hence, a lower level language directly interpretable by a machine). Second, such a language can probably be run on various types of computers. One coding (writing) in COBOL for an International Business Machines Corporation (IBM)† computer can expect to use the same or similar set of symbols to write programs in COBOL to operate on a Sperry Rand† computer. Therefore, portability from one system to another is important. Third, one instruction written in higher level languages will become multiple instructions when converted by the computer (that is, compiled) into machine-level code, which it then executes. In low-level coding one has to write each instruction necessary to conduct a task, while in higher level languages one task might require only a single line of instruction. Human programmers obviously gained speed and efficiency with this development. Finally, Sammet suggests that a high-level programming language has a notation that is problem-oriented in that its idiom is closer to "the original conceptual statement of the problem than" to machine instructions necessary to execute appropriate commands. Her four tests, first espoused in the late 1960s, have helped to delineate high-level languages from other software, some of which had language-like properties.

Since the vast majority of all programs written from the late 1960s to the present have been in high-level languages, this group of software tools constitutes an important body of programming languages. They are also the tools of data processing which have encouraged programmers to write additional applications for computers.

Programming languages have undergone four phases of development. The first constituted all kinds of instructions to machines, ranging from wires guiding looms in the early 1800s to Herman Hollerith's** card punch equipment of the 1890s down to primitive programming by Konrad Zuse** in the early 1940s. The second period, which began around the end of World War II and ended in the early 1950s, was characterized by programming in low-level code. A third commenced with the development of FORTRAN—one of the first higher level languages—in 1952 and ran to approximately the end of the 1950s. This period is considered the golden age of programming because much of today's technology first emerged in that era. A fourth phase ran from around 1960 or 1961 to the early 1970s during which many high-level languages became widely used tools contributing significantly to the expansion in the use of data processing. It was the age of COBOL and the period that saw the first use of PL/I. A final stage, less distinctive than the fourth, existed from the early to mid-1970s to the present. Many new languages appeared (LOGO and Pascal, for example) which became widely used on large computers. BASIC* came into its own as one of the dominant, if not most extensively employed, languages for microcomputers.

These periods are aribitrarily delineated and have become more difficult to define in the 1980s. Even so, they offer a frame of reference for examining changes in the structure and use of programming languages over time. The one exception was the first period, which one could argue had no programming languages. It was an important period nonetheless, for people had attempted to communicate instructions to machines long before programmers sat down to "write" programs with pencil, flowchart, and pad. Those early efforts became a formalized activity that influenced the early attempts to create languages in data processing.

Programming, that is, giving instructions to a machine, dates back to the seventeenth century when such experimenters as Gottfried Wilhelm von Leibniz** (1646–1716) and Blaise Pascal** (1622–1703) built calculating machines. When wheels and gears were correctly positioned and moved, these machines could perform simple mathematical steps such as adding and subtracting. More sophisticated was the work of Joseph Marie Jacquard** (1752–1834) who, in 1801, built a device that could do automated pattern weaving. Cards with holes were used to direct threads to the loom, creating predefined patterns of cloth. The pattern was determined by the arrangement of holes on the cards, with wire hooks passing through such holes to grab and pull through specific threads to be weaved into the cloth. The process was so successful that thousands of such looms were in use in Spain and France within one decade.

By the 1880s Herman Hollerith (1860–1929) was employing the idea of punched cards, giving machines instructions. His idea was to have each hole in a card represent a number intelligible to a card reader or card punch. The combination of such cards could be interpreted by his machines to be data or instructions. The use of cards formed a crucial background for future computers because, throughout most of the twentieth century, information (data or instructions) has been fed into or received from computers through this medium. In fact, not until the 1970s did other techniques become more widely used to take information and instructions to and from computers. When programmers first began to write software in the 1940s, such information frequently was first captured (punched) on cards, with holes representing numbers and later letters and other symbols. These were then "fed" to the computer.

By the 1930s, in addition to the use of card punch equipment by businesses and government agencies, scientists were already experimenting with the use of electrical impulses to instruct computer-like devices. These involved using the "on" and "off" characteristics of electricity in binary combinations and employing Boolean logic (algebraic expressions). Thus, a series of ones and zeros reflected the on/off of electrical behavior and became the basis of the earliest machine-readable instructions. From that beginning and the use of electrical devices in card punch equipment came the "wiring" of boards and machines so that circuits ran in a predescribed manner as the earliest form of programming. Another early example of programming, in addition to punched card devices, was the work done at the Massachusetts Institute of Technology (MIT) by Vannevar Bush** (1890–1974) with analog computers* in the 1930s. Digital computers,* which came in the next decade, employed similar techniques. These included the computers built by John Presper Eckert, Jr.,** and John Mauchly** (1907–1980) at the Moore School of Electrical Engineering† (ENIAC,* EDVAC,* BINAC,* and, to a certain extent, their UNIVAC* series built in the 1950s).

But the use of something other than hardware to do the actual programming did not first emerge until around 1945. One of the earliest serious cases of a shift to nonhardware programming involved the work of Konrad Zuse, a German inventor of computers. Zuse first began work on such devices in the 1930s and developed other machines into the 1950s. In 1945 he described a language (which was never implemented) called *Plankalkül* (Plan Calculus), which involved describing algebraic expressions usable by a machine. These could be translated into the yes/no capability of the electronics involved. His conceptual design even detailed how complicated problems might be handled. Theoretical work on solving scientific numeric problems during World War II also paralleled some of Zuse's thinking in the United States and in Great Britain. Study of the relationship between a computer's architecture and its programming culminated in the work of John von Neumann** (1903–1957). In 1945 he wrote one of the first, most complete papers on the design of computers which addressed the issue of programming as well and influenced the construction of digital computers to the present.

Thus, at the end of the first phase in the history of programming languages, practical ways existed to feed information into computer-like devices electronically and then to receive answers back. Theoretical work had been done on how to express mathematical and complex algebraic problems to such machines. Finally, early designs of primitive languages were also in various stages of development.

The years from 1945 to 1952 witnessed the emergence of new hardware, expanded function, and the birth of modern programming. It was the period during which programmers could go beyond simply wiring machines, as did Grace B. M. Hopper** with the Harvard Mark I* during World War II, to drafting with paper and pencil instructions which a computer could convert into intelligible commands that could be acted on.

The best example of this evolution was Mauchly's proposed Short Code* which was created and used at Remington Rand† in 1949–1950. The description of Short Code published in October 1952 detailed a language that was used. A two-character code was designated either a task (operation) or a variable when combined in groups of six to create UNIVAC's twelve-character length words. A = B + C thus looked to the computer as 00 S0 03 S1 07 S2. The language offered thirty different operations.

Another early example of a primitive high-level language was IBM's Speedcoding* System of 1953 for the IBM 701* computer. Like Short Code, it offered operations and the ability to trap data for numerical scientific problems—the general application area of greatest concern to developers of languages throughout the 1950s. The third early illustration was a development at MIT between 1952 and 1953 for the WHIRLWIND* computer. It was called the Laning and Zierler System* and appeared to be the first American language that allowed mathematical expressions to be written in a manner resembling normal algebraic syntax. Until 1960 other languages emerged that improved on this fundamental objective of being able to draft in normal mathematical notation languages that could be interpreted and used by the computer. These included Remington Rand's A-2* and BACAIC.*

FORTRAN, the most spectacular success in this direction, ushered in a new era in programming. FORTRAN became one of the most widely used programming languages ever, and it introduced a number of important features seen in subsequent languages. FORTRAN began as a language to support work in scientific and engineering computing, but by 1960 it was also used to write such commercial applications as payroll and sales analysis. As of 1986—over thirty years after its creation—FORTRAN was still a popular language, especially among mathematicians, scientists, and engineers.

In the early 1950s the term used to describe higher level languages was *automatic programming*. Some programmers did not like the concept, however. They believed they could do a better job and in fact were pleased with their abilities to employ and improve computer efficiencies. They maintained that the ability to write programs using minimal amounts of instructions and computer memory was critical. They feared that more progressive tools might introduce inefficiencies, which would be a serious problem inasmuch as hardware was so expensive and memory on such devices was quite limited. Computer users placed a premium on maximizing efficiencies in programming to avoid any waste of hardware resource. By the early 1980s, when hardware costs were minimal compared to the total cost of computing, such efficiencies were no longer critical. In 1952, for example, the most expensive resource was the computer, not the programmer; in 1982 the reverse was true. FORTRAN was thus born in an environment that promised to help those in need of tools to do scientific calculations, but was also constrained by the harsh economics of early programming. Figure 9 suggests the long-term trend toward greater ease of language of which FORTRAN was a part.

The original specifications for FORTRAN, developed at IBM, were ready by November 1954, and the language became operational soon after. It was the most efficient language to date. FORTRAN introduced the idea of the subroutine and, over the next several years, English-like statements and symbols to express

Figure 9
Computer Hardware Configuration ($/Hour for 3 × 10⁵ Instructions/Second)

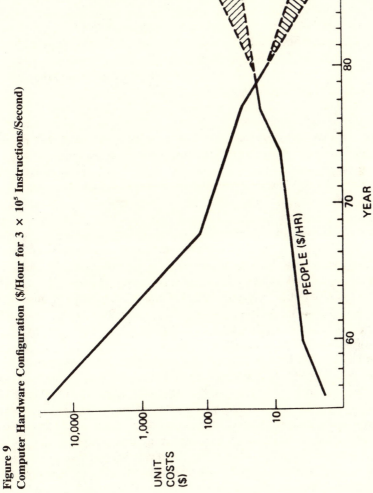

SOURCE: James Martin, *Application Development without Programmers*, © 1982. p. 3. Reprinted by permission of Prentice-Hall, Inc., Englewood Cliffs, New Jersey.

notation (e.g., GO TO, IF, STOP, DO, etc.). It proved that efficient object code could be written and compiled. Thus, programmers could become more efficient by producing more and better programs using a high-level language. Its value is demonstrated by the fact that in 1986 it was still widely used, had been enhanced over the years, and was very popular in the scientific and engineering communities to which it was originally aimed. Other languages were developed that were more appropriate or efficient for commercial applications (such as COBOL and BASIC).

During the mid- to late 1950s, while various projects were underway to create new languages (including MATH-MATIC* at Remington Rand), some attempts were made to develop English-like languages. The first of these was FLOW-MATIC,* which was also one of the first languages created exclusively for business applications. It was developed for the UNIVAC I* computer and became available in 1958.

During this period scientists working in the field of computing were beginning to develop a sense of community, encouraged no doubt by scientific conferences sponsored by universities (such as at Harvard and later at Dartmouth College) and by companies (most frequently sponsored by IBM and Sperry Rand*). A growing desire for higher level languages developed, along with the wish for a universal language that all programmers could use on all machines for both scientific and commercial data processing. By 1958 the pressure for such a language led Americans (representing the Association for Computing Machinery [ACM]†) and Europeans (within) to define the specifications for a universal language. It was basically an algorithmic language which they named International Algebraic Language (IAL) and became better known as ALGOL* 58 and 60. The result of this pressure was more research on universal languages and the development of other languages throughout the 1960s: JOVIAL,* MAD,* and NELIAC.* Better compilers were also written in the 1960s, encouraging additional research on programming languages.

The years 1958 and 1959 were significant ones, witnessing the development of ALGOL which led to the development of other widely used languages. One of the more important list processing languages, called IPL-V,* became operational. The most popular language for artificial intelligence (AI),* called LISP,* was created and maintained its favored position within the AI community over the next thirty years. COMIT,* another string processing language, also came into existence. But the most significant development of all was the initiation of efforts leading to COBOL, the most extensively employed language for commercial applications in the past quarter century. That fact alone would have justified the observation that 1958–1959 was a special period in the history of programming.

In May 1959 the Conference On Data Systems Languages (CODASYL) Short Range Committee was formed for the purpose of defining the specifications for COBOL. Its formal name was Common Business Oriented Language, suggesting the intent of the new idiom. COBOL was the result of a group effort involving

Table 60
Supply of Computers and Programmers in the United States, 1955–1975,
Selected Years

Year	Number of Computers	Number of Programmers
1955	1,000	10,000
1960	5,500	30,000
1965	22,500	80,000
1970	70,000	165,000
1975	225,000	220,000

SOURCE: James W. Cortada, *EDP Costs and Charges: Finance, Budgets, and Cost Control in Data Processing,* © 1980, p. 37. Reprinted by permission of Prentice-Hall, Englewood Cliffs, New Jersey.

computing scientists and computer manufacturers in the United States. COBOL introduced a variety of easy-to-use functions which included English-like commands, divisions in which one described data files and operations to be performed, and a system whereby both words and numerical data could be handled. This language created jobs for tens of thousands of programmers in the 1960s and 1970s, led to billions of lines of software, and, by some estimates, dominated about 85 percent of all business applications written in the United States by the early 1970s (Table 60).

COBOL was not a scientist's dream language, nor did it advance the science of computing languages so greatly, but it was the language that most contributed to the expanded use of data processing in the 1960s and 1970s. It eventually gained IBM's blessing and ran on its S/360*—the most widely used computer of the 1960s—and became a mainstay of almost all large mainframes from 1965 to the present.

The 1960s were bountiful years for programming languages. Well over 100 languages were developed during this period for scientific computing, business applications, artificial intelligence, list processing, and special applications, including numeric control languages, modeling tools (such as GPSS*), and packages for teaching programming to college students. Work was also done on database management systems* as part of the general movement to improve programming and the use of computers with software. Attempts were made to develop languages that could do both scientific and commercial applications. All of these efforts were highly successful. This success can be attributed in part to changes in hardware. Such vendors as IBM, General Electric (GE),† and Radio Corporation of America (RCA),† to mention only three, began making computers that could handle both types of transactions—a departure from the practice of the 1950s when the machines built serviced either one or the other type of applications, but usually not both within the same device.

An early effort that led successfully to a language usable by both types of users appeared in 1963–1964. Called PL/I, it was the result of a joint development effort by SHARE† and IBM. Since then PL/I has been used continuously and on a widespread basis, particularly in the United States. It was relatively efficient and used computer resources effectively. PL/I was intended for use by both scientific and commercial users because it could take advantage of new hardware and the technology available in the 1960s. This was particularly the case when employing the S/360 family of computers and disk storage. Again striving for a universal language (much as with ALGOL), some hoped that such a language might eliminate the need for both FORTRAN and COBOL as well. Instead, all three have coexisted to the present. Yet PL/I coopted the best of ALGOL, COBOL, FORTRAN, JOVIAL, and other languages of the day. It made possible operations on strings of data, concatenation, and a broader use of syntax than earlier languages. It took into account a growing requirement of programmers and users to interact with operating systems.* It was rich in commands to deal with the allocation of computer storage, the management of individual tasks, and exception handling. PL/I has been called the first of the multipurpose languages, and its arrival has been heralded as an important technological event within data processing. Many idioms developed after the mid-1960s benefited from the experiences with and functions of PL/I.

APL,* another language, offered similar benefits and enjoyed a long life. It could handle vector and matrix operations, had a large character set and significant efficiencies in the use of its broad set of operators, and required less coding than FORTRAN. Thus, for example, five lines of FORTRAN to do a specific multiplication could be coded in one using APL. Such improvements in efficiencies in the 1960s helped make its use for analysis of scientific and business data more widespread.

Various factors influenced the development and use of programming languages. One of the earliest concerned the role of the military in the United States. Beginning in the 1950s, the U.S. Department of Defense sponsored a considerable amount of research on hardware and software. It was also the largest customer of computer-related products for most vendors. Consequently, military personnel frequently influenced the course of research and development. This circumstance was evident with programming languages. Like the data processing community as a whole, they too sought universal languages and tools that the entire military establishment could use. The Defense Department, for example, was one of the strongest supporters of COBOL. In the early 1970s another effort flourished within the defense community for more modern languages. In 1975 Defense sought a language for embedded computer systems (computers within other machines) which could handle such applications as air traffic control, computation for process control, and weapons. In 1979 initial specifications for such a new language (now called Ada) emerged. The specifications were completed in July 1980, and they called for a more universal language that could be employed across a wide range of systems than originally envisioned. Work progressed on this project during the 1980s.

Several other developments were instrumental in the evolution of languages, the most obvious one involving batch versus online computing. When programming first began, it was batch oriented. This meant that a programmer sat down and wrote a program, had it punched out on cards, which in turn were "batched" together with other programs and run on the computer at a time scheduled by whoever managed the device. Thus, a programmer might write a program, turn it over to a computer operator, have it run several hours later, and then receive output (e.g., a report). This process was repeated until all errors (bugs*) were gone and it could run "clean." This programming method continued until the mid-1970s.

Still, this situation represented an improvement over earlier methods. In the 1930s and 1940s there were no programmers, but only scientists and engineers who had a great deal of knowledge about the computer they were using. In addition to the slow process of wiring and using primitive methods of instruction, they probably designed the equipment and had to fix these devices when they broke down, which occurred frequently. The introduction of programming languages in the 1950s began a process whereby programmers slowly divorced themselves from the operation of computers. It meant they could write more programs and focus on applications rather than on hardware. To be sure, they were still constrained by the limitations of computers (for example, by the size of the computer's memory), but operations (actual pushing of the buttons) became a separate job within data processing. By 1970 managers of data centers were actually prohibiting programmers from even entering the machine rooms where computers were housed, forcing them to submit their "jobs" to librarians, secretaries, and operations personnel through Dutch door arrangements and data processing mail boxes. Today there is no reason for a programmer ever to step inside a data center's machine room.

The one factor most responsible for these changes was online programming. As early as the years immediately preceding the watershed announcement of the S/360 in 1964, computer scientists had been experimenting with online processing. A programmer in this environment would sit down at a terminal and directly code programs into the system via a keyboard. Such programs would be compiled immediately, and answers appeared on the screen or were printed out within seconds or minutes. By the mid-1970s online programming or "real-time" processing became increasingly the norm. By 1986 most programmers had a cathode ray terminal (CRT) on their desks, which had become as common as a pencil holder, a coffee mug, and a telephone. Online programming sped up the process by which lines of code were written and tested. This process drove programming costs down while shortening turnaround time on projects. In general, the only batch processing left in the 1980s consisted of programs that required enormous amounts of machine time and were not dependent on timing during the normal working day as might shorter programs. Common examples of contemporary batch processing included running a Bill of Material or Material Requirement Plans (MRP) in a manufacturing plant, applications that were typically run at night.

Languages developed in the 1950s and 1960s were generally for batch processing. In the 1970s many of these older languages were released with newer versions that were online (FORTRAN and COBOL). Others were not used in an online manner until 1970 (such as PL/I and BASIC). Nearly all languages that appeared in the 1970s could be used via a terminal (such as LOGO and Pascal). In short, interactive computing had become the norm by the mid-1970s.

Hardware also contributed to the evolution of programming languages. During the late 1960s breakthroughs in the technology of hardware were becoming evident, while the capacities of memory, disk drives,* tape drives,* and their speeds increased. Hardware costs kept dropping and by 1980 could be tracked at a decline of nearly 20 percent annually. This trend meant that more power could be dedicated to running systems faster. Because online programming always required more computing power than batch did, the trend toward additional computer power continued. Computing power therefore went up as costs declined, creating the economic environment to justify more online programming and larger, more complex systems using languages that were becoming ever richer in function. Yet another economic factor influenced these changes. The cost of programmers had gone up continuously during the 1960s and 1970s to the point where by 1978 the average data processing department was spending nearly 60 percent of its budget on salaries. There was a persistent shortage of qualified programmers to handle a rapidly growing backlog of unwritten applications. These circumstances meant that new techniques were needed which increased the productivity of existing staffs while making it possible for even less technically oriented people to do programming. Online computing became part of the solution to the problem.

As languages passed through various releases (analogous to new editions of a book) during the 1960s and 1970s, additional functions appeared which increased programming productivity even further. First, macro commands were written, which made it possible for a programmer to draft one command to cover what before would have required many lines of code. Second, more "default" functions were added whereby the programming language assumed that certain conditions existed (such as how to handle files) unless told otherwise. Third, languages became more forgiving in syntax, reducing the amount of correct grammar required to make a program run. Fourth, they evolved into more English-like (natural human language) phrases. One could see the evolution dramatically by comparing a line of FORTRAN which was technical jargon when compared to a line of BASIC.

The number of languages that have appeared is difficult to calculate. In the 1950s there were probably fewer than three dozen in the United States and in Great Britain—the two most active countries in computing. In the 1960s slightly fewer than 200 have been identified. In the United States alone approximately 170 languages were always around between 1967 and 1977. At the same time between 25 and 30 died, only to be replaced by a similar number. A reasonable guess as to the number outside the United States might be an equal amount, a suspicion that is supported not with facts but is simply a heuristic feeling. Although universities were the source of most new languages, the most widely employed were those that a company developed or that were adopted by computer vendors. For example, in the United States, a language had not "arrived" in the 1960s or 1970s unless IBM introduced a compiler for it. All languages for artificial intelligence were products of universities and remained there or in commercial research laboratories. The widely used BASIC language of the 1980s emerged primarily as a result of the rapid growth in the use of microcomputers from under 1,000 in 1975 to over 2 million sold worldwide in 1985.

The subject of computing languages has attracted a rapidly growing literature and yet remains complex and ill-defined. Many observers within the data processing industry have also commented on languages and project that by the mid–1990s the next great "revolution" in technology will not be with hardware but rather the creation of programs. The anticipated breakthroughs involve new programming tools (such as better report generators and file managers, and not just more languages). Vendors also believe that they have to create the conditions to bring about this revolution quickly. Some qualities of artificial intelligence will be associated with new languages which will be part of the highly touted and anticipated "fifth generation" of technology.

The history of software development, however, indicates that not all languages have evolved from each other in the same direct, almost serial fashion as has hardware. As a consequence, predicting the future evolution of programming languages has historically been a failure. The problem is that computer technologies typically evolved from one to another (e.g., vacuum tubes to transistors to chips*). Software tools have often developed independently of each other or concurrently. Thus, COBOL and ALGOL grew up in a similar period, yet neither made a dramatic contribution to Pascal, or to today's versions of BASIC. Certain inherent aspects, such as algorithms for compilers and principles of data handling, influenced subsequent languages, but not as directly as happened with hardware. As the cost of computers and peripheral equipment dropped (especially for machine-readable storage, such as disk and tape along with computer memories), the reason for more powerful languages became compelling. Nowhere was this more obvious than with the IBM PC and Apple's† micros of the early 1980s. These machines brought programming out of the data centers to millions of people who were not programmers working in offices and at home. They did not want to learn about data processing's technology, and yet they had need of computing assistance whether for games, studying, and financial or other business analysis. As usage rose, so did demand for more power. The result was a series of product announcements by all manufacturers of micros in which more memory and storage and processing capabilities were offered without fundamental increases in the cost of such devices.

Yet certain trends and circumstances remained unchanged over the past thirty-five years. First, there was a continuous desire to make programming languages easier to use by either making them more natural (human) in syntax or by

Table 61
Cost of Processing Using IBM Computers, 1955–1981, Selected Years

First Shipped New Technology	Cost/million Instructions
1955	$40.00
1961	2.00
1965	0.40
1971	0.11
1977	0.08
1979	0.04
1981	0.02

The economic and managerial dynamics caused by such profound reductions in the cost of computing can be studied in detail in James W. Cortada, *Managing DP Hardware* (Englewood Cliffs, N.J.: Prentice-Hall, 1983): 14–26.

providing additional functions within commands (using macros usually). The urge to merge the requirements of scientific/mathematical processing with commercial needs always existed. Although major successes took place in the 1960s with APL, BASIC, and to a certain extent COBOL, only in the 1970s did the true marriage of these functions occur. The desire to integrate features in greater amounts has continued, as has most recently been evident in that group of software packages that is called productivity tools by most vendors and that does word processing, prepares graphics, and performs financial and "what if" analysis. Some argue that these are borderline examples of programming languages.

The history of programming languages is therefore the story of various successful efforts to define better ways for people to deal with and instruct computers, of their attempts to turn programming (historically considered an art by programmers) into a more logical, predescribed process with standards, and with the ease of other human tools. The cost of programming also decreased, as Table 61 suggests. Its history is one of brilliant scientists struggling with tools to teach scientific principles while working on notions of artificial intelligence. It is also a history of computer vendors developing software tools that can help sell more products. By the 1980s the position of programming languages as a carrier of change had become evident. With over 70 percent of the workforce in the industrialized world dependent on computing for their livelihood, the role of software and their programming languages has become critical. By the 1980s software (actual programs as opposed to programming languages) had become more important, with sales a multibillion dollar sector of the data processing industry. Programming efficiencies made directly possible the sale of so many products. In 1986 alone, for example, programming led to the availability of over 3,000 software packages just for use on IBM's PCs. It had made IBM's annual sales of software alone exceed $2.5 billion. The total software portion of the industry was nearly twice that in the same year. IBM, like most other vendors, pondered the real possibility that software sales could outstrip hardware sales by the late 1990s.

For further information, see: F. E. Allen, "The History of Language Processor Technology in IBM," *IBM Journal of Research and Development* 25, no. 5 (September 1981): 535–548; George Dantzig, "Reminiscences About the Origin of Linear Programming," in Arthur Schlissel, ed., *Essays in the History of Mathematics* (Providence, R.I.: American Mathematical Society, 1984): 1–11; Ellis Horowitz, ed., *Programming Languages: A Grand Tour* (Rockville, Md.: Computer Science Press, 1983); N. Metropolis et al, eds., *A History of Computing in the Twentieth Century* (New York: Academic Press, 1980); Jean E. Sammet, "History of IBM's Technical Contributions to High Level Programming Languages," *IBM Journal of Research and Development* 25, no. 5 (September 1981): 520–534, *Programming Languages: History and Fundamentals* (Englewood Cliffs, N.J.: Prentice-Hall, 1969), "Programming Languages: History and Future," *Communications of the ACM* 15 (1972): 601–610, and her "Software History," in Anthony Ralston and Edwin D. Reilly, Jr., eds., *Encyclopedia of Computer Science and Engineering* (New York: Van Nostrand Reinhold Co., 1983): 1353–1359; R. Wexelblat, ed., *History of Programming Languages* (New York: Academic Press, 1981).

PROJECT MAC. This project was undertaken at the Massachusetts Institute of Technology (MIT) during the 1960s to provide time-sharing capabilities across the entire campus while developing new technologies. It was a successful program which supported research in the field of computer science. Project MAC also represented a logical outgrowth of an earlier program at MIT called CTSS which first introduced the academic community there to time-sharing computing.

MIT had long been home for important research in the general field of computing. As recently as the 1930s Vannevar Bush** had worked with analog computers,* and in the 1940s and 1950s MIT supported the development of WHIRLWIND,* a major system for the U.S. military community. During the summer of 1961 MIT launched the Compatible Time-Sharing System (CTSS), using an IBM 7090 computer, which made a computing system available to students and faculty alike. By May 1963 it supported twenty-one typewriter users and worked well, proving that time-sharing was possible. However, the data center that supported this system ran as a batch operation, and the number of users who could use CTSS was limited. The need for a data center dedicated completely to time-sharing was recognized, and in 1962 MIT decided to establish one.

Professor Robert M. Fano** was given responsibility for creating the new center and quickly brought together a team, some of whom had worked in the development of CTSS. These included, for example, Fernando J. Corbató,** who had been a leading figure in the creation of CTSS, and Marvin L. Minsky,** who was already a recognized leader in the new field of artificial intelligence.* Fano stated the following objectives for Project MAC: to build a time-sharing system (1) that could be used by many people with public files, display terminals, and better equipment, (2) that would support research and education through the use of computing, and (3), finally, that would familiarize students and faculty in the use of computers.

Project MAC stood for two ideas: Machine-Aided Cognition and Multiple-Access Computer. Corbató later commented that Project MAC represented "the first coming of age of computer science as an organized group at MIT." It was an important point to make because, up to that time, computer projects were independent activities located in departments scattered across MIT. Part of MAC's mission was to bring some cohesion to all of these efforts without destroying their individuality. That effort contributed to the first word in its title: Project. MAC's policy board also reflected that mission since it had representatives from across the campus: Philip M. Morse, then director of the Computation Center (serving as chairman of Project MAC); Dean Gordon S. Brown from the School of Engineering; Professor Peter Elias, who headed the Department of Electrical Engineering, where many data processing projects had been undertaken over the years; Dean Howard W. Johnson, from the School of Industrial Management; Dean George R. Harrison, representing the School of Science; Professor Albert G. Hill who taught physics; Professor Carl F. Overhage, director of the Lincoln Laboratory; and Professor Fano, representing MAC as its director.

The first step in implementing Project MAC was to establish a duplicate of the system which was then operating at the Computation Center (CTSS). By November 1963 this had been accomplished and by the spring of 1964 supported 200 users in ten different departments. The network supported 100 teletypewriter terminals linked to MAC through MIT's telephone network. The system could handle twenty-four concurrent users, while other projects of a specialized nature were driven through the system in batch mode. Some, for example, were Minsky's various artificial intelligence experiments. Throughout the 1960s, hardware and software* were modified as experimentation on time-sharing systems progressed. The net result was a gradual increase in both function and users. By 1970 it had even been linked to the Advanced Research Projects Agency network (ARPAnet) which connected numerous American and European data centers. During the 1960s fundamental work was done in the development of graphics and in the use of computer-aided design systems. As early as 1963 one doctoral student, Ivan E. Sutherland, had developed Sketchpad; it was originally on the TX-2 computer within the Lincoln Laboratory† but later ran on MAC. Project MAC improved on the original cathode ray terminal (CRT) system. New television display terminals (CRTs) were designed, resulting in one of the most sophisticated and advanced graphic display systems in use during the early 1960s.

By 1965 the data center run by Fano included two sets of core memory of 32,000 characters each (which was large by the standards of the day) connected to an IBM 7094 computer. It had six sets of data channels along with a variety of disk files, two magnetic drums, magnetic tape drives,* a card reader and one punch, printers, display terminals, and a telecommunications control unit that linked into a PBX, TWX, and a Telex network.

The software needed to drive that complex configuration rapidly became outdated. In an attempt to replace CTSS in 1964, work began on the Multiplexed Information and Computing Service (Multics), a project that also involved help from Bell Telephone Laboratories† and the Computer Department at General Electric.† The system became available to MIT's users in 1969 and later emerged as a commercial offering from Honeywell Corporation.† The system proved effective and popular. By 1971, for example, it had 500 users, with 55 simultaneously capable of real-time processing. It ran seven days a week, around the clock.

Multics reduced the need to focus Project MAC's attention primarily on time-sharing systems. By the early 1970s a variety of software was also widely available within the data processing community to offer those services. So beginning in October 1969, the Multics Systems was turned over to MIT's Information Processing Center to manage. Hence, time-sharing no longer represented a frontier in computer science. Further work during the 1970s centered on narrower computer science projects, with some effort in engineering. Students and faculty who dominated Project MAC's attention came either from the Electrical Engineering School or from the Computer Science Department. Because of its narrower scope, its name was changed to MAC Laboratory and, in 1975, to its current designation: Laboratory for Computer Science.

Professor Fano, who had launched the project, served as its director until the start of the 1968 school term. By then most of Multics had been completed. The next director, J.C.R. Licklider, who picked Fano for the director's job in 1963, ran the center until 1971, giving it new focus. Edward Fredkin managed Project MAC until 1974, when, under the leadership of Professor Michael Dertouzos, the center became the large Laboratory for Computer Science.

For further information, see: R. M. Fano, "Project MAC," in *Encyclopedia of Computer Science and Technology* (New York: Marcel Dekker, 1979), 12: 339–360; Karl L. Wildes and Nilo A. Lindgren, *A Century of Electrical Engineering and Computer Science at MIT, 1882–1982* (Cambridge, Mass.: MIT Press, 1985).

Q

QUADRANT. This tool was used to perform astronomical and mathematical calculations primarily by sighting angles between stars and the earth. It existed in ancient Greece, and variants were probably used in Babylonia. It was a popular navigational tool in Europe by the beginning of the late Renaissance and hence was one of many aids to calculations contributing to the early history of computing technology.

The quadrants that proved most relevant to the history of computing were the variant European models of the sixteenth century, all of which were extremely similar in design. One of the earliest surviving models, Gunter's quadrant, dates back to the late seventeenth century. It was built by Dr. Edmund Gunter (1581–1626) who taught astronomy and mathematics at Gresham College in London. The scales on his gadget were similar to those found in another early astronomical tool, the astrolabe.* During his lifetime Gunter popularized many aids to calculation and thus quickly contributed to the increased use of quadrants. His book, *The Description and Use of the Sector*, first published in 1623, was one of the most important studies ever published in English for the improvement of navigation.

Another important quadrant, less complex than Gunter's, was developed by William Leybourn (1626?–1700), who was also a professor of mathematics in London. Like Gunter, he was known for his publications in similar fields: astronomy, mathematics, science, Napier's "bones", and slide rules.* Though his was only one of many quadrants available in the late 1600s, we know more about Leybourn's device because of his writings. It was constructed out of brass in the shape of a quarter circle with two holes along one edge for sighting the elevation of stars or anything else. The hole in the middle would have had a string serving as the plumb bob to determine the elevation of stars. This device allowed the user to take readings on the distance between scales etched on the plate. Scales on the front side of the quadrant were used primarily to perform trigonometric functions, for use in navigation and study of the stars. Lines on the face made it possible to define the radius for trigonometric scales, tangents, sines, and secants. Additional circular scales were used to establish time, locate stars, and develop horoscopes. It listed specific and important stars, signs of the zodiac, and divided the quadrant into 90 degrees.

The back of this and similar quadrants was also intended for mathematical calculations. One set of scales made possible the appropriate mathematics necessary for determining squares and square roots, cubes, and cube roots, whereas a second set made it possible to calculate the sun's position for any time during the year and to build sundials. In short, the device could be used to perform normal mathematical functions. The device provided evidence that mechanical aids to mathematics could be invented and used effectively. It existed at the same time that other Europeans were trying to design mechanical calculators. These included Blaise Pascal (1623–1662),** Wilhelm Schickard (1592–1635),** and Gottfried Wilhelm von Leibniz (1646–1716).** Theirs, however, could not perform the enormous variety of tasks possible with the quadrant, although, in fairness to the inventors, mechanical calculators had other purposes. The quadrant helped establish the notion that mathematics did not have to be done completely by the human mind. Thus, it encouraged others who sought to design computational equipment.

For further information, see: M. Dumas, *Scientific Instruments of the 17th and 18th Centuries and Their Makers* (London: B. T. Batsford & Co., 1972); David W. Waters, *The Art of Navigation in England in Elizabethan and Early Stuart Times* (London: Hollis and Carter, 1958); Michael R. Williams, *A History of Computing Technology* (Englewood Cliffs, N.J.: Prentice-Hall, 1985).

QUIKTRAN. One of the first online programming languages* devised for numerical scientific computing, QUIKTRAN was developed at the International Business Machines Corporation (IBM) in 1961 in order to create a debugging tool for users of FORTRAN.* Its developers took an existing language (FORTRAN) and modified it for use through terminals instead of its more traditional batch form. In effect, they eliminated some commands and functions within FORTRAN while adding others that would allow a batch programming language to function online and even from a remote terminal. Programs written in QUIKTRAN ran as regular FORTRAN programs in a regular batch processing system.

The head of the team that designed QUIKTRAN, John Morrissey, sought to create a tool that would enhance FORTRAN, which was IBM's major language at the time and was used by more of its customers than any other. What he actually developed was a successful debugging and terminal control program. However, his group made it too machine dependent; that is, it only functioned on IBM 7040/44 computers with 1050 terminals, which meant that only a limited number of data centers could install the software.* Despite that restriction, the language worked and was used. The first version became operational in mid-1963 and relied on the USASI Basic FORTRAN.

The language was the first online system that relied on standard computer equipment commercially available. Its compatibility with an existing and widely accepted language signaled a new era of compatibility with existing software tools. As the number of programs that a user had written, and hence invested in, increased, the need for compatibility in subsequent software became critical by the mid-1960s. From then on, languages were developed to be compatible with earlier versions in order to motivate large numbers of users to move their libraries of programs to more current or new releases of existing programming tools. Future software from QUIKTRAN was also influenced by the idea that if compatibility were not possible some conversion tool would have to be written if a vendor wanted to move the users of one language to another.

For further information, see: Jean E. Sammet, *Programming Languages: History and Fundamentals* (Englewood Cliffs, N.J.: Prentice-Hall, 1969).

R

RAMAC. RAMAC, also known as the IBM 305 RAMAC (random access method of accounting and control), was the first computing system that included direct access magnetic disk files in its configuration. The RAMAC was not a full computer but rather a unit record configuration that could handle stored programs and, of course, direct access to files stored on magnetic disk drives.* That last feature (introduced in the late 1950s with the 305) did not become a common part of computer systems until the mid-1960s and was not central to applications in general until the early 1970s. With direct access a new world of possibilities opened up, including large data files available for rapid access, online and real-time systems, and interactive computing. Until disk drives became a normal part of a computer system, most jobs were run in batch with little or no interactive computing; this was especially so with commercially available computer systems. The common magnetic media for data were drum memory systems, which were very limited in capacity and for auxiliary storage, tape that could only be read or written to sequentially. International Business Machines Corporation (IBM)† introduced a major technological leap forward when RAMAC appeared.

A typical 305 configuration included a processor, a disk file, one console for the operator to use to run the equipment, typically one printer, a card reader, and a card punch. The computational part of the configuration could store 100 instructions which came in and out of the system from a magnetic drum. The instruction set included commands to transfer data, conduct decimal arithmetic, do disk input/output (I/O), and set switches. I/O functions were performed by using plugboards with wires which were reset as needed by operators. These plugboards were located in both peripheral equipment and within the processor. To run a job the boards had to be wired and then loaded into the appropriate equipment to be used. All programs were written in machine-level language; no higher level languages were supported on the 305.

The exciting new technology associated with the 305 was the disk. This disk file had fifty magnetic disks, each of which had 100 recording tracks. Each track was capable of housing 10 to 100 characters. A fully configured 305 could handle up to 20 million characters of storage. Another innovative feature was the fact that two 305s could share the same file. Each record had an address, and the idea of organizing data on cylinders was done by programming. One could extract information off a disk file, called access time, in less than one second. That speed was profoundly important because it represented a two hundredfold increase in speed over the more conventional tape drive.*

The 305's printer was a slow "serial-strick" device; yet it could be replaced by a 407 unit record accounting machine which operated with a rated speed of 150 lines per minute. The slowness of the printer was compensated for by the ability to attach numerous such devices to the system.

RAMAC was frequently sold to customers who had been heavy users of IBM's unit record equipment and other computational products in the 1950s. They typically were commercial organizations and were comfortable in using unit record (card-based) equipment but wanted to begin using computational devices. Thus, RAMAC represented another gradual migration into the world of computers. Traditional unit record applications were converted to the 305, which obviously could process more, faster, and with greater function. Some of the key sets of applications found on the 305 included inventory control, distribution, payroll, and accounting. The 305 gave IBM one of the earliest opportunities to develop application packages, that is, programs that performed certain functions (usually in business and accounting) sold to customers to operate on the RAMAC. The company learned better ways to manage magnetic files because of RAMAC, lessons reflected in products announced during the 1960s.

IBM announced the 305 RAMAC in September 1956 and delivered its first copies in 1957. The new disk drive was called the 350. The 350 made it possible for the user to obtain information faster than on other contemporary systems. One of the earliest groups of users to realize the benefits of such a system was airline companies because now they could access customer and plane-seating files directly, making real-time airline reservation systems a reality. Of course, the 350 was much boxier in size and shape than disk drives of the 1960s and 1970s. Every disk was 2 feet in diameter, and complete stacks of such disks rose 2 feet high. There was one read/write arm for each side of a disk, and up to fifty disks were available. The arrangement has been likened to a jukebox. Since it would be several years before competitors were able to develop disk drives, IBM had a unique marketing advantage which it was able to exploit effectively. The company has introduced leading edge disk drives down to the present, taking similar advantages repeatedly. The most recent example was with the IBM 3380 disk drives of the early 1980s, which introduced technologies other competitors would not deliver on a reliable basis for nearly three more years. The lessons from RAMAC were well learned at IBM.

RAMAC remained IBM's leading edge system until the company announced the IBM 1401* in October 1959, making the earlier system obsolete. The 1401, a second-generation computer, was extremely popular, selling between 15,000 and 20,000 copies. Prior to its availability, only 6,000 electronic digital computers* had been installed worldwide. Because the 1401 was very powerful and less expensive than older IBM equipment (such as the 650*), it soon replaced many other systems, including the RAMAC.

For further information, see: C. J. Bashe et al., *IBM's Early Computers* (Cambridge, Mass.: MIT Press, 1986); Franklin M. Fisher et al., *IBM and the U.S. Data Processing Industry: An Economic History* (New York: Praeger Publishers, 1983).

RAYDAC. This computer was built by the Raytheon Manufacturing Company† in 1951 and was subsequently used by the Naval Air Missile Test Center in California. The Raytheon Digital Automatic Computer was unique among the early digital computers* in that it had a four-digit word and its arithmetic unit generated a 5-bit check for all operands and results. Thus, the machine could conduct verifications of its transactions. It also had a floating-point type of capability. In addition, it was the first machine to use plastic tapes with its tape drives.*

RAYDAC was designed and built by Louis Fem in 1951. Table 62 shows the machine's features as of July 1952, after some additional modifications were made on behalf of the Naval Air Missile Test Center. Like many U.S. digital computers of the early 1950s, this was a one-of-a-kind type built for the military community. The Navy turned to Raytheon because it had established itself during World War II as an important manufacturer of radar and related equipment for the U.S. military. Its first attempt to build a computer came with the RAYDAC; initial work began in 1947 for the National Bureau of Standards† and finally for

Table 62
Technical Features of RAYDAC

Technology	Parallel
Word length (bits)	36[a]
Instruction length	72
Instruction format	(3+1)-address
Main store size	1,024
Main store type	Delay
Backing store type	Magnetic tape
Average add time	707 microseconds
Average multiply time	868 microseconds
Basic clock frequency	4 MHz
Approximate number of valves	5,200
Approximate number of GE diodes	17,000

[a]Contained 36-bit word which included four "transfer-count" check digits.

SOURCE: Reprinted with permission from *Early British Computers*, Simon Lavington, First Edition. Copyright © Digital Press/Digital Equipment Corporation (Bedford, MA), 1980.

the U.S. Navy to whom it delivered the machine in 1951. Between 1953 and 1954 the company began a commercial version of the machine called the RAYCOM which it intended as a general-purpose computer, also digital. The project was canceled after the company decided not to venture into unfamiliar territory after all. Engineers who had worked on both of these projects subsequently were assigned to a joint Raytheon-Minneapolis-Honeywell Regulator Company (later called Honeywell)† in a joint venture called Datamatic Corporation in 1955. The result of this effort was the Datamatic-1000, later the Honeywell 800 and 400 machines, and finally the 200.

For further information, see: Franklin M. Fisher et al., *IBM and the U.S. Data Processing Industry: An Economic History* (New York: Praeger Publishers, 1983); Simon Lavington, *Early British Computers* (Bedford, Mass.: Digital Press, 1980); René Moreau, *The Computer Comes of Age: The People, the Hardware, and the Software* (Cambridge, Mass.: MIT Press, 1984).

ROBINSON MACHINES. These British computational devices were built during World War II to help decipher German military messages. Constructed at the Dollis Hill research facility of the British post office, little was known about them. The first machine was completed by April 1943 and named "Heath Robinson" after a cartoonist who drew pictures of strange machines. This particular device was electronic and relay-based, and used high-speed accumulators made out of vacuum tubes. Output appeared in printed form. Its functions were never revealed, and details concerning its construction and use remained secret. It could perform specific Boolean operations on data fed to it via paper tape* in loops of some 2,000 characters. The device was used at Bletchley Park,† home of Great Britain's computerized projects devoted to the decoding of German messages. Two subsequent and more reliable models were said to have been built called the "Peter Robinson" and the "Robinson and Cleaver," both names of stores in London.

For further information, see: Michael R. Williams, *A History of Computing Technology* (Englewood Cliffs, N.J.: Prentice-Hall, 1985).

S

SABRE. The Semi-Automatic Business-Related Environment (better known as SABRE), one of the first of the large, online, real-time applications using computers developed in the United States, was American Airlines' passenger reservations system. Orders for tickets were taken, seats on flights assigned, inventory control of all flights maintained, and fees tracked on a nationwide basis using one of the largest data communications networks of the early 1960s. SABRE ushered in a new era of computerized applications involving the use of thousands of terminals, networks, and multiple computers in critical business applications. Much of what was learned with SABRE was applied to other airlines in the 1960s and 1970s and to major systems in all industries. This application was remarkable because until SABRE was developed most business applications were batch; there was little reliance on online use of terminals by people who knew nothing about computers. In short, SABRE represented a major leap forward in technical sophistication.

SABRE grew out of American Airlines' business problems in the mid-1950s. Across the entire United States, prior to SABRE, customers telephoned in their reservations either to an airline or to their travel agent, either of whom had to rely on an airline's manual records to see whether there was room on a particular flight. In order to insure that passengers with reservations were given seats, an airline had to leave some seats unreserved and therefore unsold up to the last minute on each flight. This practice cost American Airlines millions of dollars in lost potential revenue. Although the company recognized the severity of the problem as early as 1954, correcting it took years of work. Meanwhile, the business impact remained severe. For example, in 1960 American Airlines flew 8.615 million passengers. To do that, some 26 million telephone calls had to be made by all involved across the entire United States. Manual bookkeeping could no longer keep up with the volume of business.

In 1954 International Business Machines Corporation (IBM)† and American

Airlines began studying the problem, and in November 1959 the two companies signed a contract for the development of a communications-based reservation system. The intent was to manage seat inventory for all flights and all aspects of customer relations. The final product succeeded and also allowed American Airlines to connect to other airline reservation systems via telegraphic communications in order to book seats on other vendors' flights. The two companies completed the project in 1964, although SABRE continues to change and improve. Yet by 1964 anything associated with the sale and even control of flights could be handled by a single, large system.

SABRE was a massive project in its day and rivaled major military projects such as SAGE.* Some of the experience gained by IBM in developing the military defense system known as SAGE* was applied to SABRE. This included understanding the elements necessary in a new operating system,* a great deal about the software* and hardware needed for telecommunications, and the requirements for database management.* A few statistics suggest the scope of the project. Nearly 1,200 terminals were connected nationwide across 12,000 miles of telephone lines through newly developed multiplexors and other telecommunications gear over full-duplex leased telephone lines. The system was up 24 hours per day, 365 days per year. Two major computers, IBM 7090s, were at the hub of the system. One operated the system, whereas the other served as backup to insure uptime around the clock. Each system had 32,768 characters of main memory,* which was considered large in the early 1960s. A new operating system was written for this computer, and the application program was extensive, using over 100,000 words of storage. Secondary memory for customer files consisted of 7.2 million characters of drum memory, whereas banks of magnetic disk drives* provided an additional half million characters of storage. The need for sophisticated direct access storage was essential since the records required for this system included all passenger names, information on the availability of seats, time and destination of flights, locations, type of aircraft, fees paid, and so on, with backup to tape subsystems. All such data had to be available to terminal users across the nation.

In order to manage such large quantities of information in a real-time environment, developers of the system adopted a new strategy. Because information was accessed directly, they relied on disk instead of tape as the primary technology for housing data. Databases were employed for the first time with database management software. Data communications systems also required software to manage the network and provide timely information. The system was dedicated to SABRE and so could not be used by any other application. Finally, all application programs were written so that even people who knew nothing about data processing could use them. These included travel agents and clerks behind sales counters at airports across the United States. The requirement that non-data processing personnel use the system was remarkable because in the early 1960s over 90 percent of all computer users had a good background in data processing. The ability of nontechnical personnel to use SABRE sent a clear and well-publicized message through American industry that a new era in application development had been launched. Today one cannot imagine an airline without a system like SABRE.

For further information, see: Franklin M. Fisher et al., *IBM and the U.S. Data Processing Industry: An Economic History* (New York: Praeger Publishers, 1983); David R. Jarema and Edward H. Sussenguth, "IBM Data Communications: A Quarter Century of Evolution and Progress," *IBM Journal of Research and Development* 25, no. 5 (September 1981): 391–404; René Moreau, *The Computer Comes of Age: The People, the Hardware, and the Software* (Cambridge, Mass.: MIT Press, 1984).

SAGE. SAGE was one of the largest computer-based applications undertaken in the 1950s to provide an air-defense system for the United States. Costing millions of dollars, involving dozens of organizations and companies, and thousands of people, it was a massive technological effort involving the development of software,* hardware, procedures, radar, and communications. The first SAGE center became operational on July 1, 1958, and four were still in use at least as of late 1983.

Two surprise experiences of the 1940s—the Japanese bombing of Pearl Harbor on December 7, 1941 and the Soviet Union's testing of a nuclear bomb in August 1949—led the Department of Defense to commit itself to a more modern air-defense system. In December 1949 the Air Defense Systems Engineering Committee (ADSEC), also known as the Valley Committee, became responsible for initiating what was announced to the public as the Semi-Automatic Ground Environment (SAGE) in January 1956.

The concept proposed called for radar coverage of the skies, with radar stations linked by telecommunications controlled by computers to provide nationwide coverage. It involved both immediate upgrades of existing World War II-vintage systems and their systematic replacement with new technologies. The first phase, that of replacing existing technologies, was assigned to Western Electric and Bell Laboratories.† This project was called the Continental Air Defense System (CADS). To generate technology for a more permanent arrangement, the Lincoln Laboratory† was created in 1951.

Lincoln Laboratory developed some technologies and appropriated those of other institutions. Perhaps the best known was the work done at the Massachusetts Institute of Technology (MIT) that was now merged with SAGE: the development of the WHIRLWIND* computer. This system was the first real-time control computer which, in the layperson's terms, could trap information on radar, transmit it instantaneously to someone to study, and display the information either on a terminal (cathode ray terminal—CRT) or print it out. It would become very reliable and fast. WHIRLWIND was developed by Jay W. Forrester* and a small army of engineers and graduate students at MIT who worked to build what eventually became the most expensive computer system of its time.

Next, scientists working for the Lincoln Laboratory tapped the Air Force Cambridge Research Laboratory (CRL) for communications technology. From this technology it was able to devise video-like images on CRTs of whatever radar picked up. The Cape Code System was a pilot version of the network, complete with communications and computers, built under the direction of the Lincoln Laboratory. In 1952 this project validated SAGE's technical assumptions.

Computer developments continued unabated throughout the 1950s. The follow-on to WHIRLWIND I and II was the FSQ-7 computer developed jointly with International Business Machines Corporation (IBM)† for SAGE. FSQ-7 computers were installed and accomplished a unique feat: two computers at each installation went into operation, one of them active and the other in test mode, and the second was prepared to take over the duties of the first should it fail to perform. This new computer incorporated random access memory, which, at that time, was leading edge technology. Random access memory was one of many examples of technical spinoffs from SAGE. Core memory* on the FSQ-7 emerged in a commercial product first in the form of storage on the IBM 704* in 1955, suggesting yet another byproduct of this project. The operating system for the FSQ-7 was the largest and most complicated written up to that time, requiring the work of hundreds of programmers. The management of such programming was new and led to vital experiences essential for the successful development of the large operating systems* of the 1960s and 1970s. Like those of future decades, that on the FSQ-7 had to be written by many programmers who collectively generated hundreds of thousands of lines of workable code. To make that work, management of technological projects rose to heights never before reached.

Various organizations participated in a coordinated manner. For example, IBM was responsible for building the computers used in SAGE. The Systems Development Corporation (an offshoot of the RAND Corporation for this project) took on the operational programs. Burroughs Corporation† constructed FST-2, which was a beam splitter. American Telephone and Telegraph† became the contractor for developing digital ground communications. The overall project was managed by Defense Engineering Services (ADES) under the initial direction of Colonel Richard M. Osgood. The late 1950s witnessed the integration of existing air-defense establishments and technologies with these new computer-based systems. Full deployment finally came in 1963 when twenty-three locations became operational, each with a data center responsible for a specific geographical area covered by radar and linked to each other by ground communications. The first such location was at McGuire Air Force Base near Trenton, New Jersey, which went "live" in July 1958.

The original idea was to monitor incoming aircraft and to manage manned interceptors. The introduction of a variety of new weapons in the late 1950s required that they be programmed into the network. These included not only such obvious weapons as the F-102 and F-106 and the Nike missile, but also others that were less publicized. The entire collection of centers required fifty-four computers. During the 1960s the entire system was again refurbished with new technologies, including merging communications linked to computers at bases run by the Strategic Air Command (SAC). By the 1980s new systems were replacing the FSQ-7s, although four were still operational as of 1983.

The costs of the SAGE project are still not fully understood from a technical point of view. It was the largest computer application of the 1950s, far more expensive than even the airlines reservation systems which was then just being studied. In addition, with its new vocabulary developed for the project, it had created a world of its own. SAGE represented one of the massive technical activities of the twentieth century which, like the development of the atomic bomb, showed the faith of responsible people in their ability to invent things. Moreover, it showed how existing technologies were grafted together with various disciplines to create new inventions. Finally, SAGE was another example of how military projects pushed forward the frontiers of technology, making the military perhaps the most crucial institutional element in the development of computers.

For further information, see: Morton M. Astrahan and John F. Jacobs, "History of the Design of the SAGE Computer—The AN/FSQ-7," *Annals of the History of Computing* 5, no. 4 (October 1983):340–349; Herbert D. Benington, "Production of Large Computer Programs," ibid.: 350–361; Robert R. Everett et al., "SAGE—A Data Processing System for Air Defense," ibid.: 330–339; John F. Jacobs, "SAGE Overview," ibid.: 323–329; C. Robert Wieser, "The Cape Code System," ibid.: 362–369.

SCAMP. The Scientific Computer and Modular Processor (SCAMP) was developed by the International Business Machines Corporation (IBM)† laboratory in Hursley, England, but was never made into a product. The experience gained in the development of this small binary computer was used effectively in the design of the S/360.* SCAMP represented a serious attempt by IBM World Trade Corporation to develop a fast machine that could support applications not possible on the IBM 1620* processor available in 1960. The machine was never made into a product because IBM decided to focus all of its attention on the development of what became known as the S/360 family of computers, a set of over 200 products that initially also included five computers, all compatible, and that represented the single most important and successful product introduction in the history of data processing.

During the process of selecting which laboratories would work on the S/360, Hursley acquired various responsibilities as a result of its experience with SCAMP. The most important byproduct of this computer was its control store which appeared in the S/360. Control stores (also known as control storage) were microcode instructions buried in the hardware of a computer. The control store developed for SCAMP emerged in the S/360 as the first read-only control store in a series of commercially available computers. (Today that is a common feature of all mainframes.) In addition to eliminating the need for a programmer to code additional instructions, the computer now had the capability of running programs written for use in computers with radically different architectures. Thus, an IBM could now hold instructions available only on non-IBM equipment.

SCAMP's engineers also devised instructions that would allow a user of an IBM 704* computer to run instructions (programs) on their machine. Without the work done at Hursley, IBM might well have decided not to include this feature in the S/360 machines announced in the mid-1960s.

For further information, see: Charles Bashe et al., *IBM's Early Computers* (Cambridge, Mass.: MIT Press, 1986); Emerson W. Pugh, *Memories That Shaped an Industry* (Cambridge, Mass.: MIT Press, 1984).

SCOOP PROJECT. Also known as Project SCOOP and the Scientific Computation of Optimum Programs, this effort was initiated by the U.S. Air Force to mechanize and improve planning and deployment at the end of World War II. It was an early and important step in the direction of operations research and the use of linear programming. The project was dominated by George B. Dantzig** who, in 1947, began the process of building a mathematical model of planning. His work led to the "simplex method," a process for computing best strategies in a given situation and widely used by the U.S. military down to the present. As part of the SCOOP Project, the U.S. Air Force also agreed to support the development of the UNIVAC* for use with this program in the late 1940s. Because the UNIVAC would take longer to build than originally thought, the U.S. Air Force next turned to the National Bureau of Standards† to quickly build a computer that could be used in Project SCOOP. The result was the SEAC* which began operation in April 1950.

For further information, see: Charles Eames and Ray Eames, *A Computer Perspective* (Cambridge, Mass.: Harvard University Press, 1973); Philip M. Morris, *Methods of Operations Research* (Cambridge, Mass.: MIT Press, 1951).

SEAC. SEAC was one of the first stored-program computers built in the United States and was constructed by the National Bureau of Standards (NBS)† as part of its overall program of supporting scientific work with computational equipment during the late 1940s and early 1950s. SEAC, like other computers built for or by the NBS, including the SWAC,* promoted computer technology at a time when there were less than three dozen major computer development projects in the United States, Great Britain, and France.

In 1947 NBS established the National Applied Mathematical Laboratories to centralize computational projects of the U.S. government. The laboratory's director, Dr. John H. Curtiss,** was supported by other government agencies, including the military branches (especially the Navy) in the encouragement or actual building of computers. Inspired by work done successfully at the University of Pennsylvania in the construction of the ENIAC* and EDVAC,* and work to systematize the architecture of computers by John von Neumann,** Curtiss campaigned aggressively for the construction of various devices, including the SEAC.

In May 1948, NBS decided to construct its own computer in Washington

Table 63
Technical Features of SEAC

Word length (bits)	45
Instruction length	45
Instruction format	(3+1)-address
Main store size	512 full words
Main store type	Delay
Backing store type	None
Add time (average)	864 microseconds
Multiply time (average)	2.98 ms
Basic clock frequency	1 MHz
Approximate number of valves	747
Approximate number of GE diodes	10,500

SOURCE: Reprinted with permission from *Early British Computers*, Simon Lavington, First Edition. Copyright © Digital Press/Digital Equipment Corporation (Bedford, MA), 1980.

(later known as the SEAC) and another at its Institute for Numerical Analysis at the University of California at Los Angeles (known as SWAC).

SEAC was designed under the supervision of Samuel Alexander. Concerned that the device be built quickly and be extremely reliable, his team elected to borrow from the EDVAC its design for memory* and circuitry. Thus, they used already proven mercury-delay memory technology which consisted of sixty-four eight-word lines, performing at the speed of 1 mH. The memory was later enhanced by using Williams tube memory which allowed the capacity of SEAC to grow to 1,500 words of data storage. It was also reliable and fast to work with. The arithmetic function of the computer included integer and logical multiplication and integer division. All logic depended on germanium diodes. It was the first computer to rely on solid-state components to handle all logic for the machine. This was done because vacuum tubes were too unreliable and slow, although they were used in the machine to amplify pulses. SEAC had 750 vacuum tubes (instead of the thousands common in machines built in the late 1940s), but it also had 10,000 germanium diodes. Input/output was handled by tape and punched paper.

SEAC's design and speed improved as engineers tinkered with the system over the years, but in general it could perform an addition in 192 to 1,540 milliseconds and multiply a number in 2,300 to 3,600 ms, both of which were good speeds for the early 1950s. (See Table 63 for a listing of SEAC's technical features.)

NBS's system was officially named the Standards Electronic Automatic Computer (SEAC). Its initial tests were completed in May 1950 when it performed calculations concerning skew rays in optics. NBS in Washington, D.C., used it until October 1964, when a superior computer technology replaced SWAC. Like many other U.S. government computers, it was used primarily for scientific applications that included military calculations, weather research, and work brought to it by scientists from various American universities.

For further information, see: S. N. Alexander, "The National Bureau of Standards Eastern Automatic Computer (SEAC)," *IRE Eastern Joint Computer Conference, 1951* (n.p., 1951): 84–89; H. D. Huskey, "SEAC," in A. Ralston and C. L. Meek, eds., *Encyclopedia of Computer Science* (New York: Petrocelli/Charter, 1976): 1239; P. D. Shupe, Jr., and R. A. Kirsch, "SEAC—A Review of Three Years of Operation," *IRE Eastern Joint Computer Conference, 1953* (n.p., 1953): 83–90; R. J. Slutz, "Memories of the Bureau of Standards' SEAC," in N. Metropolis et al., eds., *A History of Computing in the Twentieth Century* (New York: Academic Press, 1980): 471–477.

SELECTIVE SEQUENCE ELECTRONIC CALCULATOR (SSEC). With this machine International Business Machines Corporation (IBM)† made its first important foray into the computer arena. First announced in January 1948, it ushered in a new era in the history of IBM, one that would result in the rapid decline in the introduction of card punch equipment which had dominated the company's business volumes over the previous three decades. The SSEC was also the first device to combine the concept of the stored program with electronic computation and to operate on its own instructions as data. Although it was overshadowed within a few years by the introduction of the UNIVAC I* by Remington Rand Corporation† and the IBM 650,* for a while SSEC was the only large computer able to compute complex and bulky problems. The SSEC generated for IBM a major patent in the area of the stored program—a feature that would exist in all computers by the early 1950s. The experience of building the SSEC gave a generation of engineers within the company valuable experience prerequisite to launching IBM's future computer projects, particularly the IBM 650 which became the most widely used computer of the vacuum tube era.

Engineers at IBM, particularly at Endicott, New York, did some research in the 1930s on electromechanical devices, and, between 1937 and 1944, IBM sponsored research at Harvard University under the direction of Professor Howard H. Aiken** on what became known as the Harvard Mark I.* Despite a falling out between Thomas J. Watson, Sr.,** president of IBM, and Aiken (caused by Aiken's reluctance to give IBM as much credit for work on the project as Watson wanted), company engineers continued their interest in computational devices. Early byproducts of their concern included the IBM 603 Electronic Multiplier, introduced in September 1946. This machine relied on electronics more than its predecessors and broadened the engineering department's skills in the area of computational electronics.

Watson ordered the building of a high-speed calculator that would surpass the technology of Aiken's Mark I. Frank E. Hamilton, an experienced IBM engineer, was given responsibility for the project, and Robert R. "Rex" Seeber joined him. Seeber had worked at Harvard on the Mark I and had a good grasp of the theoretical aspects of computer architecture, whereas Hamilton understood how to manage large machine-building projects. During the fall of 1945, the computer was designed by various groups within IBM. Construction took place throughout 1946. By the end of that year, the relay and vacuum tube components had been assembled, leaving cabling and testing to 1947. That second set of tasks was completed by mid-1947, and the machine was finally dedicated in New York at IBM's headquarters on January 27, 1948. It rose from floor to ceiling, looking very much like a machine from a science fiction movie. It was packaged in metal casing with sleek cabinets. Lights and tapes blinked and moved, making it the caricature of what a computer should look like for a generation of cartoonists. It was given wide press coverage, thereby bringing the public's attention to a new technology.

The SSEC was a large system occupying 1,500 square feet of floor space. It contained 21,400 relays and used about 12,500 vacuum tubes. Its programs were modifiable and could operate under their control. It relied on a large table-lookup facility, and it could branch at any instruction. These two features alone helped to make this machine far easier to use than older computers, especially with complex problems that depended on intermediate results in order to continue computation. A multiplication of fourteen digits by fourteen other digits could be performed in one-fiftieth of a second, division in one-thirtieth of a second, and either addition or subtraction only 1/3,500 of a second. The heart of the system was an arithmetic unit. It also had a vacuum tube storage unit of 160 digits. Data moved back and forth from memory to the arithmetic unit by way of switching circuits. Much of how mathematics were conducted had already been developed for the IBM 603 and were simply replicated in the SSEC. It used standard IBM card equipment for input/output along with a tape subsystem for additional memory. It could also send output to printers.

The SSEC rivaled such existing machines as the Mark I and the ENIAC* because of its large memory. Moreover, its ability to change programs made it possible to solve larger problems than could other existing machines. Because its programs could change, it could perform calculations on the results of calculations already performed by programs in progress. No other machine at that time could do that. While it functioned, scientists from various laboratories throughout the United States used the machine. The experience gained on the project allowed IBM's engineers to develop the IBM 650 which became the most widely used computer of the 1950s (over 1,800 were built). Neither the SSEC nor the IBM 650 represented giant steps forward in technology, yet each was capable of doing more than its predecessors and proved very reliable by the standards of the day. The SSEC was shut down in July 1952.

For further information, see: Charles J. Bashe, "The SSEC in Historical Perspective," *Annals of the History of Computing* 4, no. 4 (October 1982): 296–312 and Charles J. Bashe et al., *IBM's Early Computers* (Cambridge, Mass.: MIT Press, 1986); John C. McPherson et al., "A Large-Scale, General-Purpose Electronic Digital Calculator—The SSEC," *Annals of the History of Computing* 4, no. 4 (October 1982): 313–326; B. E. Phelps, "Early Electronic Computer Developments at IBM," ibid., 2, no. 3 (1980): 253–267.

SHORT CODE. One of the first higher level programming languages* ever developed, Short Code was written for use on the UNIVAC* computer around 1949. The first important description of this language appeared in 1952. It apparently was the brainchild of John Mauchly,** the developer of UNIVAC and earlier ENIAC* with others at the Moore School of Electrical Engineering† at the University of Pennsylvania. Short Code was originally to run on the BINAC* computer in the late 1940s. Like other very early programming languages, this one was designed to do numerical scientific problem-solving.

This language had a two-character code that designated operations or variables. A programmer would use six of these to total the twelve characters in a word used on the UNIVAC. One could select from thirty different operations, including floating-point arithmetic, bracket indicators to evaluate expressions, find roots, test sizes, and do mathematical functions and input/output. It did not have a compiler and in fact was interpretive.

Short Code had limited importance and impact. Soon after World War II computer scientists were already at work on creating high-level programming languages. It was part of a collection of computing successes associated with a series of computers beginning with the ENIAC and that carried through to various UNIVACs in the 1950s. Short Code probably brought experience to various engineers at Remington Rand† and later Sperry† in the design of programming languages. It did not appear to have been very effective or widely used.

For further information, see: Jean E. Sammet, *Programming Languages: History and Fundamentals* (Englewood Cliffs, N.J.: Prentice-Hall, 1969).

SIMULA. This programming language was based on ALGOL* and was intended primarily for use as a software* simulator. Simulation Language relied on ALGOL 60 and first appeared in 1965, supported on Univac's 1107 computer. It consisted of a group of programs (called processes) designed to operate in parallel fashion. The processes individually executed functions in groups which the language's developers called active phases and events. Each process would consist of data and performed specified actions. This modeling tool was developed in Norway and, because it emerged in the early 1960s, was one of the first simulation languages in Europe where support for ALGOL as a programming language was greater than in the United States. Its use of classes of data also contributed to the design of data-handling mechanisms in other languages as a forefather of database management* software.

SIMULA was the creation of Kristen Nygaard** and Ole-Johan Dahl,** both of whom worked for the Norwegian Computing Center (NCC) in the early 1960s. Like their counterparts in the United States who developed such modeling packages as GPSS,* they wanted a software tool to do simulations. In 1961 they began to formulate their concept of what such a package should look like. The result was SIMULA I. NCC and Univac negotiated an arrangement whereby the NCC would acquire one of Univac's computers, while the computer manufacturer would fund development and later the marketing of SIMULA. The first SIMULA I compiler was completed in January 1965.

The language was developed in four stages. Between the summer of 1961 and the autumn of 1962, it was designed using "discrete event network" ideas. The designers also relied on mathematics to describe functions. Between the late summer of 1962 and September 1963, these ideas were merged with ALGOL's syntax, resulting in a more refined design. From then until March 1964, the design was modified to take advantage of storage management procedures designed by Dahl. The decision to implement the language was also taken. During the fourth stage, from March through December 1964, the NCC implemented the language by writing its compiler. In May the originators of SIMULA I published the first user's manual.

Those familiar with the tool perceived that SIMULA was a "system description language" that remained simple. It was accepted owing to Univac's support for the compiler both by funding, which sustained the development of SIMULA, and later by selling copies of the language to its customers. The two authors also contributed to the success of the language by spending the majority of their time designing and implementing it rather than lecturing and writing about their work. Added help came from use of SIMULA I at the NCC where at an early date it was adopted for use in its various research projects. After Univac began selling the compiler as a program product, interest expanded. Then in 1968, as proof of this growing interest, the Burroughs Corporation† brought out a version that ran on its B5500 computers.

Yet as happened with most new programming languages,* use created the need for additional function. By early 1966 users recognized that the language had a clumsy collection of element and set concepts and that other functions were also awkward (e.g., its inspect mechanism for remote attribute accessing). Because of its design, which relied too heavily on its role as a simulation package, the language could not be used for other applications. The authors also noted that the software did not use computer storage efficiently. These concerns led to their development of SIMULA 67.

In the new package, the concept of classes of data and subclasses of processes would be developed effectively. Classes of data, by being implemented efficiently, showed many designers of computing languages how to handle data in new and effective ways. The net result was a giant step forward toward the creation of database management* packages which formed the backbone of many large applications written in the 1980s. SIMULA 67 was very instructive for other computer scientists.

What eventually emerged as SIMULA 67 has been described by its developers, K. Nygaard and O.-J. Dahl, as "a general, high-level programming language and a system description language." It was developed and tested in the early months of 1967 with support from Control Data Corporation (CDC).† Versions kept being drafted, and finally, in 1969, a complete release of SIMULA 67 appeared. A Univac version was presented in March 1971, designed to run on its 1100 series of computers—then the company's main line of large processors. A compiler also appeared for use on International Business Machines Corporation's (IBM's)† System 360* line of computers.

The language enjoyed considerable popularity in Europe because of its convenience and support from Univac, CDC, and the NCC. In the United States its use on Digital Equipment Corporation (DEC)† equipment also increased its acceptance, particularly in the mid-1970s. By the end of the decade, SIMULA 67 had active followers in both Europe and the United States, although it is not known how large of a community these users represented.

For further information, see: K. Nygaard and O.-J. Dahl, "The Development of the SIMULA Languages," in Richard L. Wexelblat, ed., *History of Programming Languages* (New York: Academic Press, 1981):439–488.

SIMULTANEOUS EQUATION SOLVER. This machine was built by John V. Wilbur at Massachusetts Institute of Technology (MIT) in the 1920s and used by various researchers to perform large calculations by 1930. It appeared like a row of wooden columns, which together extended over 6 feet in length and more than 3 feet in height. To these were mounted tilting steel plates to represent unknowns, while steel plates represented equations to be solved. The contraption could work on nine equations simultaneously. The machine received some publicity when an economist at Harvard University, Russian-born Wassily W. Leontief, used it in the early 1930s to perform necessary calcuations required to predict the consequences of any economic decision. By the time this economist had devised a methodology, it became obvious that the mathematics involved would require 450,000 multiplications or two years to do. Wilbur's device cut that time dramatically and, as Leontief recalled later, "You could really change the coefficients slightly by simply sitting on the frames" of the Simultaneous Equation Solver.

For further information, see: Charles Eames and Ray Eames, *A Computer Perspective* (Cambridge, Mass.: Harvard University Press, 1973); John V. Wilbur, "Mechanical Solution of Simultaneous Equations," *Journal of the Franklin Institute* 222 (December 1936): 715–724.

SLIDE RULE. This device for performing mathematical calculations was used in the Western world for over 300 years. By the early 1900s hardly any student of mathematics or engineering was without one. It was invented because of the need to simplify the application of logarithms in arithmetic calculations. It continued in use until the wide availability of the digital hand-held calculator made it obsolete in the 1970s. But while in use, it was the world's single most important mechanical aid available to calculate multiplication, division, square, and cubical expansions. The wide acceptance of this device was testimony to its practical use. In turn, such acceptance encouraged mathematicians and inventors to continue the search for computational aids, which ultimately, although indirectly, led to the development of the modern computer.

The histories of the slide rule and of logarithms are closely intertwined. The requirement for mathematical aids to calculate other than simple addition and subtraction, particularly in the study of astronomy, inspired a Scotsman, John Napier,** to develop logarithms. He began work on logarithms in 1594 and had worked out his ideas by 1614. In effect, he defined the geometric progression of powers. Each progression was relative to the number one, and each power had a number assigned to it called a logarithm. Although it would be left to Henry Briggs,** professor of geometry at Oxford University and Napier's contemporary, to develop logarithmic tables per se, the Oxford don thought correctly that such tables would speed up calculations. One could use the tables to look up the two numbers to be multiplied or divided and then their logarithms would either be added or subtracted to reach the answer. Performing squaring and cubing could also be facilitated by a similar process.

Briggs's and Napier's work was instantly accepted as a major step forward in mathematics. By 1610, the tables were widely used. Yet the need to automate or ease their use led several inventors and mathematicians to experiment with what eventually became known as the slide rule. Historians credit William Oughtred, an English minister, with the invention of the first practical slide rule. In 1621 he made a device that consisted of two pieces of wood, so attached that one would slide over the other, and with numbers imprinted on each from end to end on a scale separated relative to their logarithms, much like modern slide rules. In order to multiply, for example, a user would align the multiplicand opposite the number one and then search for the multiplier. The number on the board opposite the multiplier was the answer. To perform division, one simply reversed the procedure. The slide rule was fast and was immediately received as a new and useful productivity tool. Even though it was not extremely accurate, it was acceptable, particularly by mathematicians willing to track where decimal points belonged in an answer. All answers were approximations and yet close enough to insure the popularity of the invention for some 300 years.

The slide rule underwent various modifications over that period of time. By 1630, for example, a circular slide rule was available; circular devices continued to be developed until the early 1900s. In 1650 a spiral slide rule was built. The calculations performed with slide rules also became increasingly sophisticated. For instance, Sir Isaac Newton in 1675 did cubic equations using three parallel logarithmic scales. By 1722 square and cube scales were commonly available. Three-part and inverted slide rules were developed in England during the eighteenth century. Runners down the middle of a slide rule were added to various models during the late 1700s and became a standard design feature of all models by 1850. During the nineteenth century, scales were expanded and specialized versions were built to cater to the needs of particular scientific disciplines, while the evolution of the device took place all over Europe, not only in Great Britain.

Although many firms were manufacturing slide rules, in the late 1800s the German firm of Keuffel and Esser (K&E) had become a major supplier in Europe and, by the early 1900s, in the United States. By the 1920s the best and most popular slide rules were K&E's. Hardly any engineering students between the early 1930s and the late 1950s was without a K&E slide rule. The firm stopped making them in 1972 when it concluded that sales were dropping too fast to justify continued manufacturing. The decline was caused by the introduction of hand-held calculators in the United States. Although slide rules are still manufactured by other firms (particularly out of plastic, in Asia), the era of the slide rule ended symbolically when K&E gave the Smithsonian Institution some of the equipment it had used to manufacture the device.

The slide rule was a precise piece of manufacturing. The best were always made out of mahogany because of the attributes of the wood. The basic strip of wood and the runner in the center that moved back and forth were always made from the same piece of mahogany. This insured that expansion and contraction of the wood in a particular slide rule would always be exactly the same. The moisture content was supposed to be relatively the same so that the wood would not swell up so much that the runner would get stuck and hence be momentarily unusable. In the last versions (from the 1920s forward), slide rules usually had a glass and later plastic plate that moved from end to end. Each had a black line perpendicular to the length of wood to help mark exactly where answers were on the slide rule. Until the 1960s, when etching techniques allowed manufacturers to produce very exact black lines, such markings on the glass or plastic were black spider webs. Indeed, until the 1960s slide rule and other instrumentation manufacturers raised spiders for the webs, using the strands for markings.

In the parlance of computers and data processing, the slide rule was an "analog" device because it measured analogically. That is, the slide rule measured a physical property analogously to a number; hence, it never counted. An absolute number therefore could not be derived from a slide rule as it could from a "digital" calculator which rendered a discrete number for an answer. To this day, computers are either analog* or the more widely used digital.* The distinction between analog and digital is important because each represents a line of research and historical evolution down to the modern computer. The use of the abacus* (a digital device) and the slide rule (analog) clearly marked out for mathematicians and inventors of calculating devices the existence of two paths of work as far back as the late 1600s. Scientists preferred digital devices. Among the early workers on calculators were Johann Kepler,* Wilhelm Schickard,** and Blaise Pascal** in the early 1600s—all worked on digital-type units. In the late 1600s they were followed by Baron Gottfried Wilhelm von Leibniz** who in 1673 built his own calculator. Even in the nineteenth century, the bias was for digital. During the early 1800s Charles Babbage* designed machines that would give absolute answers.

The use of electricity in sophisticated calculators during the early twentieth century marked the beginning of significant technological developments in analog devices. The most dramatic early example of such equipment was the differential analyzer* built at Massachusetts Institute of Technology (MIT) by Professor

Vannevar Bush** during the 1930s. Yet even that device did not supplant the slide rule as a practical alternative. It would take the electronics-based hand calculator of the 1970s (a digital device) to sound the death knell of the slide rule. Ironically, electronic analog computers appeared as early as the 1940s. They were used primarily in process control applications, usually in scientific and manufacturing facilities, operations which slide rules could not handle.

For further information, see: E. M. Horsburgh, ed., *Handbook of the Napier Tercentenary Celebration or Modern Instruments and Methods of Calculation* (Edinburgh: Royal Society of Edinburgh, 1914; reprinted Los Angeles: Tomash Publishers, 1982).

SNOBOL. SNOBOL, one of the first string and list processing languages developed in the United States, was widely used in Bell Laboratories† and at many American universities during the 1960s and 1970s for research on the theory of automata, graph analysis, and manipulation of formulas. List languages emerged when computer scientists needed to maintain tables for different kinds of information for which an end user had no idea how much computer storage to allocate. Therefore, not knowing in advance how many elements would be involved in such tables, the user needed a tool that could operate in this environment. This requirement led to the development of *list* languages. Later, they were used to write compilers for other languages, prove theorems, manipulate formal algebraic expressions, and do other work in artificial intelligence.*

SNOBOL consisted of the original language and three other major revisions called SNOBOL2, SNOBOL3, and SNOBOL4. The earliest work done to develop what would become the first release of SNOBOL at Bell Labs took place in 1962 within the Programming Research Department located at Holmdel, New Jersey, by Ralph E. Griswold, David J. Farber, and Ivan P. Polonsky, with some assistance from M. Douglas McIlroy. It took them three weeks to develop the earliest version of the language, and by the summer of 1963 it was in wide use within Bell Labs. The earliest users found it convenient to handle algebraic expressions and their evaluations, to convert descriptions of networks into FORTRAN* doing statistical analysis on them, and to write a compiler for FORTRAN. Within a year some users had also employed it to do text processing, graph analysis, and simulations of automata. Members of Bell Labs used the language by the mid-1960s to generate other programs and even to develop experimental compilers.

The question of giving this language a name posed a problem for its authors, much like that faced by the creators of other programming languages.* One of its earliest names was SEXI (String Expression Interpreter) but for obvious reasons was not serious enough for the tool. Eventually, the developers settled on String Oriented Symbolic Language (SNOBOL). It followed the fashion of the time of developing names that were clever descriptions. The humor was extended to messages in the earliest releases of SNOBOL that were not serious:

"ALL OUT OF SPACE, YELL FOR HELP" and "NO END STATEMENT, WHISPER FOR HELP." Subsequently, other programmers carried the humor further with their SNOBOL-related software tools and enhancements: ICEBOL (1968), SNOFLAKE (1970), SPITBOL (1971), FASBOL (1971), ELFBOL (1973), SNOBAT (1976), and SLOBOL (1977), to mention the most prominent.

Because the original release of SNOBOL was primitive and had limited capabilities, its developers continued to enhance and refine it with new functions and corrections to earlier problems. The biggest need was to provide built-in function mechanisms which were standard procedures (e.g., the subroutines needed to do simple arithmetic without having to write that capability each time a programmer sat down to solve a problem in SNOBOL). SNOBOL2 was developed and first used at Bell Labs by April 1964. SNOBOL3, released in July 1964, provided the capability for a programmer to define functions. Like older versions, it ran on an IBM 7090/94 and was well received. But its lack of ability to manipulate structures (names usually of aggregates of data), along with the evidence that hardware technology was changing with a shift to larger, faster machines with online programming, made the developers of SNOBOL see the need for a major restructuring of the language.

By 1965 Bell Labs was also in the process of installing GE 645 computers so that a conversion of the language would be necessary anyway to move it from the International Business Machines Corporation (IBM)† computers. Serious consideration was given to a total rewriting of the language rather than simply a revision. A larger staff than before was assembled to work on what would become SNOBOL4 in February 1966. The language took several years to develop and proved far more extensive and sophisticated than earlier versions. It evolved through the release of versions of SNOBOL4 throughout the late 1960s. The earliest version was operational in April 1966; by August it ran on an IBM 7094, and by the end of the year on an IBM 360.* Subsequently, all development focused on 360 varieties because of the rapid acceptance of that family of computers within Bell Labs and across American industry in general. By 1968 fifty-eight copies of the language had been installed on IBM 360 computers. A major revision of the language, called Version 2, complete with full documentation, appeared in December 1968, and Version 3—the last to come from Bell Labs—came out in November 1969. Other researchers made various and sporadic enhancements to the language during the early 1970s, but for all practical purposes, its great period of development had ended by the start of the new decade.

Thus, the final version of the language—SNOBOL4—became the edition used throughout the 1970s on various types of computers. It differed substantially from earlier versions. The simplicity of the language was dropped in SNOBOL4 in exchange for more powerful functions and additional capabilities required by a community of programmers technically more sophisticated than they had been in the early 1960s. The last version of the language had more semantic content than before. But more fundamental, SNOBOL4 emerged as a general-purpose language, whereas SNOBOL3 and earlier releases were simply string manipulation languages. SNOBOL4 could handle a larger number of data types, had more capability to work with patterns such as those of data objects and types, tables, and arrays, while offering functions conveniently.

The language settled down in the late 1960s into one used primarily for the production of other compilers and in doing work with algebraic expressions. It was employed on a large number of IBM's systems (7090/94, 360, 1620*, 370*), on a variety of PDP* minicomputers, and on computers of the 1960s from General Electric (GE)† and Radio Corporation of America (RCA).†

SNOBOL's technological impact was limited. Within the small group of string and list processing languages, it played the important role of introducing the ability to establish strings and substrings, along with a series of standardized mathematical functions that could process such strings. The language provided methods for recognizing patterns conveniently and remained an excellent vehicle for matching patterns and solving problems with string handling. It was also one of a very few programming languages that was widely available on many types of computers with full documentation not offered by a data processing vendor. Lack of a vendor's support frequently killed a language; SNOBOL obtained support and maintenance from Bell Labs, however. After Bell's interest in the language declined, support came from people who had worked on SNOBOL at Bell Labs but who were not at various universities. In the final analysis, SNOBOL was a useful language within a narrow range of programming problems.

For further information, see: D. J. Farber, R. E. Griswold, and I. P. Polonsky, "SNOBOL, A String Manipulation Language," *Journal of the ACM* 11, no. 1 (1964): 21–30; R. E. Griswold, "A History of the SNOBOL Programming Language," in Richard L. Wexelblat, ed., *History of Programming Languages* (New York: Academic Press, 1981): 601–657; W. D. Maurer, *A Programmer's Introduction to SNOBOL* (New York: American Elsevier, 1976); Jean E. Sammet, *Programming Languages: History and Fundamentals* (Englewood Cliffs, N.J.: Prentice-Hall, 1969).

SOFTWARE. A formal definition of software would include programs, rules, any procedures, and documentation related to the operation of a computer. In practice, the term refers to programs written to instruct a computer what to do. These programs are always written using a programming language. The term originated during the early days of digital computing, in the 1940s, when equipment was called hardware and thus anything else, particularly programs, was called software. By 1960 it was a widely used term within the data processing industry. Thus, with first- and second-generation computers, a distinction was made between equipment and instructions. With the creation of programming languages* in the 1950s and especially FORTRAN* in the early 1950s, the distinction became clearer. Software consisted of programming languages, the actual programs written in these languages, operating systems,* and database management systems.*

For further information, see: S. Rosen, "Software," in Anthony Ralston and Chester L. Meek, eds., *Encyclopedia of Computer Science* (New York: Petrocelli/Charter, 1976): 1283–1285; Nancy Stern and Robert A. Stern, *Computers in Society* (Englewood Cliffs, N.J.: Prentice-Hall, 1983).

SPEEDCODING. This was one of the first programming languages* developed for numerical scientific computing. John Backus** of International Business Machines Corporation (IBM),† along with John Sheldon, began work on this language in January 1953 in an attempt to create a language for use on the IBM 701.* By September they had a system ready. The language had two sets of operations: *OP1* and *OP2*. The first had three addresses, and the second, one. The format used was a standard card layout, which meant a programmer or mathematician could use one of each type of operation. The language had forty-five different operations under *OP1*, including ten arithmetic functions and an additional five for mathematics. The rest were for input/output. *OP2* handled such things as testing, modification of addresses, and control transfer instructions. It had operations for programmed index registers. The entire system was considered interpretive.

Speedcoding was one of the earliest attempts at IBM to develop a programming language. In the case of John Backus, it was an important experience since he would play a key role in the development of all of IBM's programming languages throughout the 1950s and into the 1960s, including the creation of FORTRAN* which became the most widely used scientific programming language in the history of data processing.

For further information, see: J. W. Backus and H. Herrick, "IBM 701 Speedcoding and Other Automatic Programming Systems," *Symposium on Automatic Programming for Digital Computers* (Washington, D.C.: Office of Naval Research, Department of the Navy, 1954): 106–145; and J. W. Backus, "The IBM 701 Speedcoding System," *Journal of the ACM* 1, no. 1 (January 1954): 4–6.

SPRINT. This list and string processing language of the early 1960s should not be confused with the telephone communications service of that name. The objective of the language was to be machine independent; in fact, it ran on International Business Machines Corporation's (IBM's)† 7094 and 7040 computers. Little is known about this language, and it was hardly used by anyone other than its developer, C. A. Kapps, as an experimental language.

For further information, see: C. A. Kapps, "SPRINT: A Direct Approach to List-Processing Languages," *Proceedings of the SJCC* 30 (1967): 677–683.

STRELA. This early Soviet computer was constructed by Y. Y. Vasilevsky in 1953. By 1956 this machine had evolved into enhanced versions, the last of which was the STRELA3, the Russian word for arrow. The machine used 43-bit words. Multiplication took 500 microseconds to perform using Williams tubes for main memory.* That memory had a capacity of 1,024 words. Access time was 20 microseconds in memory, which made the machine slower than its contemporaries in the Soviet Union, the BESM 1* at 6 microseconds, for example. It was built with 8,000 tubes and for input employed paper tape.* Output went to punched cards.

For further information, see: René Moreau, *The Computer Comes of Age: The People, the Hardware, and the Software* (Cambridge, Mass.: MIT Press, 1984).

STRETCH. This was the code name given to a computer developed by International Business Machines Corporation (IBM)† with the help of the Los Alamos Scientific Laboratory between 1955 and 1961. Also known as the IBM 7030, this computer provided twice the computing power of existing machines. When actual design was begun, the targets set for speed and performance were raised even further. In the end, eight copies of the computer were constructed. The first was delivered to Los Alamos in 1961 and was shut off in 1971. Some of the technological innovations of this machine were eventually incorporated into other second-generation computers and into the S/360.* STRETCH was IBM's first "supercomputer." Most important, the packaging of components in this machine (transistors, magnetic core storage, circuitry) taught IBM engineers a great deal that influenced the development of other computers such as the 7000* series and the 1401.*

As early as 1954, company officials had recognized the need to develop transistor-based computers at its development facilities in Poughkeepsie, New York. This computer had to reflect the manufacturing and processing lessons learned with the IBM 650* and the 700 series, while incorporating many of the real-time characteristics of the SAGE* online system developed for the military branches of the U.S. government. The initial performance target was to build a machine with 100 times better performance than the 704* for large scientific applications. New technologies were required because existing components could not do the job, hence the code name Project STRETCH.

By relying on transistors instead of vacuum tubes, IBM had committed itself to abandon the basis of all first-generation computers and to move into the new world of the second generation—an arena of undefined technologies in some instances and of unknown costs and manufacturing procedures. The benefits of some of the technology were already appreciated in the early 1950s. Transistors had greater reliability than vacuum tubes, second-generation computers would have smaller components and faster circuits, while expenses would drop and the consumption of electricity would decline as a whole. Yet the use of such new technology also required rethinking about the architecture and function of computers in general. To spread around the anticipated expense and risk of development, IBM signed a contract with Los Alamos. IBM anticipated development costs to be about $15 million. The firm signed a contract with Los Alamos to deliver a machine at $3.5 million and expected the first computer to cost $4.5 million in addition to the development costs. IBM managers thought that the funds from Los Alamos would help defray costs for creating what they thought might be a final production run of some twenty machines. If that many were built, then the entire project would be profitable.

In April 1961 the new machine was completed and at a considerable loss to the company. Estimates vary, but it appears that development expenses for the entire project were $25.4 million and that in the end the entire project cost $40.7 million in excess of receipts for the sale of some machines. As a consequence, the company learned to be more cautious when setting expectations for new technologies in future systems. Scientists throughout the data processing industry and especially within IBM concluded that the unquantifiable benefits of the new computer outweighed the cost overruns that so irritated IBM's top management.

Perhaps the most important benefit from this system was the understanding gained in how to develop what became known as standard modular systems (SMS) component packaging, which became the basis of second-generation computer architecture for the company. Engineers also learned a great deal more about how to manufacture in a cost-effective manner printed circuit cards while enhancing the quality of back panel wiring. IBM learned how to size a semiconductor project and then to manage costs which were reflected in pricing structures for individual products. The lessons were learned well because profits with all second-generation products expanded throughout the 1960s.

The computer itself had functions that were not available in earlier machines which became standard fare in all future systems. The 7030, the most powerful computer of its time, had more function. It could, for instance, have six instructions executing concurrently at various stages of completion instead of one at a time. This feature alone improved throughput for the system as a whole. In addition to this great speed, the system had parallel floating-point arithmetic which was necessary for scientific computing and, within the same machine, serial, variable-length, fixed-point arithmetic which, along with character processing, were typical functions of a commercial machine. Prior to STRETCH, only one or the other set of functions were available in a computer, not both in the same machine.

Words common to data processing today were introduced into the vocabulary with STRETCH. The most widely used of all was *byte*, the term used to describe a binary character (e.g., a single letter or digit) and the smallest piece of information in a computer's addressable storage. A byte is made up of bits (in hexadecimal notation the series of zeros and ones that serve as codes for decimal numbers and alpha-characters). The binary-addressing capability of the machine allowed words to be of various lengths (bits into bytes) as long as each address was of a length defined as powers of two. From STRETCH down to the present, all computer systems had their addresses and storage defined as powers of two, hence the odd numbers used to describe size: 64 bits or 2,048 bytes of storage, for example. The computer could handle both binary and decimal arithmetic which made the acceptance of such vocabulary convenient.

Another common phrase was *I/O interface*, which emerged as the channel in the computer that could allow up to thirty-two different devices to be part of the computer system. The I/O interface had responsibility for all I/O instructions. Finally, the term *architecture*, which today is widely used in the data processing community to describe the fundamental design characteristics of a device, came into the language. Such features as multiprogramming, storage protection, and interrupts, were now described as part of the computer's architecture. Variations of the term were also evident. Thus, one might have said that the system was ''architected'' to allow both commercial and scientific computing.

Although many technological innovations were introduced with STRETCH, the project was scrubbed after eight machines because of cost overruns. A ninth device, called Harvest, was built for the unique purpose of handling data streaming and table lookup for special applications. The work was done on a byte-for-byte basis and represented a unique machine. Only one Harvest (also called the IBM 7951) was ever built and that for the U.S. National Security Agency.

For further information, see: C. J. Bashe et al. ''The Architecture of IBM's Early Computers,'' *IBM Journal of Research and Development* 25, no. 5 (September 1981): 363–375; W. Buchholz, ed., *Planning a Computer System (Project Stretch)* (New York: McGraw-Hill, 1962); Franklin M. Fisher et al., *IBM and the U.S. Data Processing Industry: An Economic History* (New York: Praeger Publishers, 1983); René Moreau, *The Computer Comes of Age: The People, the Hardware, and the Software* (Cambridge, Mass.: MIT Press, 1984).

SWAC. SWAC was one of several computers built in post-World War II American laboratories under the sponsorship of the National Bureau of Standards (NBS)† in an attempt to consolidate and coordinate all work done by U.S. government agencies on computers. Excited by the success of the ENIAC* at the University of Pennsylvania, and at a time when there were no commercial ventures doing significant work on computers (with the possible exception of the Eckert-Mauchly Computer Corporation), NBS seized the leadership in computer development projects. SWAC became one of the most successful of these efforts. It began with NBS's decision to construct a machine for its own use in Washington, D.C.

The decision to build the machine was made in late 1948; construction began in January 1949 and ended in July 1950. Engineers were able to work quickly by relying on existing computer technology with some enhancements. The project was under the direction of Harry D. Huskey,** of the University of California, a scientist who knew about American and British computer work during the 1940s. Originally called the ZEPHYR, the computer was renamed the National Bureau of Standards Western Automatic Computer (SWAC). A staff of eleven engineers built SWAC in California where it stayed for its entire career. They made it out of thirty-seven cathode ray tubes (CRTs, or large TV-like screens) (for memory*), 2,600 tubes, and 3,700 crystal diodes (for the arithmetic logic unit).

This computer system, like all such devices of the 1940s onward, had three parts: memory, arithmetic unit, and input/output devices. Memory relied on the Williams tube run in parallel, which meant that data could move in and out of the computer's tubes at the same time, thereby increasing the speed with which information was handled. At the time the machine first became operational, 16,000 three-address additions could be done per second. In the same time, 2,600 multiplications were possible, which made this computer's memory extremely fast by contemporary standards. It allowed 256 words, each made up of 37 binary bits, to be stored on each of 37 CRTs. Memory cycle time was 16 microseconds (us). SWAC's engineers added extra storage in the form of a magnetic drum which could house 4,096 words with an access time of 500 us instead of the 16 us possible with CRTs.

The arithmetic unit relied on 37-bit registers, all based on a binary counting system. Hence, electronic impulses were either positive or negative, and the combination of such charges indicated the value of a particular number. SWAC could perform such mathematical functions as addition and multiplication.

Input/output (I/O) reflected commonly available technologies of the day. Typewriters and punched paper tape* were the workhorses of the system. Because of the slowness of such devices and their heavy dependence on maintenance, they were soon replaced with an IBM 077 collator and a IBM 513 card punch to handle I/O. Now cards were read at 240 per minute and punched at 80 per minute. The system could read 78 bits (or two words each made up of 37 bits each) off each row on a card. Each card had eleven rows.

Operators controlled the SWAC with eight instructions to handle addition, subtraction, multiplication rounded, product, comparisons, extractions, input, and output, with variations in command. Floating-point capability was later added to enhance the mathematical functions of this machine through the use of SWACPEC, an interpretive floating-point system for programming.

The computer was mounted in three cabinets measuring 12 by 5 feet, and 8 feet high—small by contemporary standards. WHIRLWIND,* the largest computer of that era, took up over two floors at the Massachusetts Institute of Technology (MIT), while other devices consumed a healthy portion of any single-floored laboratory at other universities and government agencies. In addition to the I/O already mentioned, SWAC had a console and power generator, all housed in an air-conditioned room of 40 by 30 feet. The system was built and maintained on the campus of the University of California at Los Angeles (UCLA).

SWAC performed relatively well. By mid-1953 during 70 percent of the time when it was run, outputs were useful. SWAC was then also averaging some fifty-three hours of work per week. All applications on the SWAC consisted of complicated scientific calculations, one of which took 453 hours to complete. Other applications included linear equations, Fourier synthesis of X-ray diffraction patterns of crystals, work on Mersenne primes, differential equations, and various other numerical analyses for both government and private engineers and scientists. Air circulations around the earth were also studied using the

SWAC. NBS continued to use this device until 1954 when UCLA took over management of the system. UCLA in turn operated it until December 1967. Jobs run in the 1950s for government and scientific applications represented the most common use of computers at that time, although by the late 1950s various companies, such as International Business Machines Corporation (IBM)† and Sperry Rand† (with its UNIVAC*) were selling machines for commercial applications.

For further information, see: H. D. Huskey, "Characteristics of the Institute for Numerical Analysis Computer," *Mathematical Tables and Other Aids to Computation* 4, no. 30 (1950): 103–108, "The National Bureau of Standards Western Automatic Computer (SWAC)," *Annals of the History of Computing* 2, no. 2 (April 1980): 111–121, "SWAC," in A. Ralston and C. L. Meek, eds., *Encyclopedia of Computer Science* (New York: Petrocelli/Charter, 1976): 1382–1383, and "The SWAC: The National Bureau of Standards Western Automatic Computer," in N. Metropolis et al., eds., *A History of Computing in the Twentieth Century* (New York: Academic Press, 1980): 419–431.

T

TAPE DRIVES, IBM. These primary auxiliary data storage devices were used with computers in the late 1940s and throughout the 1950s. By the late 1960s they had been displaced by disk drives* as the primary medium for storage of machine-readable information. From that time forward, tape storage was used primarily for archival purposes. During the 1940s it served as a major input medium for new, often experimental computers. But by the end of the 1950s they were standard components of computer configurations. Because International Business Machines Corporation (IBM)† wrestled technical and marketing leadership away from Remington Rand† (later known as Sperry†) by the end of 1956, its products became major elements within the data processing industry. As with many other IBM products, particularly disk drives, technological standards were set by the large vendor. This circumstance was evident by the time IBM introduced the 3420 tape drive, the peripheral that dominated tape drive sales in the industry in the 1970s despite severe competition from other vendors (such as Telex), which standardized on the technology and specifications of this tape drive.

Tape had been used since the dawn of modern computing. Paper tape* had first appeared in the 1850s for use with telegraphic equipment. By the end of the 1930s Konrad Zuse** had employed exposed photographic film for his Z series* of computers. In the 1940s tape was used with the ENIAC,* EDVAC,* BINAC,* and in such other equipment as the WHIRLWIND,* the Mark* series from Harvard University, and in every commercial system of the 1950s, including the celebrated UNIVAC.* In the early 1950s magnetic tape displaced punched cards as the medium of choice for storing large files of machine-readable data. Tape made it possible to process information at faster speeds than with cards while relying increasingly on programs to manage data.

With the early and immediate marriage of tape to computers, it was expected that when IBM entered the computer business it would rely on tape as part of a system's configuration. Beginning with its engineering activities in the early 1950s, IBM began to introduce magnetic tape equipment. The first product of any significance, shipped initially in 1953, was a relatively high-performance digital magnetic tape drive for use with the IBM 701.* It used a vacuum-column design and served as the basis of IBM's offerings during the 1950s—the great era of tape drives in general. Officially known as the IBM 726, it employed a non-return-to-zero encoding (NRZI) form and recorded data on seven tracks (all parallel), using a half-inch plastic tape that was magnetic. By 1960 tape was being coated with magnetic iron oxide. During the decade, the density at which data were recorded on IBM's tape went from 100 to 800 bits per inch (bpi), while the speed at which tape worked grew from 75 to 112.5 inches per second (ips). Engineers improved the speed at which start/stop functions were performed in order to reduce the amount of tape wasted (much like unused runway that an airplane flies over in landing). That unused portion, known as the interblock gap (IBG), was constantly a subject of attention and measurement. In the 1950s the IBG hovered at 0.75 inch. Engineers also worked on expanding the cluster of data together (called physical block sizes) to improve both the efficiency of tape and the effective rate at which data were handled.

Because tape evolved in an environment first dominated by card punch equipment, data were grouped together much as they had been on cards. Operations were "batched" together, hence the term *batch jobs*. Information could be read or written only sequentially; direct access would have to wait for disk drives, although a number of vendors attempted to develop accessing via tape in the 1950s. They failed to develop useful, direct access tape systems. As long as computers were slow, sequential accessing was not a severe problem, only an irritant. In 1956 IBM introduced the 777 Tape Record Coordinator as one of the first tape control units to manage data, much like data channels would in computers in the 1960s. It worked with the IBM 705 II processor, buffering data so that the computer could overlap tape operations and processing. (See Table 64 for a listing of the technical features of IBM's tape drives during 1953–1966.)

Between 1963 and 1966 engineers interested in tape focused their attention on products for inclusion with the S/360* family of computers, ultimately developing two tape drives. The first was a new version of the IBM 7340 Hypertape drive with very dense packing of 3022 bpi. It relied on one-inch tape mounted in a cartridge, but because it was incompatible with the widely used half-inch tape of the day, the IBM 7340 failed commercially. The IBM 2401, however, was a real success. It employed a nine-track format, was compatible with existing tapes, and was offered for use on the S/360 as its basic tape drive. It recorded at 800 bpi and later appeared with the ability to record at a density of 1600 bpi. To meet the requirements of customers who had both nine- and seven-track formatted tape libraries, IBM introduced several models of the 2401. The interlock gap dropped to 0.6 inches on nine-track tape but could not be

Table 64
Select Technical Features of IBM Tape Drives, 1953–1966, Selected Years

Feature	726	727	729-III	729-IV	2401-6
Year introduced	1953	1955	1958	1961	1966
Data rate (in./s)	75	75	112.5	112.5	112.5
Characters/in.	100	200	556	800	1,600
Characters/s	7,500	1,500	62,550	90,000	180,000
Terminal velocity[a]	75	75	112.5	112.5	112.5
Characters/s/ $/Monthly rent	17.6	27.3	69.5	94.8	216

[a]Rates effective acceleration.

lowered on the seven-track beyond the standard 0.75 (if such tape was to remain compatible with those of other vendors).

After 1967 tape became less important than disk storage but remained a critical component of any major system. In a commercial environment, for example, payroll applications were traditionally run using tape drives as the primary housing for data. The same held true for very large data files run against manufacturing applications. Not until the late 1970s and early 1980s did disk storage become so cheap that it made sense to migrate even these applications to disk. Yet back in the 1960s tape drives became faster in order to keep up with the increased processing speeds of computers and to serve as backup units to ever faster disk drives.* Typically, software* delivered by a vendor came on tape. That tape, as with most reels, was in 2,400-foot lengths. Large libraries of tapes mushroomed by the early 1960s which, by the 1970s, were frequently massive, containing tens of thousands of tapes, particularly for insurance companies and government agencies. Even companies with intermediate-sized installations would shelter libraries containing several thousand reels. Almost all recovery and backup systems were recorded on tape, along with back copies of data files.

Beginning in 1967 and throughout the 1970s, IBM introduced new tape drives that caused increased density of data, going from 1,600 to 6,250 bpi (Table 65). The velocity of tape rose from 112.5 to 200 ips, while start/stop functions were improved. Such improvements first appeared in the IBM 2420, which was initially shipped in 1969. This device had an ips of 200 within 2 ms from a complete stop. Speed and function were related to customers in much the same fashion as automobiles were described because these two features made tape more or less attractive. Furthermore, by the early 1970s one could buy equipment from IBM in various colors: IBM blue was popular along with red and canary yellow. Read/write errors occurred less frequently than on disk; therefore

Table 65
Select Technical Features of IBM Tape Drives, 1971–1973

Feature	3420 Model 7	3420 Model 8
Year Introduced	1971	1973
Data rate (in./s)	200	200
Characters/in.	1,600	6,250
Characters/sec.	320,000	1,250,000
Terminal velocity	200	200
Characters/s/$/Monthly rent	478	1,440

reliability, while always a factor, was less important with more proven tape technology.

The centerpiece of IBM's tape offerings was the IBM 3420 which was first shipped in 1973. It proved to be a highly reliable, long-lasting product that is still the industry's standard today. The IBG dropped to 0.3 inches, start/stop time went to under 1 ms, and bpi grew to 6,250 (Table 66). The data rate (the speed at which information could be read) decreased to 1.25 Mbps. In 1979 a servo-controlled reel-to-reel tape drive became available on the IBM 4300 and the IBM 8100 systems.

In the early 1980s IBM jumped ahead of the industry in tape by announcing the 3480, a tape drive that represented an important improvement over older devices. Instead of using a reel-to-reel format, it returned to the concept of a tape cartridge. This offered much higher levels of reliability, greater control through programming, and more density, while reducing the physical size of reels of tape and their associated tape drives. For example, instead of being 6 feet tall, boxey affairs, the new drives were table-high, used less electricity, and required substantially less air conditioning. The cost of storing data on tape had declined once again, as usual in this industry.

For further information, see: Charles J. Bashe et al., *IBM's Early Computers* (Cambridge, Mass.: MIT Press, 1986); Franklin M. Fisher et al., *IBM and the U.S. Data Processing Industry: An Economic History* (New York: Praeger Publishers, 1983); J. P. Harris et al., ''Innovations in the Design of Magnetic Tape Subsystems,'' *IBM Journal*

Table 66
Data Rates on Select IBM Tape Drives, Early 1980s

Drive	At 1,600 BPI	At 6,250 BPI
3420/4	120	470
3420/6	200	780
3420/8	320	1,250

All data rates are in Kilobytes.

of Research and Development 25, no. 5 (September 1981): 691–699; Emerson W. Pugh, *Memories That Shaped an Industry: Decisions Leading to IBM System/360* (Cambridge, Mass.: MIT Press, 1984).

THOMAS ARITHMOMETER. This device was one of the earliest successful calculators of the nineteenth century for doing mathematical calculations. As early as 1671, Gottfried Wilhelm von Leibniz** had worked on a machine that would do multiplications by way of successive additions. His first machine appeared in 1694. It did not work well, not because of design errors, but because of the poor machining techniques of the seventeenth century. The ideas embodied in the device were kept alive by various inventors. The first practical version of this technology appeared around 1820 when Charles Xavier Thomas de Colmar built a machine later known as the Thomas de Colmar Arithmometer. During the four decades following its invention, the device was well known among inventors of aids to calculation. Charles Babbage,** then working on his various engines, was aware of Colmar's device and may have been influenced by it. Colmar's machine was the one other inventors sought to improve by the 1860s in order to develop a viable product.

The challenge was to make a machine that was at least as easy to use as this one. Colmar's was easier to use than older devices. For one thing, the operating motion during multiplication was reversed for the result registers, which meant that the result registers shifted to the left and above the stepped-gear mechanism. This solved the earlier model's problem of having to move the entire stepped-gear mechanism under the result registers. Better quality manufacturing also insured a more reliable device. Thus, it worked well and rarely broke down. Its disadvantages were its heaviness and size (it took up an entire table). This last disadvantage was a real problem until the Baldwin class came along with models one-fourth the size and weight of the Arithmometer.

No figures are available on the number of copies of the Arithmometer in use. The device was made one at a time, primarily in France, and only limited numbers were available in the 1820s and 1830s. By the late 1860s, however, others were building and selling variations of this technology in Western Europe and the United States. By the early 1880s, more sophisticated machines were available, including devices manufactured by Burroughs Corporation† and the National Cash Register Company (NCR).† Copies of the machine were familiar fixtures of many laboratories until World War I.

It varied in design, with some models possessing six or seven or even eight figures in the setup registers and as many as twice the number of any setup register in the results register. In addition, some very large models could handle sixteen digits in setup and thirty-two in the output or results register. Probably up to two dozen firms built these machines both in Europe and in the United States. The Tate Arithmometer, a British model of the late nineteenth century, had movable markers that looked much like the bass and treble controls on a modern stereo and a button above all ensconced in a wooden case. Another, called the Edmunson Circular Calculating Machine, resembled a portable roulette wheel and was also in a handsome wooden box. But all arithmometers were bulky and awkward to use compared to more modern devices. Thomas dominated the market with his machine during the early nineteenth century because his was produced at a time when greater amounts of data had to be processed. By the late nineteenth century many varieties of calculators and other mechanical devices for handling data would be available.

For further information, see: M. d'Ocagne, "Vue d'Ensemble sur les Machines à Calculer," *Bulletin des Sciences Mathématique, 2ᵉ Série* 46 (1922): 102–144; E. M. Horsburgh, ed., *Handbook of the Napier Tercentenary Celebration* (Edinburgh: Royal Society of Edinburgh, 1914), reprinted, Los Angeles: Tomash Publishers, 1982); F. Reuleaux, *Die Thomas'sche Rechenmaschine* (Leipzig: n.p., 1892).

TIDE PREDICTORS. These represented a class of analog aids to calculation in the study of tides. The need for equipment to predict the rise and fall of tides had existed from earliest times. For seamen, failure to appreciate when the tide was high upon entering or leaving a port posed the risk of tearing the bottom out of their vessel. To avoid that disaster, sailors went to great lengths to monitor the behavior of tides. But visual evidence and experience were not enough, and so from the earliest days of sailing, men sought more precise mechanisms and techniques to help in the process—an exercise that has continued to the present.

The application is both simple and complex. It is simple in that tables can easily be compiled suggesting the ebb and flow of tides for a particular port; in a relative sense, people have been doing that for centuries. Navigational aids published in Europe frequently contained both astronomical data to guide direction and tidal tables for various ports. The application is difficult in that certain tides and obviously complex estuaries still defy precise definition. Computation by Fourier sequences relying on records of past tides improved the situation. This method of calculation works relatively well, but because the sun and moon influence the characteristics of tides, a process not yet fully understood, calculations remain inexact. Moreover, these calculations are cumbersome and can be time-consuming; hence, the need for aids to speed up the process while making the results more accurate.

In the fifteenth century a machine existed which could predict tides for a few European ports; it is now in the Bodleian Library at Oxford University. Other machines were probably in use in the same years, but only the British version has survived. The Oxford device has two pointers, like those of a normal clock, one with a picture of the moon and the other of the sun. They are directed toward marks indicating the positions of these two bodies. The user would peek through a hole in the center, looking for a light-colored part that varied from the thin crescent to the moon's full circle. If what he saw on the device was the same as what was in the sky, as seen through the hole, the gadget had been set correctly. Then, appreciating the "aspect" of the particular port in question, the user could read on the device approximately when high tides might be expected.

This method was used for hundreds of years until the mid–1800s when the quality of tide predictors was improved through the use of mathematics. By then harmonic analysis had progressed to the point where a collection of cosine coefficients existed, which made it possible to create a formula good enough to predict tides. The formula required was developed by Lord Kelvin (1822–1892), a Scottish physicist:

$$y = A \cos(u) + B \cos(v) + C \cos(W) + \dots$$

Once Kelvin had developed the formula, he began to build a machine to use it.

Lord Kelvin's tide predictor was made up of a collection of lower and upper gears. The upper gears turned a certain amount proportional to a coefficient (e.g., A, B, C). A rod connected to the upper and lower gears moved up and down as they turned, the rod serving as a mechanical analog of the cosine function. Amplitude of movements was established by gear ratios. Thin wires connected rods to pulleys so that when the rod moved up or down, their pulleys also rose or fell. Yet another wire threaded through all the pulleys and attached to one end of the machine was, at the other end of the wire, connected to a pen that recorded on chart paper. As the device turned, the wire loosened or tightened, causing the pen to rise or fall on the paper—in effect, tracing the function that was twice the sum of the cosine functions generated by the gears. A user had to understand the gear ratios needed for a particular location to be studied (such as a particular port) and would then calculate future tides by turning the machine's drive wheel. The results were recorded on the chart. These were in turn were worked up into publishable form.

The British government adopted his machine. In the United States similar devices were built at the end of the nineteenth century and during the early years of the twentieth century. The most important of these machines was constructed by the U.S. Coast and Geodetic Survey which was in charge of navigational work for the American coastline. That agency began constructing a machine in 1905 and put it into production in 1911. It was quite large: 7 feet high, 11 feet long, and nearly 2,500 pounds in weight. It was far more advanced than Kelvin's machine in that it could calculate the height of tides to the nearest 0.10 feet for every minute of the year. It predicted tides for the U.S. government until the mid–1960s.

The device was finally replaced with an IBM 7094 which was programmed to do the job. The amount of data used then was far greater than before. The 7094 used a cosine subroutine 20 million times to develop all the predictions for one location for one year. Because the application was now programmed and could therefore use a general-purpose computer, it became relatively easier to move on to newer computers in the 1970s and 1980s than to the first processor.

For further information, see: Edward Roberts, "Tide-predicting Machine," in E. M. Horsburgh, ed., *Handbook of the Exhibition of Napier Relics and of Books, Instruments, and Devices for Facilitating Calculation* (Edinburgh: Royal Society of Edinburgh, 1914, reprinted as *Handbook of the Napier Tercentenary Celebration or Modern Instruments and Methods of Calculation*, Los Angeles: Tomash Publishers, 1982): 249–252; Michael R. Williams, *A History of Computing Technology* (Englewood Cliffs, N.J.: Prentice-Hall, 1985).

TRAC. Also known as the Text Reckoning and Compiling language, TRAC was a string and list processing programming tool created in the early 1960s. Its designer intended TRAC to provide an aid to handling text interactively by users who would communicate with computers by typing commands on a keyboard. It was one of the earliest software* languages to use nested functions or macro facilities, functions that became relatively common by the early 1970s.

Calvin Mooers began to design TRAC in 1960 while he was at the Rockford Research Institute, and four years later, Peter Deutsch made it run on a PDP–1. Later, the language operated on other PDP* computers (models 5, 8, and 8S), on General Electric's (GE's)† Datanet–30, and on an IBM System 360* Model 67. It was reportedly also functioning in the 1960s on two European computers, a SAAB D–21 and an ICT 1202.

Mooers originally wanted a language that could create strings with sequences of actions (also known as functions) and nest them. Nesting was the process of incorporating structures of a particular kind within the exact same kind of structures. He intended for TRAC to evaluate nesting beginning with the innermost level to that most outward in the program, reading, as in English, from left to right as the means of executing a program. To control a program written in TRAC, a user employed two fundamental commands called *idling* procedures. **PS** stood for print and **RS** for read.

TRAC was not a widely used language in the 1960s. It spawned one other piece of software, however, called General Purpose Macrogenerator (GPM), which was developed in the mid–1960s for the same reasons as TRAC but in a considerably different format. For example, GPM handled data strings differently. It more frequently mixed its specific executable operations with data strings than did TRAC in its original form.

For further information, see: C. N. Mooers and L. P. Deutsch, "TRAC, a Text Handling Language," *Proceedings of the ACM 20th National Conference* (1965): 229–246; C. N. Mooers, "TRAC, a Procedure-Describing Language for the Reactive Typewriter," *Communications, ACM* 9, no. 8 (March 1966): 215–219.

TRANSISTORS. See CHIP

TREET. This programming language* was designed to do string and list processing and was developed by E. C. Haines as his master's thesis in 1964 for the Massachusetts Institute of Technology (MIT). TREET relied heavily on another language, LISP,* for syntax and structure. It was designed to display trees (hierarchies of files with single data paths) and was one of the earliest languages to allow a user to perform queries. As a research tool, it underwent a considerable number of changes throughout the 1960s and ran on various

International Business Machines Corporation (IBM)† computers: STRETCH* (7030) and various System/360s.* TREET has been used for data management and text processing on these machines.

TREET consisted of functions defined by a programmer or user using English-like commands and notation. Its intent was to offer the same functions as LISP, only in a notation that was easier to use and read. Differences developed over the years, making much of TREET's syntax independent of LISP, although it could be translated back into LISP release 1.5.

For further information, see: E. C. Haines, Jr., *The TREET List Processing Language*, Information System Language Studies Number Eight SR–133 (Bedford, Mass.: MITRE Corporation, April 1965) and his *The TREET Programming System: IBM 7030 Implementation*, MTP–58 (Bedford, Mass.: MITRE Corporation, March 1967).

TURING MACHINE. The concept for this computing device was developed by Alan M. Turing** in 1936 and published in a paper in 1937. Much like John von Neumann** and the Von Neumann Machine,* the concept of the Turing Machine provided a useful abstract of what a computational device might look like and offered a description of the functions it should perform. Unlike von Neumann's conception (developed in the mid–1940s), Turing's was essentially the first of any significance in the twentieth century. Furthermore, his was an attempt to deal with a specific class of mathematical problems; von Neumann's was a direct statement of how to design computers.

Turing's machine was made up of three parts. The first was a *control unit* which could consist of various forms. The second was a *tape* (probably of finite length) on which discrete pieces of information could be stored on predefined sectors by way of a previously established set of notations or symbols. Finally, Turing conceived the idea of a *read-write head* which might travel above a tape extracting information from it for use by the control unit (computer) or which would deposit information back on tape. His machine would have operated like a batch processor, that is, computing a sequence of specific steps or jobs. The symbol being read off the tape by the read-write head would determine the action of the machine, as would the nature of the control unit itself. A program in his model was thus a combination of symbols and state-of-being of the control unit, both working together to generate the action of the machine.

Turing's conception was both simple and complex in describing the fundamental components of a computer. It had a finite program, storage of information, and specific modes of operation. His approach proved that any computer could be simulated using his concept. Hence, anything that could not be computed in his model could not be computed at all. This concept was particularly important because, in the 1930s, Turing was interested in problems that could not be solved. That concern led to his devising the Turing Machine (as it came to be called) as a way of determining whether a problem could be solved in the first place.

Turing's paper describing his concept, "On Computable Numbers with an

Application to the Entscheidungsproblem'' (1937), was a remarkable publication published when he was only twenty-six years old. He described computers before any existed, and he provided a useful approach to the study of problems that apparently had no mathematical solutions. It was immediately recognized as an important paper in the field of mathematics and soon after was read by others interested in what came to be known as computer science. Since publication, it has been the subject of continuous debate and study. Variations of the Turing Machine have been created conceptually since the late 1930s.

For further information, see: J. Hartmanis, *Feasible Computations and Provable Complexity Properties* (Philadelphia: Society for Industrial and Applied Mathematics, 1978); Andrew Hodges, *Alan Turing: The Enigma* (New York: Simon and Schuster, 1983); Alan M. Turing, ''On Computable Numbers with an Application to the Entscheidungsproblem,'' *Proceedings, London Mathematical Society*, Series 2, 42 (1937): 230 and reprinted in Martin Davis, ed., *The Undecidable* (Hewlett, Long Island: Raven Press, 1965); Sarah Turing, *A. M. Turing* (Cambridge: Heffer and Sons, 1959); J. H. Wilkinson, ''Some Comments from a Numerical Analyst,'' *Journal of the ACM* 18, no. 2 (1970): 137–147.

U

ULTRA. The British gave this code name to their effort to break German codes during World War II. The British were able to decode almost every military message sent by the Germans during the war covering every important campaign of the conflict. At the start of the war, the British government called together at Bletchley Park,† an estate near London, a number of scientists, mathematicians, and chess champions to work on various intelligence projects. One of these projects involved the decoding of German messages originally coded in Germany or Western Europe on the ENIGMA* machine.

The British were fortunate in that in 1938 a Polish engineer named Richard Lewinski, who had knowledge of the ENIGMA, offered to help the British construct a similar machine. Scientists working at Bletchley could now begin to understand how the Germans coded and decoded messages. In July 1939 the Polish government, aware that war was imminent, decided to share their intelligence-gathering capabilities with the British and French. Since the early 1930s, as it turned out, the Poles had regularly been breaking German codes. At the meeting held on July 25, 1939, ENIGMA, along with other intelligence apparatus, was explained.

Over the next several years, Winston Churchill and other Allied leaders were fed reports on German communications on a regular basis. The data provided were so valuable that even when it was learned that Coventry was slated for bombing by the Germans no precautions were taken for fear of exposing the existence of ULTRA. The result was loss of life and the destruction of one of England's most beautiful churches.

After the war, many of those who worked on ULTRA went on to develop computers in Great Britain. The most notable of this group was Alan M. Turing.** While working on German codes, this group constructed a computer-like device called COLOSSUS,* giving them additional experience with the technologies that would be used to make computers after the war. As of this writing, the whole story of ULTRA, ENIGMA, and COLOSSUS remains shrouded in secrecy. The files on them remain closed to the public in Great Britain and unavailable to historians. The story of these projects was unknown until the 1970s when comments and then memoirs began to be published by participants.

For further information, see: Jozef Garlinski, *The Enigma War* (New York: Charles Scribner's Sons, 1979); Joel Shurkin, *Engines of the Mind: A History of the Computer* (New York: W. W. Norton, 1984); Gordon Welchman, *The Hut Six Story: Breaking the Enigma Codes* (New York: McGraw-Hill, 1982).

UNICODE. This programming language was developed by Remington Rand† to solve scientific numerical problems using its 1103A and 1105 computers. It was designed between 1957 and 1958, appeared much like MATH-MATIC* in form and function, and served the same purpose as International Business Machines Corporation's (IBM's)† FORTRAN.* UNICODE's fate was tied to that of the two computers; as long as they sold, customers considered using UNICODE, and when they became unattractive, the language passed into oblivion. These machines were more powerful than those on which the earlier Remington Rand language (MATH-MATIC) had relied, making UNICODE functionally more advanced in that it could handle larger problems.

In many ways the language replicated functions found in MATH-MATIC's compiler several years earlier. At the same time, it took advantage of concepts that were appearing concurrently in various releases of FORTRAN. For example, variables beginning with I, J, K, L, and M were designated as fixed point; all others were floating point. In MATH-MATIC there was no floating-point arithmetic. UNICODE kept MATH-MATIC's concept of using numerical subscripts as a useful tool. The language was actually a hybrid of MATH-MATIC and FORTRAN, and so contributed nothing to the evolution of programming languages.* Furthermore, its failure to win wide acceptance could be traced directly to the demise of the 1103A and 1105 caused by the success of IBM's products—including FORTRAN—in the late 1950s.

For further information, see: Jean E. Sammet, *Programming Languages: History and Fundamentals* (Englewood Cliffs, N.J.: Prentice-Hall, 1969).

UNIVAC. UNIVAC was one of the first widely used general-purpose digital computers* in the data processing industry. Although it was not the first computer made for commercial customers, it did become the most widely used machine in the early 1950s. The word UNIVAC became almost synonymous with computers and characterized the public's image about what computers looked like. Indeed, the UNIVAC introduced in the early 1950s made it appear that the manufacturer dominated the fledgling computer industry. After UNIVAC was accepted, many scientists, government officials, and business executives began to foresee commercial uses for computers. This computer was to mark a new chapter in the history of computers.

The UNIVAC was a descendant of earlier machines made by John Presper Eckert, Jr.,** and John W. Mauchly** during the 1940s. During World War II, while at the University of Pennsylvania's Moore School of Electrical Engineering,† they constructed the ENIAC,* proving that an electronic digital computer could be built. Then they designed the EDVAC,* which showed the value of the stored-program concept in the late 1940s while the BINAC* proved it. Then, in 1951, came the UNIVAC, also designed by these two engineers, incorporating their earlier experiences with a new machine designed as much for commercial as for scientific applications.

To supplement contracts and funding from the U.S. government for additional computer design, Mauchly persuaded the Prudential Insurance Company to support development work on the UNIVAC with an option to buy one, cementing the agreement with a contract signed on December 8, 1948. It called for the construction of a computer for $150,000, due to be delivered on September 15, 1950. The A. C. Nielsen Company also wanted a UNIVAC. Thus, by the end of 1948, Mauchly had signed up several customers for UNIVACs: the National Bureau of Standards† for the Census, Watson Labs/Teleregister, Prudential, and then Nielsen. These were in addition to contracts with other U.S. government agencies.

Additional financial support for the Eckert Mauchly Computer Company (EMCC) came from the American Totalisator Company which made pari-mutuel machines for posting odds and recording race results. The new chairman of the board of the EMCC was Henry Straus, a vice-president of the American Totalisator Company. Through his support work could continue on the UNIVAC, and the BINAC could be completed. By the end of 1949 EMCC had over 130 employees and contracts for six UNIVACs worth $1.2 million. After Straus's death in an airplane accident on October 25, 1949, American Totalisator lost interest in the UNIVAC project, a loss of support that accelerated EMCC's race toward insolvency.

Financial worries drew attention away from development work on the UNIVAC and threatened to bring the whole project to a close. Finally, on February 1, 1950, Remington Rand† bought the company, turning it into a subsidiary in exchange for stock ownership and cash. The new division was headed by General Leslie R. Groves, who had managed the Manhattan Project and was now charged with building the UNIVAC. With the acquisition of the EMCC, Remington Rand entered the computer industry as an important member. Eckert continued development work on the UNIVAC, although Mauchly later left the firm. Work on the computer progressed, and, finally, on March 31, 1951, the first UNIVAC was delivered to the Census Bureau.

The UNIVAC I, as it was called, was designed to be manufactured in quantity. In fact, forty-six were built—the single largest number of the same type of computer ever made up to that time. Most were initially shipped to government agencies, but General Electric (GE)† in 1954 became the first commercial customer to acquire one. As most manufacturers of computers in the 1950s did, both the EMCC and Remington Rand grossly underestimated market demand for such machines. Mauchly, for example, believed he could place fifteen at a cost of about $1 million each. Despite Remington Rand's failure to build an aggressive marketing force or programs to push the product, over forty were sold.

The first UNIVAC represented a continued refinement over its predecessors. Mercury delay-line technology was once again employed for memory,* while the processor operated at a fast 2.5 megahertz pulse rate, like the BINAC. Memory could house 1,000 twelve-digit characters. Speed was critical to UNIVAC's success, and it was fast by the standards of the day: additions or subtractions were performed in less than 600 microseconds, whereas multiplications took 2,500 microseconds and division 4,000 microseconds. The machine was 14.5 feet by 7.5 feet by 9 feet, and relied on a magnetic tape system for storing data and passing them to the computer quickly and in quantity. Data transmission via tape represented a significant improvement over earlier machines that had relied either on punched paper tape* or cards, both of which were much slower. UNIVAC's configuration had a card-to-tape converter, a uniprinter, and a console. It was also one of the first computers to be housed in frames that made it look like an attractive machine rather than a piece of lab equipment. The verifier, for example, looked like a desk with drawers, the tape drives* were in ample refrigerator-sized housing, whereas the high-speed printer had metal covers to hide its inner workings (Table 67).

Reliance on tape made the system even more effective than other computers. Instead of using BINAC's plastic tape, metal called the UNITAPE was used. It was a half-inch wide with 1,200-foot lengths per reel. Later, the standard for the entire industry would be 2,400-foot reels. One hundred to 120 decimal digits could be stored per inch of tape; thus, one reel of tape could store just over a million characters of data. To do the same on cards would have required tens of thousands which were bulky, expensive, and slow to feed into a computer or to punch as output. With tape, records could be longer than eighty characters; tape was also durable and easy to store.

The total configuration represented what a typical mainframe system would look like throughout the 1950s and 1960s. First, there was the computer itself (now known as the processor or mainframe), with subsystems to manipulate data and memory to store information. Other devices could be added to the system: Supervisory Typewriters and printers. Tape drives* were the primary input/output units, and they operated at a very fast rate of 7,200 characters per second. They operated as quickly as if one were reading ninety eighty-column IBM cards per second. Additional card I/O peripherals were added, and the configuration included a tape reproducer that operated at 10,000 decimal digits per second. All programming was at first done in machine language code; later, language compilers were added. With the UNIVAC the shape of the modern computer

Table 67
Technical Features of UNIVAC I

MANUFACTURER: Remington Rand Division, Sperry Rand.
OPERATING AGENCIES: Army Map Service, Washington, D.C., AEC
    Computing Facility, New York University, Radiation
    Laboratory, University of California at Livermore,
    Department of the Air Force, HQUSAF, Washington, D.C.,
    HQ Air Material Command, Wright-Patterson AFB, Ohio.
    (Commercial users not included in this list.)
GENERAL SYSTEM: General-purpose applications, mapping,
    geodesy, research and services in mathematical sciences,
    AEC reactor design problems, weapons development, Air
    Force programming, logistical business-type problem
    solutions, inventory control, payroll, billing, planning.
    Timing   Synchronous
    Operation   Sequential
NUMERICAL SYSTEM: Internal number system: Decimal; decimal
    digits per word: 11 plus sign 12 alphanumeric; decimal
    digits per instruction: 6; decimal digits per instruction
    not decoded: 1; instructions per word: 2; total number of
    instructions decoded: 63; total number of instructions
    used: 45; arithmetic system: fixed point; instruction
    type: onde address code; number range: -1 to +1; floating
    point performed by subroutines supplied with computer.
    Words may be made up of alphabetic, numeric, and typewriter
    characters. (AMS and USAF report 64 and 46 instructions
    decoded and 40 and 46 instructions used, respectively.)
ARITHMETIC UNIT: Add time (excluding storage access): 120
    seconds. Multiply time (excluding storage access): 1,800
    microseconds. Divide time (excluding storage access):
    3,600 microseconds. Construction: vacuum tubes. Number of
    rapid access word registers: 4. Basic pulse repetition rate:
    2.25 megacycles/second. Multiply and divide times depend on
    numerical value of multiplier, dividend, and divisor,
    respectively. All quantities processed by the computer are
    in units of 11 digits plus a sign. Time includes simultaneous
    computation in duplicate circuits and comparison or results
    for identity.

STORAGE:

Media	Words	Digits	Microsec. Access
Acoustic Delay			
line-Hg	1,000	12,000	400 maximum
Magnetic Tape	120,000	1,440,000	

1,500 foot magnetic tapes used.

INPUT:

Media	Speed
Magnetic Tape	12,800 characters/ second read-in speed 100 inches/second
Metallic Tape:	1/2 inch wide in lengths of 100, 200, 500, or 1,500 feet, recorded at densities of 20, 50, 120, or 128 characters/inch.

Input media are prepared by: Unityper I Keypunching.
Records at 20 character/inch. Loop controlled. When

Table 67 *(cont.)*

used with Printing Unit produces printed copy. Unityper
II Keypunching: records at 50 characters/inch. Printed
copy produced simultaneously. Card-to-Tape-Converter:
240 cards/minute instantaneous conversion. 80-column
punch card input 120 characters/inch. 90-column Card-to-
Tape Converter: 240 cards/minute. Tape Operated Unityper
II (prototype) 6-10 characters/second. Converts 5-channel
punched paper tape to 7-channel (plus sprocket channel)
magnetic tape recording. Punched Paper Tape to Magnetic
Tape Converter: 200 characters/second; high-speed
conversion. Magnetic Tape Recording of Unityper II
verified by: Verifier Keypunching which verifies original
recording; provides for correcting mistakes on original
recording; produces printed copy simultaneously with the
other two functions. Magnetic tape recording of Card-to-
Tape Converters verified internally.

OUTPUT:

Media	Speed
Magnetic Tape	12,800 characters/second with speed 100 inches/second. Recorded at 128 characters/inch.

Output Equipment using Magnetic Tape input:

Uniprinter	10-11 characters/second. Converts recording on magnetic tape to desired printed format.
High-Speed Printer	600 lines/minute, adjustable to 200 and 400 if desired. 120 characters/line; 130 characters/line maximum with repetition of characters.
Card Punching Printer (Delivery: October, 1956)	Will print on both sides of a card and will punch the card.
Tape-to-Card Converter	120 cards/minute. Converts magnetic tape recording to 80-column punched cards. Detachable plugboard provides for field rearrangement.
Tape to 90-Column Card Converter	Converts to 90-column punched cards; otherwise similar to Tape-to-Card Converter.
Rab Lab-Buffer Storage	Hg 3,500 microseconds/60 words

USAF-AMC-Typewriter not used for normal input-output.

SOURCE: Specifications are from Martin Weik, BRL Report No. 971. Reprinted with permission from *From Eniac to Univac*, Nancy Stern, First Edition. Copyright © Digital Press/ Digital Equipment Corporation (Bedford, MA), 1981.

had at last been formed. The only peripherals it lacked which are common to today's computers were disk drives* for storage of data, but that technology was added to future models of the UNIVAC.

Although technological leadership in the design of computers passed from Remington Rand to International Business Machines Corporation (IBM)† through the lack of aggressive product development in 1954–1955, other UNIVACs were designed. The UNIVAC I soon experienced competitive pressures from IBM, particularly because of its slow memory after tape subsystems were available on competitors' systems. UNIVAC II finally appeared in 1957 after long delays caused by a lack of resources for development, uncertainty within the company on how to manage product development, and marketing. In 1960 the UNIVAC III and the UNIVAC 1107 were announced. UNIVAC III was compatible neither with earlier models nor with other UNIVAC devices, which meant that if a customer wanted to use one of the new machines all programs running on other UNIVACs would have to be converted—not a small or inexpensive task. Like the UNIVAC II, it also came on the market late and, therefore, was technologically not state-of-the-art. Yet, about 100 were ultimately sold. UNIVACs (I through III) and other computers made by Remington Rand and later Sperry* insured that this manufacturer would remain a major force in the computer industry from the early 1950s to at least through the 1970s.

The most important of all these machines was the UNIVAC I. It was technologically advanced for its day. It revolutionized the cost and effectiveness of computing, and was the one machine most responsible for the widespread use of data processing by commercial enterprises during the early 1950s when the modern age of data processing was born.

For further information, see: Franklin M. Fisher et al., *IBM and the U.S. Data Processing Industry: An Economic History* (New York: Praeger Publishers, 1983); Herman Lukoff, *From Dits to Bits: A Personal History of the Electronic Computer* (Portland, Oreg.: Robotics Press, 1979); Nancy Stern, *From ENIAC to UNIVAC: An Appraisal of the Eckert-Mauchly Computers* (Bedford, Mass.: Digital Press, 1981).

V

VON NEUMANN MACHINE. This term refers to the theoretical design of digital computers* described by John von Neumann** in a paper circulated after June 1945. Although controversy exists as to who actually developed many of the concepts he describes in the paper, particularly the notion of the stored-program computer, the fact remains that collectively these ideas have been labeled the Von Neumann Machine. They in turn formed the backbone of almost all computers designed since the mid–1940s.

Von Neumann was a scientific consultant to the U.S. government during World War II, playing an influential role in the development of computers. The most dramatic example of this role was the help he gave to the engineers at the Moore School of Electrical Engineering† at the University of Pennsylvania who were building the ENIAC,* the first electronic digital computer,* during 1944–1945. It became operational in 1946. He wrote the paper described above in collaboration with Herman H. Goldstine** and Arthur W. Burks,** both of whom were involved with the ENIAC, the first as a military liaison and the second as an engineer on the project. But only von Neumann's name appeared on the title page of the paper, probably encouraging the use of his name to describe the class of machine it detailed. The paper was entitled ''First Draft of a Report on the EDVAC,'' and it would become the most influential document ever written in the field of data processing. It was initially circulated to 101 people in an unpublished form, and it was later published.

The basis of the paper was the concept of the stored program, an idea that called for all data and instructions to be stored together within one memory* system in the computer, intermixed and not separated as before. This approach would allow a computer to compute at the speed of electronics, while both data and commands could be read or written under the control of programs, not peripheral equipment or by people (human intervention along a job stream). Memory would thus house data which, in itself, could be information or instructions, distinguished only when they moved into the processing part of the computer where they were either data for a program or instructions that did something to data. He also described how this could be managed, focusing on the logical modification of an instruction or data to instruction in a system. To this day the idea remains central to the design of a computer.

The paper also introduced the idea of a program counter, which was a register used to indicate the location of the next instruction to be executed. That task would be done each time an instruction was fetched from memory. That register would then update its record of addresses in memory where an instruction was housed.

Von Neumann's document described the components of a computer's system: memory, a processor in which the actual calculations took place, the concept of input/output components such as printers, card punches and readers, tape, and other storage media. It also discussed how these components might interact. Conceptually, and in his logical scheme, a system began with input of data (and equipment to do that), went on to a central processing unit (CPU) to be acted on (calculations done), and then to an output device eventually to become output data.

The EDVAC* was the follow-on computer to the ENIAC, the ostensible reason for writing the paper in the first place. Regardless of who actually deserves the credit for the stored-program concept, von Neumann must be recognized for presenting the rationale through which other scientists would attempt to design computers. Nancy Stern concludes that the concept of the stored program "was not the invention of von Neumann alone." Rather, he most effectively, and earliest of all, "elucidated" (Goldstine's word) the idea. Another early computer scientist, the Englishman Maurice V. Wilkes,** later noted in his memoirs that von Neumann "appreciated at once the possibilities of what became known as logical design" and the potential benefits of using stored programs. As a highly respected scientist, von Neumann gave respectability to the idea of designing computers, particularly the concept of the stored program.

For further information, see: Nancy Stern, *From ENIAC to UNIVAC: An Appraisal of the Eckert-Mauchly Computers* (Bedford, Mass.: Digital Press, 1981); John von Neumann, *The Collected Works of John von Neumann*, 6 vols., ed. A. H. Taub, especially vol. 5 (New York: Macmillan, 1963): 34–79.

W

WHIRLWIND. This computer, developed at the Massachusetts Institute of Technology (MIT) in the 1940s and early 1950s, was the first electronic digital computer* with "real-time" capability, meaning it could respond instantaneously to data fed to it. It became the first computer to employ random access magnetic core memory,* and the first to use synchronous parallel logic in its design, thereby allowing it to process more than one transaction at a time. This first-generation computer could thus track and measure events as they happened and was usable in both scientific and commercial applications involving instantaneous feedback of information. Thus, it advanced the technology of computer design significantly over previous machines.

The project resulting in the construction of WHIRLWIND I dates back to the early days of World War II when the U.S. Navy had to rapidly train flight crews for bombers. By 1943 the need for ground training equipment employing flight simulators was compelling. Captain Luis de Florez of the U.S. Navy approached MIT with a proposal to build such a device. He also believed that, along with the benefits of faster training, such equipment might reduce training costs. In 1944 the U.S. Navy and MIT discussed a project that would satisfy these demands. The device planned was called the Airplane Stability and Control Analyzer (ASCA).

MIT was a logical place for such a complex project, for it had been performing research for the government for decades and already had a fine reputation for scientific work. Captain Luis de Florez, having graduated from MIT, was familiar with existing trainer-analyzer work done at the school and by others, such as by Bell Laboratories.† Special facilities were also available at MIT. For example, Jay W. Forrester** and Gordon Brown had established the Servomechanisms Laboratory at MIT in 1940 which had done useful work on fire control and legitimately claimed particular strength in naval radar systems. The U.S. Navy's Office of Research and Inventions therefore believed that key engineers and scientists at MIT could assume this new complex project. This lab was part of a broader group within the Department of Electrical Engineering which, like Forrester, had employees and students eager for more government work.

Negotiations over ASCA and modifications to that project ultimately led to an agreement with MIT on December 18, 1944, calling for the construction of a new trainer-analyzer. The device was never built, for the task was too ambitious for the times. What emerged instead was an important computer of broader scope.

Forrester, head of this new task, became intrigued by the engineering challenges posed by the contract. Robert R. Everett** soon joined the project. For the next several years these men dominated WHIRLWIND. Forrester was an engineer by training and had only recently completed his undergraduate work at the University of Nebraska in 1939. He did his graduate work at MIT and in the early years of World War II gained experience with radar and military projects requiring practical use of integrated systems and servomechanisms. Everett, a New Yorker educated at Duke University, completed his undergraduate work in 1942 and also came to MIT for graduate work. He went to work for Forrester in the Servomechanisms Laboratory.

The two men began to formulate the overall design for the device which would be a computer, while pulling together a staff of engineers during the winter and summer of 1944–1945. Within two years initial design considerations made it obvious that an analog machine, borrowing from recent experiences at MIT and Bell Labs, could not work with real-time systems that required machines to provide specific and instant feedback on flight actions. Forrester and his engineers decided that instead a digital-based device would have to be designed. Although they had studied the ENIAC* which had been built and completed at the Moore School of Electrical Engineering† by 1946, they felt that many improvements would have to be made beyond the achievements of existing electronic digital computers. However, from the experience of the ENIAC they realized that an electronic digital machine would be a better path to pursue than the analog. Forrester believed that the cost of a digital computer might be less at time of construction, trouble determination would be easier to design, and ultimately such a device could be used for other applications. Although construction costs might be less, the expenses anticipated in development would be greater. That realization required going back to the Navy for additional funding. Furthermore, the time to complete the project would extend beyond what the original contract called for.

At the end of the second year of work, designing therefore shifted from analog to digital and at a time when Congress was cutting military spending. Despite this problem, fundamental design characteristics emerged in 1946–1947. For example, Forrester's team decided to install a parallel processor rather than a serial machine in order to better control costs and produce a much faster device. The machine would therefore have multiple processing capabilities to handle more than one set of transactions at the same time. These men also decided to use 16-bit word lengths. Significant enhancements to current memory technology were designed in 1947 calling for specially designed tubes which made possible the development of commercial computers in quantity and reliability during the 1950s—a fundamental technological innovation. The device would have thirty-two registers composed of toggle switches and an additional five made up of flip-flops to house data. The entire system would be a stored-program machine: that is, programs to run in the machine would be housed within its memory and be called to use by the machinery itself instead of being fed in manually as needed. Finally, more so than any other computer of its day, WHIRLWIND was massive, taking up more than two floors of a building.

Instead of a trainer-analyzer, the team decided to build a general-purpose computer because the problems associated with the first were too great to solve using analog methods and the opportunity to design a better device of wider application became exciting to this band of engineers. The delays caused by the shift in goals annoyed the Navy and some leaders at MIT between 1947 and 1949. Forrester persisted nonetheless. In 1949 the device already occupied 2,500 square feet in Barta Building at MIT and was still growing. However, testing of storage portions had begun, and additional components became operational in 1950 and 1951. The computer was by now called by its final name, WHIRLWIND I. Other computer projects sponsored by the Navy were nicknamed Hurricane and Tornado.

By the fall of 1949 the computer consisted of 3,300 tubes, over 8,900 diodes, and by June 1950 could perform error-free for one hour. In March 1951 it was operating in a normal fashion for thirty-five hours a week. Two years later magnetic drum memory replaced electrostatic memory, and a magnetic tape subsystem was added for additional memory (data storage in machine-readable form). By the end of 1954 WHIRLWIND was a very large computer with 12,500 vacuum tubes and 23,803 diodes. The entire system consisted of a central processing unit, console, and cathode ray tube (CRT) terminals, all on one floor of Barta Hall. Data communications and drum storage were on the floor below. The basement housed power and cables, and air conditioning units stood on the roof. Power consumption approximated 150 kW.

Despite its huge dimensions, WHIRLWIND was an operational electronic digital computer using a 16-bit parallel, single-address binary architecture. Each word of instruction or data took up 16 bits of memory. Operational code handled multiplication and division in thirty-two operational functions. Early in its life, the computer had automatic marginal-checking capability. It could check the rate of voltage anywhere within itself and regulate power as the designers wanted. Components thus could be monitored to determine which were headed for failure. That feature allowed preventive maintenance to take place or, during development, redesign to improve the system's overall reliability and availability (also known as up-time). Input/output consisted of very large CRTs, Flexowriters, and photoelectric tape readers. CRT output could also be microfilmed.

The machine was capable of performing 20,000 transactions per second in the beginning and, with redesign and the use of magnetic core memory, that figure rose to 40,000—a major breakthrough that would influence the design of other computers in various companies and universities for the rest of the 1950s (Table 68). WHIRLWIND had a batch operating system and could spool to offline printing. Its parallel functions were particularly attractive to other scientists, particularly to those who designed International Business Machines Corporation's (IBM's)† 700* series and machines at Digital Equipment Corporation (DEC),† the future home of many veterans of the WHIRLWIND project.

The software* developed for this machine was as significant as the hardware. The MIT group pioneered in the creation of a coding system for interpretive algebra. Through this system a programmer could write familiar mathematical expressions using programming language capabilities and functions that looked almost the same if someone were writing normal mathematical notation. Machine-level language was not heavily used, thereby making it possible to encourage people to write "programs" for this computer. Coding at a higher level than machine language was real progress, accomplished in 1952–1953 when many developments in programming languages* were occurring. Thus, software developments with WHIRLWIND were dramatic and major.

WHIRLWIND was the most expensive computer built up to that time. Total development costs were approximately $5 million. In comparison, Harvard's Mark II* had come in at $695,000, the ENIAC at $600,000, EDVAC* at $470,000, ORDVAC* at $250,000, and the UNIVAC* I at nearly $500,000. Despite the Navy's frustration over cost, the longer the project lasted the more difficult it became to terminate it prematurely.

While in its final stages of development, the government used WHIRLWIND as the prototype for the SAGE* Air Defense System. In order to provide instantaneous feedback for that application, the memory was upgraded to the new technology of core storage. This change meant that the system would be far more reliable and faster than before. In addition, WHIRLWIND now had greater capabilities than before, if for no other reason than it could handle vast quantities of information within a shorter period of time. The SAGE project eventually absorbed some of the engineers who had worked on WHIRLWIND and thus were able to use their experiences on new computer-related projects of the late 1950s.

WHIRLWIND made many technological contributions: (1) Easily the most influential technical impact on computing technology in general came from the use of random access, magnetic core memory. Such technology became a common feature of many future computers because it was faster and more reliable than earlier vacuum tube affairs of the 1940s. (2) The ability to self-check for failing components (called marginal checking at the time) was extremely important in improving reliability. (3) The extensive use of cathode ray tubes (CRTs) was new and refreshing; today the practice is standard on all computer

Table 68
Technical Features of WHIRLWIND

Central Processing Unit (CPU)	Single address, parallel, binary CPU	35,000 operations per second
Main memory	Magnetic core=2,048 16-bit words (add=24 microseconds, multiply=40 microseconds, control transfer=16 microseconds	
Secondary storage	ERA Magnetic DRUM, 1 x 2,048 words, 60 rps	31,000 wors/sec.
	Raytheon Magnetic TAPE, 4+1 units at 125,000 words each	390 words/sec.
Input	Ferranti photoelectric 7-hole READER, 2 units	205 characters/sec.
	Flexowriter mechanical 7-hole READER	10 characters/sec.
Output	Raytheon magnetic TAPE for later printing or punching 2+1 units at 53,000 characters each	133 characters/sec.
	Flexowriter printer, 1 direct, 2 from tape	8 characters/sec.
	Flexowriter punch, 1 direct, 2 from tape	11 characters/sec.
	16-inch SCOPE with visible face and 16-inch SCOPE with computer-controlled Fairchild CAMERA.	6,200 points/sec. 200 digits/sec. 1,200 digits/sec.
Word	5-bit operation+11-bit address=Sign+15-bit fraction	
Character	6-bit representation of the 50 keys on a Flexowriter in arbitrary, teletype-like code.	

Performance Characteristics (1959)

Operating time usable	96.5%
Average time between failures	10.6 hours
Scheduled maintenance time	1.25 hours/day
Size of system	13,000 tubes

SOURCE: Adapted from tables in Robert R. Everett, "WHIRLWIND," in N. Metropolis et al., eds., *A History of Computing in the Twentieth Century* (New York: Academic Press, 1980): 381. Reprinted with permission.

systems. (4) It was a real-time system and not simply batch. (5) It had extensive programming and software in order to respond to data from radar. (6) It had more communications capabilities than any earlier device; only in this way could it become a real-time defense system. (7) It had synchronous parallel logic, which in the layperson's terms means that the machine could transmit electronic pulses (otherwise known as digits) simultaneously rather than one at a time (sequentially) while continuing to maintain control and logical coherence. No computer had done that before. Today that is a normal feature of all computational equipment, including the hand-held calculator. This feature therefore introduced the possibility of dramatic increases in processing speed.

One member of the project, Kenneth Olsen, went on to form DEC, one of the most important computer manufacturing companies within the data processing industry. Other engineers designed computers for different firms or did research at universities and within government agencies during the 1950s and 1960s. As a result of their work on WHIRLWIND, they introduced significant improvements in the sophistication and quality of computer technology.

This computer lingered on long after first-generation computers had passed into history and third-generation machines were being installed. Although use by the government technically ended in 1959, during the 1960s the Wolf R & D Corporation used it until the early 1970s. Today portions of the machine are at the Smithsonian Institution and at the Computer Museum in Boston.

For further information, see: R. R. Everett, "WHIRLWIND," in N. Metropolis et al., eds. *A History of Computing in the Twentieth Century* (New York: Academic Press, 1980): 365–384; Herman H. Goldstine, *The Computer from Pascal to von Neumann* (Princeton, N.J.: Princeton University Press, 1972); Kent C. Redmond and Thomas M. Smith, *Project Whirlwind: The History of a Pioneer Computer* (Bedford, Mass.: Digital Press, 1980); T. M. Smith, "Project Whirlwind: An Unorthodox Development Project," *Technology and Culture* 17, no. 3 (July 1976): 447–464.

Z

Z COMPUTERS. This series of German computers was built by Konard Zuse** between the late 1930s and the late 1950s. One of his machines was operational by December 1941, which made him the first engineer who ever built a usable digital computer.* His computer functioned even before the ENIAC,* which is usually called the first electronic digital computer, constructed at the Moore School of Electrical Engineering† during the last days of World War II and the early period following. Some historians, however, argue that the Z3 was an automatic calculating machine rather than a digital computer.

There were several Z computational devices. The Z1 was an electromechanical machine used as an aid in mathematical computation, built between 1934 and 1938 in the living room of Zuse's parents. A second device was apparently built, although it no longer exists—perhaps a victim of an Allied bomb. The Z3, however, was completed in 1941 and worked. It employed a binary numbering system and floating-point arithmetic. Zuse added a relay adder which allowed him to use four relays to develop the sum of two binary digits. The Z3 was destroyed by Allied bombing at the end of the war. Zuse continued to work on computers in postwar Germany. The Z4, also constructed during the war, survived and became public in 1949 when it was installed at the Eidgenössische Technische Hochschule in Zurich. That event called attention to his nearly solitary and unknown efforts. In 1954 the Z4 was moved to another research facility near Basle where it functioned until 1959.

In 1945 the Z series also led Zuse to develop a *Plankalkül*, a programming language* that was algorithmic. Characterized by a matrix form of notation, it handled numerical and nonnumerical problems. In time, other languages functioned on the Z series. To capitalize on his work, Zuse formed a company in 1949, called ZUSE KG, to market his machines. The firm's first important product was the Z11, used in geodetical and optical applications, and was a relay processor. The next device, named the Z22, relied on vacuum-tube technology. Its successor, the Z23, used transistors. All of these machines appeared during the 1950s. The Z22 was first shipped in 1958, and eventually some fifty copies were built. The same year saw the introduction of the Z64, also known as the Graphomat, a computer-driven plotter. In 1969 Siemens AG acquired ZUSE KG, together with responsibility for marketing and maintaining the Z series of computers. But Zuse no longer built machines, thereby ending the history of the Z series.

A closer look at the Z3 suggests why it was the most important of the series. It was reconstructed in 1961 and is now housed at the German Museum of Technology. It consisted of two cabinets holding relays and an operator's panel made up of a keyboard and a display. Perforated movie film was read into the machine (as the data entry vehicle) from a device to the right of the operator's console. The Z3 was equipped with two racks of relays used as memory* to house sixty-four numbers. The entire device was physically small; each cabinet for memory was only 6 feet high by 3 feet wide. Zuse used nearly 2,600 telephone relays for his computer. Some 1,800 were dedicated to memory and another 600 to arithmetic. The last 200 were devoted to handling the film reader, keyboard, and operator's display panel.

The entire device was built without the aid of research institutes or universities. Zuse and his friends tinkered with the device on their own time. One historian, Paul E. Ceruzzi, estimates that Zuse must have spent approximately $6,000 to $7,000 on materials (or, according to Zuse, about 25,000 Reich Marks). No other computer built in the same period cost that little. Given the parts he used— all proven technologies of the 1930s and most of the relays, even second-hand— they were generally very reliable and thus allowed him to operate the device without significant time being devoted to repairs. That stood in sharp contrast to large American computers built in the 1940s which rarely ran for more than a few hours without requiring some amount of attention.

How important was the Z3 historically? Ceruzzi argues that "it stood at the gateway to the computer age." Another historian, Michael R. Williams, calls Zuse's work a stepping stone "toward the electronic stored program computer." The Z3 was scarcely known to have existed until the late 1940s and even then to only a few individuals, and thus Zuse worked in a parallel stream to many other computer projects but not in their mainstream. By the mid–1940s computers were being developed, and by the late 1940s they were being used routinely in the United States and in Great Britain. These originated from work done in those two countries, at the same time Zuse worked on the Z3. The American and British work was widely publicized, and their findings were shared with other engineers, thereby influencing computer research, particularly in the United States.

The Z4's history reflected wartime conditions. It was a full-scale relay computer started in 1942, based largely on the Z3 and yet using a 32-bit word length. The Z3 employed a 22-bit binary word length. Expanding the word length made it possible to have a seven-decimal place answer, taking calculations up to a range of 10^{20}. Memory was expanded eightfold over the Z3. Yet because of Allied bombing of Berlin, especially in 1944, Zuse had to move his work three times to different places in the city. On April 6, 1945, his Z3 was destroyed in a bombing raid while he was again moving the Z4. In the early months of 1945 he moved the Z4 out of Berlin to Göttingen and then to an underground military installation in the Harz Mountains. Next he transferred his device to a village deep in the Bavarian Alps. He kept it packed at that location until 1950 when he re-assembled it again for transfer to Zurich and the Federal Technical Institute. While there, engineers could claim that the Z4 was the only computer functioning in Europe. Although it was well exercised, it was not as significant an influence on data processing as other machines in Europe, particularly in Great Britain.

The Z5, built for the Leitz Optical Company, played a greater role. Although similar to the Z4, it was the basic machine later known as the Z11, forty-two of which were constructed. That led to the Z22, already mentioned, and that in turn to a transistorized version and 200 copies of it.

Before leaving the story of the Z series, one additional footnote involves what Zuse did while in the Bavarian Alps with his Z4. Zuse could not work on this or any other machine, for parts were not available and living conditions were harsh. Instead, he turned his attention to his "Plan calculus," better known as the *Plankalkül*. It was during this period that he designed what amounted to a programming language,* one of the first such efforts attempted in the history of computers.

For further information, see: F. L. Bauer and H. Wössner, "The Plankalkül of Konrad Zuse: A Forerunner of Today's Programming Languages," *Communications of the ACM* 15 (1972): 678–685; Paul E. Ceruzzi, *Reckoners: The Prehistory of the Digital Computer, From Relays to the Stored Program Concept, 1935–1945* (Westport, Conn.: Greenwood Press, 1983); Michael R. Williams, *A History of Computing Technology* (Englewood Cliffs, N.J.: Prentice-Hall, 1985).

Appendix:
Chronology

ca. 3000 B.C.	Abacus first developed, probably in Babylonia
ca. A.D. 800	Zero—probably imported from India—began to be used in China
ca. 1000	Gerbert described the use of an abacus with "apices"
ca. 1200	Earliest minted jetons appeared in Italy
1200s	Ramon Lull conceived of a logic machine and possibly built part of one
ca. 1300	Modern wire-and-bead abacus began displacing Chinese calculating rods
ca. 1500	Quadrant went into wide use in Europe
1580	Rabbi Judah ben Loew built an automata in Prague, called Joseph Golem
ca. 1600	Modern wire-and-bead abacus was introduced into Japan
1614	John Napier published his *Canon of Logarithms*
1617	John Napier published *Rabdologia*, in which he described "Napier's bones" and "Multiplicationis Promptuarium"
ca. 1620	Sector was in wide use in Europe
1620	Robert Napier published John Napier's *Constructio*
1622	William Oughtred invented the circular slide rule
1623	Wilhelm Schickard completed invention of the first mechanical calculator
1624	Henry Briggs published his first set of modern logarithms
1628	Adrian Vlacq published the first complete set of modern logarithms
1630–1633	William Oughtred and Richard Delamain developed and introduced the slide rule in Europe
1642	Blaise Pascal invented an adding machine
1644–1645	Pascal completed work on his calculator
ca. 1650	Sliding-stick–type slide rule came into existence
1666	Gaspard Schott published *Organum Mathematicum*
1672	Samuel Morland published *The Description and Use of Two Arithmetic Instruments*

1673	René Grillet described his adding machine
1674	Gottfried Wilhelm von Leibniz completed construction of a calculating machine
1738	Jacques de Vaucanson built an automata in the form of a duck
1786	J. H. Muller published a description of an automatic difference engine
1801	Joseph-Marie Jacquard built a loom programmed with punched tape, which subsequently became an industry standard for the manufacture of textiles
1820	Thomas de Colmar invented a calculator called the Arithmometer
1822	Charles Babbage began work on his difference engine
1826	Babbage published tables of logarithms developed with the partial use of mechanical devices
1834	Babbage began to design the analytical engine
1842	British government stopped its financial support of Babbage's difference engine
1847	George Boole published *The Mathematical Analysis of Logic*—the source of Boolean logic
1850	Amedee Mannheim developed the Mannheim Slide Rule
1853	Pehr and Edvard Scheutz built the first automatic difference engine called the Tabulating Machine
1854	George Boole published *The Laws of Thought*
1876–1878	Lord Kelvin constructed a harmonic analyzer and tide-predictor devices
1878	Ramón Verea received a patent for a calculator that could perform direct multiplication and division
ca. 1880	The true variable-toothed gear was invented, making possible complex calculating machinery
ca. 1890	Mechanical disk-sphere cylinder integrators reached a practical level of proficiency
1890	U.S. Census Bureau used card punch and tabulating equipment developed by Herman Hollerith during the 1880s, thereby launching the card punch phase of data processing's history
1891	Genaille-Lucas rulers were described
1893	The first truly efficient four-function calculator was invented and called the Millionaire
1900–1910	Mechanical calculators became widely used in business and by scientists
1906	Lee De Forest invented a three-electrode tube (triode), the basis of the vacuum tube used in first-generation computers
1908	Percy Ludgate proposed his design for an analytical engine
1910–1913	Bertrand Russell and Alfred North Whitehead published their monumental *Principia Mathematica*
1911	U.S. Coast and Geodetic Survey installed a tide predictor, which it continued to use until the 1960s
1913	Leonardo Torres y Quevedo designed an electrified arithmometer
1919	W. H. Eccles and F. W. Jordan published one of the earliest papers on flip-flop circuits
1920	Torres y Quevedo publicly exhibited an electrical arithmometer

1923	Moore Schoo! of Electrical Engineering was established at the University of Pennsylvania
1925	Bell Labs, organized in 1924, officially came into existence on January 1 as a consolidation of various research activities within American Telephone and Telegraph (AT&T)
1928	International Business Machines Corporation (IBM) introduced the standard eighty-column computer card, in use for the next six decades; ENIGMA coding machine put into regular military use in Germany; Punched card equipment came into use as output devices on commercially available mechanical calculators in Germany; Punched card machines were used by L. J. Comrie to calculate the moon's motions
1929	IBM established the Columbia University Statistical Bureau to develop scientific uses for punched card equipment
1930	Vannevar Bush built his first differential analyzer—an early analog processor—at the Massachusetts Institute of Technology
1931	L. J. Comrie converted a National Accounting Machine into a difference engine
1932	A computer-like device at Bell Labs was used for the first time to reproduce music; C. E. Wynn-Williams used large-scale electronic counters to count events in a physics experiment
1934	A Bush differential analyzer became operational at the Moore School of Electrical Engineering
1936	Konrad Zuse applied for a patent on his design for mechanical memory
1937	John V. Atanasoff developed the design for his ABC device; George R. Stibitz experimented with relays to do calculations at Bell Labs; IBM agreed to help Howard H. Aiken build the Harvard Mark I; Alan Turing published his paper "On Computable Numbers"
1938	Chester Carlson, future founder of Xerox, made his first copy of a document on October 22; Claude E. Shannon published *A Symbolic Analysis of Relays and Switching Circuits*; Konrad Zuse completed construction of the Z1, considered the first binary calculating machine
1939	Engineers at Bell Labs began building the Complex Number Calculator; IBM and Howard Aiken began assembling the Harvard Mark I; Konrad Zuse completed construction of the Z2
1940	The Bell Labs Complex Number Calculator became operational on January 8 and was demonstrated to attendees at the September 11 meeting of the American Mathematical Association via a remote terminal—the first example of remote job entry (RJE)
1941	On December 5 Konrad Zuse completed construction of the Z3, considered the first fully operational calculator with automatic control of its operations
1942	John W. Mauchly published "The Use of High Speed Vacuum Tube Devices for Calculating"—an early proposal for what would become the ENIAC, considered the first electronic digital computer; John V. Atanasoff's ABC device was abandoned due to World War II
1943	Harvard Mark I became operational in January at IBM's laboratory in Endicott, New York; U.S. Army and Moore School began initial discussions leading to the construction of the ENIAC; Heath Robin-

son'' device and later COLOSSUS became operational at Bletchley Park to solve code-breaking problems; Engineers at the Moore School began building the ENIAC; Relay Interpolator became operational at Bell Labs; Project WHIRLWIND began at MIT as an analog flight simulator

1944 Engineers at the Moore School began to explore the concept of the stored-program computer; Harvard Mark I was moved from IBM's laboratory at Endicott, New York, to Harvard University; Bell Labs' Model III calculator became operational in June; In July two accumulators of the ENIAC became operational; John von Neumann visited the Moore School for the first time in September to inspect the ENIAC; U.S. Army extended its contractual support of the ENIAC to include the EDVAC, a stored-program computer; IBM built the Pluggable Sequence Relay Calculator for the U.S. Army

1945 Model IV calculator became operational in March at Bell Labs; In spring operation of the ENIAC commenced; John von Neumann completed writing his ''First Draft on a Report on the EDVAC,'' the first full description of a stored-program computer; Alan Turing began working for the National Physical Laboratory (NPL); ENIAC became fully operational in November; Engineers at MIT decided to convert Project WHIRLWIND from analog to digital

1946–1959 First generation computers predominated

1946 ENIAC was dedicated on February 16; Alan Turing reached an advanced stage of his design for the ACE processor, and John Wilkinson joined NPL also to work on the ACE; J. Presper Eckert and John Mauchly decided to build commercial versions of a digital computer and established the Electronic Control Company; John von Neumann began working to establish a digital computer project at the Institute for Advanced Study (IAS) at Princeton, New Jersey, and soon after was joined by Herman H. Goldstine and Julian Bigelow; Bell Labs' Model V machine became operational in July; Moore School lectures on computer science were held—a major event in disseminating information on computers; M. V. Wilkes began designing the EDSAC at Cambridge University; Computer Laboratory was established at Manchester University, and two months later F. C. Williams and Tom Kilburn joined the laboratory; F. C. Williams applied for a patent on his electrostatic memory tube (the Wiliams Tube), the most popular memory system of the late 1940s; Arthur W. Burks, H. H. Goldstine, and John von Neumann wrote ''Preliminary Discussion of the Logical Design of an Electronic Computing Instrument''

1947 Harry Huskey began working at NPL; Moore School made delay-line memory for the EDVAC operational in March; IAS computer's electronics were redesigned to nearly final form; In July Harvard Mark II became operational; Transistor was invented at Bell Labs, in one of the most important developments in electronics in the twentieth century; M. V. Wilkes began building EDSAC at Cambridge University; ENIAC was converted into a stored-program computer through the use of function tables

1948   IBM introduced the Selective Sequence Electronic Calculator (SSEC) in January; A. D. Booth tested a functioning magnetic-drum memory; On June 21 Manchester University's first digital computer went into limited operation—this computer ran the first fully electronic stored program; On June 30 AT&T made public the development of the transistor; Development of the SEAC began at the National Bureau of Standards (NBS)

1949   Development of the SWAC began at the Institute for Numerical Analysis of the NBS; NPL began construction of the Pilot ACE; In May EDSAC became fully operational at Cambridge University; Construction of LEO I, a duplicate of EDSAC, began; The "relocating loader" was soon after incorporated into EDSACs on order; National Research Development Corporation (NRDC) came into existence to build computers in Great Britain; BINAC, the first stored-program computer built in the United States, was tested and delivered to Northrop Aviation; In September Harvard Mark III became operational; Manchester University's digital computer became operational in October

1950   WHIRLWIND became partially operational; SEAC became operational at the NBS in April; On May 10 Pilot ACE became a functioning system at the NPL; In July SWAC became operational at the Institute for Numerical Analysis; Bell Labs' Model VI (last in the series) became operational in November; Konrad Zuse reconstructed and installed the Z4 at the Federal Technical Institute in Zurich, Switzerland; Isaac Asimov devised his "Three Rules of Robotics"; Alan Turing published his paper "Computing Machinery and Intelligence," in which he described the Turing Test

1951   U.S. Census Bureau took delivery of the first UNIVAC on March 31; Australia completed construction of its first stored-program computer, CSIRAC; IAS computer went into limited operation; LEO I became fully operational; First computer-animated movie was made at MIT; M. V. Wilkes et al. published *The Preparation of Programs for an Electronic Digital Computer*; Construction of WHIRLWIND was completed, making available the largest digital computer of its time and the first to do real-time processing; William B. Shockley invented the junction transistor; RAYDAC was delivered to the NBS; Grace Hopper developed the concept of the compiler programming languages

1952   MANIAC became operational in March at the Los Alamos Scientific Laboratory, primarily for use in atomic energy applications; ORD-VAC became operational at the Aberdeen Proving Grounds in Maryland; In May IBM announced the IBM 701, its first full-function digital computer; IAS computer became operational on June 10; Construction of the ILLIAC I was completed in September, a classic example of the von Neumann digital vacuum tube computer; University of Toronto installed a Ferranti Mark I computer; First flexible disks appeared; SHORT CODE became available as a programming language on the UNIVAC I; Construction of the EDVAC at the Moore School was completed; Harvard Mark IV became operational; Core memory was installed on WHIRLWIND I and on the ENIAC

1953   IBM shipped its IBM 701 to the Los Alamos National Laboratory; Engineering Research Associates (ERA) shipped the first 1103 computer; General Electric (GE) built the OARAC computer; S. A. Lebedev built one of the first Soviet computers, the BESM1; Rand Corporation installed the JOHNNIAC, an IAS look-alike computer; Y. Y. Vasilevsky built the STRELA computer in the Soviet Union; MIT successfully tested Jay W. Forrester's magnetic-core memory

1954   DYSEAC computer became operational at the U.S. Army Signal Corps; U.S. Navy took possession of the NORC system; Texas Instruments (TI) announced it had produced the first commercially available silicon transistors; Antonin Svoboda completed building the first Soviet fault-tolerant computer, SAPO; English Electric built the DEUCE, using the Pilot ACE as its design model

1955   ENIAC was powered off for the last time; IBM began designing STRETCH; EDSAC acquired index registers

1956   John McCarthy of MIT introduced the phrase "artificial intelligence"; IBM began shipping the IBM 705 computer; Manchester University and Ferranti Ltd., began building the ATLAS, a large processor

1957   IBM began shipping the first disk drive, the IBM 350, and also the IBM 650 RAMAC system; IBM introduced FORTRAN, the programming language most widely used by engineers and scientists through the 1970s; UNIVAC II was announced; Electrodata introduced the DATATRON 220, a vacuum tube computer

1958   IBM introduced second-generation computers, the IBM 7070 and the IBM 7090; Philco began shipping the transistorized TRANSAC S-2000, one of the first of this type; FLOW-MATIC, a business programming language, became available for use on the UNIVAC; Initial pieces of the SAGE system became operational; First AN/FSQ-7 computer became operational as part of the SAGE system; Jack Kilby of TI built one of the first integrated circuits (ICs); Jean Hoerni developed the planar process for manufacturing transistors

1959–1964   Second-generation computers predominated

1959   United Nations Educational, Scientific, and Cultural Organization (UNESCO) sponsored its first international conference on information processing, held in Paris; John Backus introduced the Backus-Naur Form (BNF), defining programming notation; IBM announced the IBM 1401 and 1620 computers in October; IBM and American Airlines agreed to develop the SABRE airline reservations sytem; First major programming language for commerical applications became available from IBM, called the Commercial Translator; First ERA 1101 was delivered to the U.S. Bureau of the Census; First major description of COBOL was completed—the language became the most widely used programming language for commercial applications in the 1960s and 1970s; PEGASUS series of British computers became available; NPL built an ACE computer; Xerox introduced the first office copier, the 914; Harvard Mark I was powered off for the last time; Fairchild Semiconductor and TI simultaneously developed the IC; Robert N. Noyce of Fairchild applied for a patent on the chip; Kurt Lehovec designed an integrated circuit whose components were isolated with pn junctions; Noyce developed a planar integrated circuit, making possible mass production of reliable ICs in the 1960s; KIEV, an early Soviet scientific computer, became operational

1960   PDP-1 was introduced by Digital Equipment Corporation (DEC) as a minicomputer; LARC, one of the first supercomputers, became operational at the Lawrence Radiation Laboratory in Livermore, California; IBM began shipping the 1400 system

1961   IBM established the Spread committee, which recommended production of the S/360 as a state-of-the-art family of computers; MIT created the first computer time-sharing system in the United States; Los Alamos National Laboratory took possession of the first Stretch computer; TI built its first computer based on the IC

1962   PDP-4 minicomputer was first delivered; Manchester University installed the first ATLAS computer

1963   Burroughs introduced its popular B5000 system; An early form of the BASIC programming language became available at Dartmouth College; ILLIAC II became operational; DEC announced its first true minicomputer; Bell Punch Company of Great Britain began marketing electronic calculators built out of discrete components

1964–1970   Third-generation computers, hallmarked by the IBM S/360, were predominant

1964   On April 7 IBM announced the S/360 family of compatible systems with some 150 different products, ushering in a new era in data processing technology; Control Data Corporation (CDC) began shipping its 6600 processors in September; RCA announced its Spectra series of computers

1965   DEC shipped its first minicomputer to be built with integrated circuits, called the PDP-8; IBM began shipping the S/360

1966   IBM announced DL/I, its database manager

1968   First microcomputer was announced by Viatron Computer Systems Corporation in October; National Cash Register Company (NCR) announced its Century series of computers; Intel Corporation introduced the 1K random-access memory (RAM), the first in the world

1970 to present   Fourth-generation computers were predominant

1970   In June IBM introduced the S/370 to replace the S/360 family of computers; IBM announced the 3420 tape drive in November, the industry standard for nearly two decades thereafter; IDMS, a popular database management program of the 1970s, was announced by Cullinane Corporation

1971   In March CDC announced the CYBER 70 family of computers; IBM first used bipolar memory in the S/370 Model 145 and began shipping the IBM 3330 disk drive—the standard of the industry for the rest of the 1970s; Intel manufactured the microprocessor and introduced the 4004 (4K) chip; Mass-produced pocket calculators became available in the United States

1972 In January Hewlett-Packard Company (H-P) announced its first hand-held calculator, called the "electric slide rule" or the HP-35; In March TI introduced the 8-bit chip; TI also introduced the DataMath, an early hand-held calculator; In April Intel introduced the 8008 microprocessor, an 8-bit chip; "Pong," the first video game, was shipped by Atari Corporation

1973 Last ATLAS was shut down at the Atlas Laboratory in Great Britain; IBM first used MOSFET memory in its S/370 Models 158 and 168; Most computer vendors were selling computers based on the integrated circuit

1974 IBM announced MVS, the operating system for its largest processors; TI introduced a series of hand-held calculators, including the SR-50 and SR-51; *Radio-Electronics* published an article on how to build a "personal computer"

1975 ILLIAC IV, one of the earliest supercomputers, became operational in November; TI introduced the first digital watch; AT&T announced the Dimension Private Branch Exchange (PBX) system; Altair introduced its first microcomputer

1977 In March IBM announced the 303X large computers; Tandy Corporation announced its first microcomputer, the TRS-80, in August; TI introduced the TI-58 and TI-59 hand-held programmable calculators; Apple II personal computer was introduced

1978 IBM announced the 8100 distributed processors in October

1979 In February IBM announced the 4300 family of medium-sized computers; On May 30 Tandy announced the TRS-80 Model II, a state-of-the-art microcomputer

1980 IBM announced the IBM 3380 disk drive, the most advanced storage device in the industry

1981 Osborne 1 microprocessor was introduced; IBM introduced the Personal Computer (PC), its first microcomputer, on August 12

1982 Lotus 1-2-3 spread-sheet software became available for microcomputers; Pac Man, a popular video game of the early 1980s, was introduced

1983 First fiber-optics telephone line between cities was opened in February by AT&T between Washington, D.C., and New York; IBM introduced the PC jr

1984 Apple Computer introduced the Macintosh protable microcomputer on January 24; IBM made public its one-million-bit RAM

1985 IBM announced the 3090 large MVS/XA computers in February

Index

About the Author

JAMES W. CORTADA is Senior Marketing Programs Administrator for the IBM Corporation. He is the author of several books on the history and management of data processing, including *EDP Costs and Charges, Managing DP Hardware*, and *An Annotated Bibliography on the History of Data Processing* (Greenwood Press, 1983), as well as two companion volumes to the *Historical Dictionary of Data Processing: Technology* which cover organizations and biographies in the history of data processing. Dr. Cortada has also published numerous articles in a variety of journals.